Unit Operations in Winery, Brewery, and Distillery Design

Unit Operations in Winery, Brewery, and Distillery Design

David E. Block

Konrad V. Miller

CRC Press
Taylor & Francis Group
Boca Raton London New York

CRC Press is an imprint of the
Taylor & Francis Group, an **informa** business

First edition published 2022
by CRC Press
6000 Broken Sound Parkway NW, Suite 300, Boca Raton, FL 33487-2742

and by CRC Press
2 Park Square, Milton Park, Abingdon, Oxon, OX14 4RN

© 2022 Taylor & Francis Group, LLC

CRC Press is an imprint of Taylor & Francis Group, LLC

Library of Congress Cataloging-in-Publication Data
Names: Block, David E. (Chemical engineer), author. | Miller, Konrad V., author.
Title: Unit operations in winery, brewery, and distillery design / David E. Block and Konrad V. Miller.
Description: Boca Raton, FL : CRC Press, 2021. | Includes bibliographical references and index. |
Summary: "This text focuses on equipment and facility design for wineries, breweries, and distilleries and fills the need for a title that focuses on the challenges inherent to specifying and building alcoholic beverage production facilities and the equipment therein. The book walks through the process flow of grapes to wine, grain to beer, and wine and beer to distilled spirits, with an emphasis on the underlying engineering principles, the equipment involved in these processes, and the selection and design of said equipment. Written at a level accessible to both engineers and non-engineers, this textbook is aimed at students, winemakers, brewers, distillers, and process engineers"-- Provided by publisher.
Identifiers: LCCN 2021035514 (print) | LCCN 2021035515 (ebook) | ISBN 9780367563875 (hbk) | ISBN 9780367563899 (pbk) |
ISBN 9781003097495 (ebk)
Subjects: LCSH: Distillation apparatus. | Wineries. | Breweries. | Distilleries.
Classification: LCC TP159.D5 B55 2021 (print) | LCC TP159.D5 (ebook) |
DDC 663/.10284--dc23
LC record available at https://lccn.loc.gov/2021035514
LC ebook record available at https://lccn.loc.gov/2021035515

ISBN: 978-0-367-56387-5 (hbk)
ISBN: 978-0-367-56389-9 (pbk)
ISBN: 978-1-003-09749-5 (ebk)

DOI: 10.1201/9781003097495

Typeset in Times
by SPi Technologies India Pvt Ltd (Straive)

Contents

Preface for Unit Operations in Winery, Brewery, and Distillery Design

This textbook has been prepared over many years as a text for VEN 135: Winery Systems at UC Davis. This is a required course for our students pursuing a Bachelors or Masters degree in Viticulture and Enology and focuses on winery equipment design, as well as more generally on winery design. More recently, it has been expanded by Dr. Miller to serve as a text for VEN 140: Beverage Distillation Processing. In turning this into a textbook, we have expanded the scope to include beer and distilled spirits production, which follow extremely similar processes and often use identical equipment. While other courses in our curriculum focus on process science, and why certain processing steps might be used, this text is more focused on how the equipment for this processing works and the engineering principles that are an integral part of these operations.

Prior to my taking on this course in 2001, VEN 135 was a 3-week short course that covered some engineering fundamentals and was a precursor to a graduate course in winery design. In 2001, we decided to expand the course to cover all equipment from the grapes coming into the winery to the product and waste leaving. It became a 4-credit course that meets five times per week during a 10-week quarter. This includes three lectures per week, along with a weekly demonstration of a piece of equipment in the winery and a problem-solving session/recitation to reinforce lecture material and prepare students for homework and exams. For the first few years that I taught the class (initially with Prof. Roger Boulton and then by myself), I would get feedback from student evaluations of the course commenting that students were unsure of why they had to understand engineering principles for a career as a winemaker. "Couldn't I just hire an engineer to do the calculations for me?" was a common question. After giving this much thought, I changed the way that I taught the course in 2005. Now, I have a competition in the first week of class to see who can come up with the best concept for a new winery, including a back-story, wines to be produced, and labels. After a small group of likely wine purchasers helps me narrow down the entries to the strongest ideas, the class votes on which concept they like, and we then design that winery throughout the quarter during our recitation time. That is, if we are discussing fermentor design in the lecture, we design the fermentors for the new winery concept that week as well. By the end of the quarter, we have designed the entire winery including a layout with some simple scale drawings of the equipment layout and building elevations. This seems to really tie in the engineering concepts to the winery design and winemaking for students and seems to be enjoyable for everyone—including me as the instructor.

I do not expect students reading this text, or taking the associated courses, to be experts in engineering (unless they started as engineers to begin with!). However, I do set out to achieve four objectives for the students:

- be familiar with winery, brewery, and distillery equipment, and process flow from grapes to wine, grain to beer, or wine and beer to distilled spirits.
- understand the idea of sanitary design and its application to plant operation and design
- be able to understand critical equipment parameters and design issues for specifying/purchasing and operating alcoholic beverage systems
- understand how plant design can influence product "style" and product "style" can dictate plant design.

My key goal is that students finish the course with the ability to converse effectively with suppliers of equipment that they may be purchasing for their production facility and have fundamentals that will allow them to operate, maintain, and troubleshoot their equipment.

This text follows the flow of the course I teach at UC Davis. It starts with some introductory ideas on sanitary design that are important for all equipment in a winery, along with sizing pipes/hoses and pumps that will be used throughout the winery. It then follows equipment through wine processing from grape receiving through fermentation, post-fermentation processing and packaging, and winery utilities. In each of the chapters, there are examples of calculations in the text (derived from wineries that I have designed in this class over the last 15 years), along with problems at the end of each chapter that are representative of exam problems that I have given since the inception of this course. In addition, as the book progresses, I sometimes include what I would call "Integrative Problems," which require understanding of multiple engineering concepts or unit operations to fully solve. These are typically real-world types of problems that help stretch understanding past what is covered in my lectures. At the end of the text, I include a description of the various wineries that I have designed in class to date. I would like to thank many talented UC Davis V&E students over the years that have contributed the ideas on which these designs have been based. Perhaps the most important winery that I have had a major role in designing is the Teaching and Research Winery at UC Davis, the most advanced and most sustainable winery in the world. There are certainly examples in this text that describe concepts from this winery for consideration of students and industry alike.

A note of special thanks is in order here to Prof. Roger Boulton. Roger has been a mentor to me since I started in the Department of Viticulture and Enology at UC Davis in 1996. I like to joke that he and I make up about 40% of the engineers working on wine in the world—which is not too much of an exaggeration. Therefore, we need to stick together! Roger contributed his notes and insights from the precursor short course, along with helping me teach the course for a couple years while I learned about winery equipment. My background previous to joining UC Davis had been in fermentation process development, facility design, and manufacturing in the biopharmaceutical industry, so I welcomed his guidance. There are certainly some of his words and ideas scattered throughout this text.

I also need to acknowledge my co-author, Dr. Konrad Miller, here. Konrad completed his doctoral studies in chemical engineering in my laboratory at UC Davis in 2019 modeling phenolic extraction in red wine processing. Some of his work is presented here in the chapter on fermentor design. He is the kind of student that every faculty member dreams of having in their lab at least once in their career—bright, creative, productive, and just pushy enough to get a steady stream of work out the door including this textbook. From his invaluable engineering experience in industry (working in wine, beer, and distilled spirits production) to his teaching in my course and being responsible for other courses at UC Davis (including Distilled Spirits Production and Wine Stability), his additions to this text are extremely important. These include the sections or chapters on mass and energy balances, brewing upstream processing, settling and centrifugation, and distilling, among others. It has been a true joy to work with him on this project—well, as much as writing a textbook can be a true joy!

With a text like this one, developed over 19 years, there are a lot of people to thank—mostly my VEN 135 students (probably over 700 by now!). I would also like to thank the Gallo family for the support of my program through the Ernest Gallo Endowed Chair in Viticulture and Enology—along the way it supported trips to wineries and research centers around the world, along with funds to make sure this project came to fruition. Special thanks are due to former students Carson Nye and Victoria Chavez for helping us to get permission to use all of the figures and graphs in this book that were not our own, along with useful comments to strengthen the text. We also gratefully acknowledge Minami Ogawa for her amazing work developing many of the figures and tables—undoubtedly, the most beautiful artwork in the book is due to her talents! Of course, my biggest thanks goes to my family (my wife Karen, and children Andrew and Jessica) who put up with me staying up late writing and correcting exams, trying out ideas on them, and telling stories about them in class!

Prof. David E. Block
Ernest Gallo Endowed Chair in Viticulture and Enology
Professor and Marvin Sands Department Chair
Department of Viticulture and Enology & Department of Chemical Engineering
University of California
Davis, CA

Prior to coming back to school for my doctorate, I worked as a process engineer in the distilled spirits, wine, and beer industries. When performing process design, equipment installation, or troubleshooting, this was academically akin to being alone in the wilderness. Textbooks for petrochemical and bioprocess equipment design exist: I have well-tabbed copies of Kister's *Distillation Design* and *Distillation Troubleshooting*, Hall's *Rules of Thumb for Chemical Engineers*, and Blanch and Clark's *Biochemical Engineering*. I pursued my master's degree on the topic of beverage distillation and my Professional Engineer's (PE) license while working in an attempt to further my process engineering knowledge. Still, no good engineering textbook or reference manual existed for alcoholic beverage processing equipment. When I was a doctoral student, I took a course in Winery Systems Engineering under

Dr. David Block, and it poleaxed me: here, in one place, was all the information and rules that I had spent the better part of a decade cobbling together from different sources and my own experience.

When David invited me to be an author on his textbook, it was a fantastic honor, and I saw several opportunities to expand his Winery Systems Engineering reader for a more general audience. First, my experience as a process engineer has shown me that many problems can be addressed via mass and energy balances, so I expanded that section accordingly. My experience in the brewing industry drove us to add beer fermentors to our discussion of fermentation, and to add a whole new chapter on beer upstream processing, as it is quite distinct from wine. My experience in the bioprocess industry has made me appreciate the importance of centrifuge separations and crossflow filtration, which are rapidly gaining traction in the beverage industry.

Finally, my experience with distillation gave rise to a whole new section of this text: the engineering that goes into distilled beverage production. Every undergraduate chemical engineer is convinced that they know the secrets to distillation, but there are substantial and non-trivial challenges that arise in alcoholic beverage distillation. While this text will serve as a primer to non-engineers on the concept of distillation, engineers will also find this chapter illuminating. This text served as the reader for my Beverage Distillation Processing course at UC Davis.

I'd like to thank the students in the courses I've taught, who have in turn taught me how to communicate high-level engineering concepts to an audience. I'd also like to thank Prof. David Block, for his tutelage over my doctorate and post-doctorate work, and for the incredible honor of asking me to be an author on this textbook. And of course, I'd like to thank my wife, Caroline Miller, for all her love and support.

Reader, I hope this book helps you in your pursuits, be you a winemaker, brewer, distiller, or befuddled process engineer. The knowledge synthesized in this book represents over a decade of industry experience, over 25 years of research and teaching experience, and quite a bit of skull sweat. Good luck!

Prof. Konrad Miller PhD, PE
Director of Process Development
Solugen, Inc.
Assistant Adjunct Professor
Department of Viticulture and Enology
University of California
Davis, CA

Authors

Prof. **David E. Block** is Marvin Sands Department Chair in Viticulture and Enology at UC Davis and holds the Ernest Gallo Endowed Chair. Since joining UC Davis, he has conducted research on various topics, from fermentation optimization methods to metabolic engineering of yeast for improved wine production, as well as more recently working on single-plant resolution irrigation sensing and control. He played a key role in designing the UC Davis LEED Platinum-certified Teaching and Research Winery. Dr. Block has received the Distinguished Teaching Award from the UC Davis Academic Senate, the highest teaching award given for teaching alone on the UC Davis campus. Prior to joining UC Davis, he worked for Hoffmann-La Roche, Inc., working on biopharmaceuticals, both in process development and in manufacturing. Dr. Block holds a BSE from the University of Pennsylvania and a PhD from the University of Minnesota, both in Chemical Engineering.

Prof. **Konrad V. Miller** is the Director of Process Development at Solugen, Inc., a Houston-based biotechnology company; and an Adjunct Professor at the University of California at Davis. He holds a PhD in Chemical Engineering from the University of California at Davis, an MS in Chemical Engineering from the University of Southern California, a BS in Chemical Engineering from the University of California at Berkeley; and is a licensed Professional Engineer. After earning his bachelors, Dr. Miller worked as a process engineer in the wine and spirits industry at E&J Gallo, the beer industry at Anheuser-Busch, and in the biotechnology industry at Amyris. He returned to school to study under Dr. Block, earning a PhD focused on the development of reactor-engineering models of red wine fermentations. His expertise is in process development and design in the food and bioprocess industries.

1 Engineering Principles for Winemaking, Brewing, and Distilling

1.1 MAKING WINE, BEER, OR SPIRITS: THE INPUTS

There are three main inputs into a finished wine that determine its sensory attributes and chemical composition. These three inputs are the vineyard and viticultural practices, the climate, and enological practices used on the harvested grapes (see Figure 1.1). Enological practices can be thought of as having two components. These are (1) the equipment present in a winery and (2) the winemaking techniques and order of operations performed with this equipment to achieve the final product. These enological practices can vary widely from winery to winery, depending on the equipment/tools available to the winemaker and his/her style.

Similarly, beer quality is a function of the type and quality of grains used (i.e. the "mash bill"), as well as additives like hops or flavorings. These raw materials can change based on how they are grown and on the climate in which they are grown. However, here too, the equipment and processes used by the brewer or distiller can also have a significant impact on the final product.

Distilleries almost always include a winery-like facility (fruit brandy, rum, tequila) or a brewery-like facility (whiskey, vodka), as these facilities generally produce the material to be distilled (wine, cider, beer, etc.). While we discuss distillation separately in Chapter 8, all winery and brewery discussions are also pertinent to distillery operations. It also clear that the type of distillation equipment and processes used by distillers will have a major impact on the sensory characteristics of the final product.

In this textbook, we will be focusing more on the equipment than on the processes, though there are many times where these are inextricably linked. For some parts of the process, the effect of a processing step will be well understood or differences in wine or beer characteristic due to equipment design can be predicted, but this is certainly not a general rule. Therefore, it is important to have as much information as possible on enological or brewing practices to make good choices on the equipment purchased and how it is used. One way to do this is to break processing down into a series of smaller steps.

1.2 THE CONCEPT OF THE UNIT OPERATION

Chemical engineers often describe each of these steps in a series of operations as a "unit operation." Each unit operation (Figure 1.2) has an input process stream with certain attributes and an output process stream with defined, desired attributes that

DOI: 10.1201/9781003097495-1

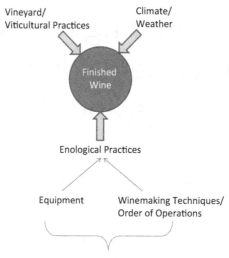

FIGURE 1.1 The three main inputs into a final wine are illustrated here. This text will focus on the equipment and systems available to winemakers, brewers, and distillers to control the final sensory and chemical characteristics of their respective end products.

must be achieved by the piece of equipment used for the unit operation. In this text, we will use this concept and discuss each unit operation separately. There will be unit operations for juice or wort preparation, fermentation, filtration, bottling, and wastewater treatment, just as examples. Some of these, such as bottling, can be broken down into even smaller units (e.g. glass dumping, rinser, filler, corker, foiler, and labeler). Many times, the unit operation can be accomplished by several different pieces of equipment. In this case, all of the options will have to be evaluated to choose one, including cost, availability, ease of use, or other considerations.

FIGURE 1.2 The unit operation, defined. A unit operation is a piece of equipment or system that transforms an input stream (e.g. wort) into an output stream (e.g. beer) with a different set of attributes. An example would be a fermentor, where yeast consume the sugars in wort to produce the ethanol and carbon dioxide in beer.

1.3 EXAMPLE: A GENERIC WINERY BLOCK DIAGRAM

If we string a whole series of unit operations together, we have a winery (or brewery or distillery, as the case may be). Figure 1.3 illustrates an example of a generic winery block diagram that describes the flow of grapes into the winery through the product and waste leaving the winery—all as a series of unit operations. In this diagram, we have grape delivery followed by juice preparation as one of the major inputs to the wine. There are several unit operations that could be used in juice preparation such as destemming, sorting, crushing, pressing, settling, and draining (and many other possibilities in various sequences). The other major input is the inoculum. Together, these streams are fed into the primary fermentor, which can also take on different shapes and forms. From the primary fermentor, a secondary fermentation can occur either in the same or different vessel. After the secondary fermentation, a series of post-fermentation processing unit operations can be chosen such as fining, cold stabilization, filtration, ethanol/VA removal, blending, and bottling, all leading to the final product. In addition to the fermentation and post-fermentation processing unit operations, several other unit operations are also required for successful winery operation. These include cleaning, refrigeration, HVAC (heating, ventilation, and air conditioning), and liquid and solid waste removal. All of these unit operations require

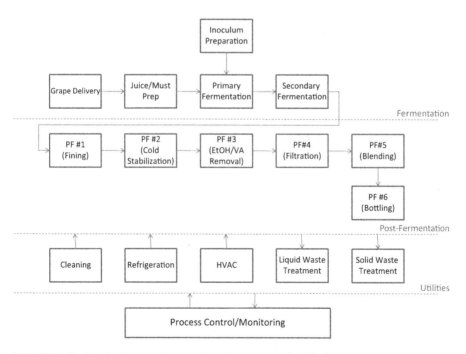

FIGURE 1.3 Block diagram of a generic winery—note that this is nearly identical for beer, with juice/must prep replaced by mashing and lautering. A collection of unit operations in series defines a process.

some sort of process control and/or monitoring. After some introductory topics related to engineering units and unit conversions, sanitary design and valves, pipes, and pumps, we will follow the process flow shown in this diagram, discussing each new unit operation as we get to it. We will start with fermentation processing equipment, then move into post-fermentation processing, followed by all of the associated winery utilities.

1.4 CHOOSING EQUIPMENT FOR A NEW OR EXISTING FACILITY

There are many **design factors** that determine the specification and purchase of equipment for a facility (Figure 1.4). One important factor is the capital available—you cannot purchase equipment with money that you do not have or cannot borrow. Another important factor is your plant's business plan. The planned production capacity for the plant is an important determinant of equipment purchased, as is the anticipated or actual "style" and varieties of product to be produced. All of this will be determined in conjunction with a marketing plan. Finally, regulations, including international, federal, and state regulations, must be taken into account in deciding what equipment is to be purchased.

The design factors will naturally lead to a series of **design choices**. The design choices are the actual pieces of equipment that will be specified and purchased, along with how they are arranged in the plant. Much of the remainder of this text will focus on giving you the information necessary to make these choices—or at least to work with engineers and vendors to help make the right decisions for your facility.

Finally, all design choices will have both expected and unexpected **design consequences**. Design consequences include impact on operating costs, such as raw material costs, energy costs, personnel costs, and waste treatment costs. One example of this is that purchasing an automated cleaning system will require more capital, but is

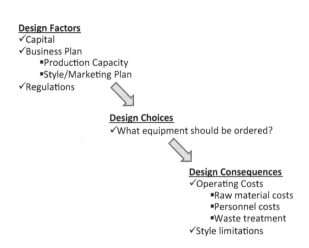

Design Factors
✓Capital
✓Business Plan
   ▪Production Capacity
   ▪Style/Marketing Plan
✓Regulations

Design Choices
✓What equipment should be ordered?

Design Consequences
✓Operating Costs
   ▪Raw material costs
   ▪Personnel costs
   ▪Waste treatment
✓Style limitations

FIGURE 1.4 Choosing equipment for a new or existing winery, brewery, or distillery. Design factors determine the choices made in purchasing new equipment. All purchasing choices have intended and unintended consequences.

likely to reduce cleaning material costs, personnel costs, and waste treatment costs, along with potentially improving product quality and consistency. Design consequences will also impact the flexibility and style of a facility and potentially limit future stylistic options. An example of these consequences would be purchasing a press sized based on crushed or machine-harvested fruit, thereby precluding its use for whole cluster pressing, except for very small lots.

2 Fundamental Concepts in Beverage Engineering

2.1 ENGINEERING UNITS AND UNIT CONVERSIONS

Nearly all numbers in the practical world have units associated with them, and these units are critical to understanding the context of the numbers. For instance, saying that the temperature is 32 outside has no meaning without units. If this were 32°C, it would be quite warm, while if it were 32°F, it would be freezing. If it were 32 K, the nitrogen in the air would be condensing! While this is a simple example, it illustrates the point that all numbers need units with very few exceptions (we will get to some later in the next chapter).

The four basic units that we will be working with in this text are units of temperature, length, mass, and time. All other units will be combinations of these four basic units. For instance, force will have units of mass × length/time². Energy or work will have units of mass × length²/time². Using units on every number can help with checking calculations as well. As an example, if we are calculating a force and we know that $F = ma$ by Newton's Second Law where "m" is the mass of an object and "a" is the acceleration of that object, then "m" with units of mass and "a" with units of length/time² would yield a product of:

$$F = ma \left[= \right] \text{mass} \times \frac{\text{length}}{\text{time}^2} = \frac{\text{mass·length}}{\text{time}^2} \tag{2.1}$$

which are units of force ([=] stands for "in units of"). If we had accidentally substituted a velocity for the acceleration (and included the units), we would have incorrectly calculated a force with units of mass • length/time, which does not make sense. Thus, always keeping track of units will help us to avoid calculation errors and incorrect substitutions into formulae. Every number in your calculations should have the units written next to it. This is important. Why is this such a big deal? Over many years of teaching, we have observed a strong correlation between an absence of units in a student's problem solution and calculation errors. Missing points on an exam is one issue, but perhaps not as big an issue as ordering the wrong pump or tank in a real job because units were ignored. Make it a habit now so it becomes second nature. You will thank us later.

There are two main systems of units in use in industry internationally, English Units and metric or SI units. It would be convenient if all manufacturers and suppliers all used just one of these systems. Unfortunately, they do not. Therefore, in this text, we will see both systems of units. This is especially the case in the US, where grapes and grains arrive in tons, juice flows into tanks with capacities listed in gallons, wort and beer is measured in barrels (31 gal), and wine leaves the wineries in 750 mL

TABLE 2.1
Unit Conversion Table for Common Engineering Units

Length

1 inch	2.54 centimeter
1 foot	30.48 centimeter
1 meter	100 centimeter
1 meter	39.37 inch
1 micron	10^{-6} meter

Mass

1 pound*	16.0 ounces
1 pound*	453.6 grams
1 pound*	7000 grains
1 ton (short)	2000 pounds*
1 kilogram	1000 grams
1 kilogram	2.205 pounds*

* Avoirdupois

Volume

1 foot³	28.32 liters
1 foot³	7.481 US gallons
1 U.S. gallon	3.785 liters

Power

1 kilowatt	56.87 Btu per minute
1 kilowatt	1.341 horsepower
1 horsepower	550 foot-pounds force per second
1 horsepower	0.707 Btu per second
1 horsepower	745.7 watts

Density

1 gram per centimeter³	62.43 pounds per foot³
1 gram per centimeter³	8.345 pounds per U.S. gallon
1 gram mole of an ideal gas at 0°C and 760 mm Hg is equivalent to 22.414 liters	
1 pound mole of an ideal gas at 0°C and 760 mm Hg is equivalent to 359.0 feet³	
Density of dry air at 0°C and 760mm Hg	1.293 grams per liter = 0.0807 pound per foot³

Pressure

1 pound per inch²	2.04 inches of mercury
1 pound per inch²	51.71 millimeters of mercury
1 pound per inch²	2.31 feet of water
1 atmosphere	760 millimeters of mercury
1 atmosphere	2116.2 pounds per foot²
1 atmosphere	33.93 feet of water
1 atmosphere	29.92 inches of mercury
1 atmosphere	14.7 pounds per inch²

Heat, Energy, and Work Equivalents

	cal	Btu	kWh	Joules
cal	1	3.97×10^{-3}	1.162×10^{-6}	4.1840
Btu	252	1	2.930×10^{-4}	1055
ft · lb	0.3241	1.285×10^{-3}	3.766×10^{-7}	1.356
kWh	860,565	3412.8	1	3.60×10^{6}
hp-hr	641,615	2545.0	0.7455	2.685×10^{6}
Joules	0.239	9.478×10^{-4}	2.773×10^{-7}	1
liter-atm	24.218	9.604×10^{-2}	2.815×10^{-5}	101.33

Constants

e	2.7183
π	3.1416

Gas-law constants:

R	1.987 (cal)/(g mol) (K)
R	82.06 (cm³) (atm)/(g mol) (K)
R	10.73 (lb/in.²) (ft³)/(lb mol)(°R)
R	0.730 (atm) (ft³)/(lb mol) (°R)
R	1545.0 (lb/ft) (ft)/(lb mol) (°R)
R	32.17 (ft) (lbm)/(s) (s) (lbf)

bottles. For this reason, it is important to be familiar with both types of units and to be able to convert readily between them. Table 2.1 is a unit conversion table that lists units for common parameters in both systems of units, along with conversion factors. Conversion between units is achieved easily by writing the original number with its units multiplied by the conversion factor written as a fraction and canceling out units. For example, if the acceleration of gravity is 9.8 m/s² in metric units and we would like to have this same acceleration in English units, we can use the following equation:

$$g = 9.8\frac{m}{s^2} \times \frac{3.28\,ft}{m} = 32.2\frac{ft}{s^2} \tag{2.2}$$

If more than one unit conversion is required for a particular number, we can just multiply by each conversion factor and cancel out all of the units that appear in both the numerator and denominator of the resulting equation. An example of this would be converting atmospheric pressure in English units to metric:

$$P_{atm} = 14.7\frac{lb_f}{in^2} \times \left(\frac{12\,in}{ft}\right)^2 \times \left(\frac{3.28\,ft}{m}\right)^2 \times \frac{4.45\,N}{lb_f} \times \frac{kPa}{1,000\,N/m^2} = 101.3\,kPa \tag{2.3}$$

As we will be using pressure quite a bit in this text, it is important to note a quirk of the English system units for mass and force. The unit for mass in the English system is lb_m; the unit for force is lb_f. An object that has 1 lb_m of mass will weigh 1 lb_f. However, if we look at Newton's Second Law for an acceleration equal to the acceleration of gravity, g, there is a problem with units:

$$F = m \times a = m \times g = m \times 32.2\frac{ft}{s^2} \tag{2.4}$$

If we plug in 1 lb_f for F and lb_m for m, the two sides are not equal. The equation we need to use, therefore, is:

$$F = m \times \frac{g}{g_c} \tag{2.5}$$

where g_c is a constant:

$$g_c = 32.2\frac{lb_m \cdot ft}{lb_f \cdot s^2} \tag{2.6}$$

This way, all units are now consistent. We can use the same Equation (2.5) with metric units, but in this case g_c is 1 without any units. This also works out, because, for instance, a 1 kg object would have a weight of 9.8 kg m/s² or 9.8 N.

2.1.1   Sugar Concentration Units in Winemaking, Brewing, and Distilling

An important idiosyncrasy of the food industry, especially the juice and fermented beverage industry, is the use of a specialized scales for the measurement of sugars dissolved in a water solution. For historical reasons, these scales are measured in the wine and juice industry as °Brix, and in the brewing industry as °Balling, or occasionally as °Plato. These scales are functionally identical, such that a value reported in °Brix is the same in Balling or Plato.

These scales are used to express the concentration of sucrose in a water solution: 1 °Brix is defined as 1 g sucrose in 100 g of solution. This is an extremely convenient scale, as a 20 °Brix juice has double the sugar concentration per unit volume of a 10 °Brix juice.

Brix or Balling can be directly measured for a sugar solution my measuring density (or specific gravity) at a fixed temperature, and using Table 2.2.

A major drawback of sugar measurement in this way is that it assumes all solvent is water and all dissolved solids are sugar. For example, a saltwater solution with a density of 1.03 g/mL would indicate a °Brix of 7.55, leading a naïve winemaker to believe the saltwater had 7.55 g/100 g sugar, when it had none. While this is usually not an issue in fruit juices, it can become a considerable source of error in high gravity wort with low fermentable extract conversion (see Section 5.3), or when ethanol is present, as ethanol is substantially less dense than water.

An alternative density scale, often encountered in Australia and New Zealand, is degrees Baumé (°B or °Bé). 1°B is the density of an 18 g/L sugar solution, which should yield roughly 1% alcohol by volume (abv) post fermentation. As such, measurement of °B for a sugar solution yields a rough approximation of final %abv. At 20°C, the relationship between the specific gravity of a denser than water liquid and °B is given by $SG = 145/(145 - °B)$, while the relationship for a fluid less dense than water is given by $SG = 140/(130 + °B)$.

It is important to note that density is always a function of temperature. Just like gasses, liquids are denser when cold and less dense when hot. Temperature correction tables are typically provided by instrument vendors.

2.2   MASS AND SPECIES BALANCES

In Chapter 1, we considered the notion of a unit operation: a step in the process where material enters, is somehow affected, and then leaves. Central to the analysis of a unit operation is the concept of a *mass balance*. A mass balance is an accounting of materials. Material enters a unit operation, material leaves an operation, and the amount of material in the unit may increase or decrease—or stay the same if the mass entering and leaving the unit operation do so at the same rate.

As a trivial example, consider a potable water storage tank for use in a brewery. Water enters the tank, perhaps as output from a reverse osmosis (RO) filter. Water enters the tank at 500 pounds per minute (lb_m/min). Water is pulled from the tank for use in the mash tun at 50 lb_m/min, for a bottling line at 150 lb_m/min, and for use in brewery cleaning at 200 lb_m/min. The rate of water accumulation in the tank is then 500 lb_m/min − (50 lb_m/min + 150 lb_m/min + 200 lb_m/min) = 10 lb_m/min accumulating in the tank. This is a mass balance.

TABLE 2.2
Specific Gravity Versus Brix/Balling/Plato at 25°C

Degrees Balling/Brix/Plato to SG Conversion Table

Specific Gravity	Balling/Brix/Plato	Specific Gravity	Balling/Brix/Plato	Specific Gravity	Balling/Brix/Plato
1.000	0.00	1.044	10.94	1.088	21.10
1.001	0.26	1.045	11.18	1.089	21.32
1.002	0.51	1.046	11.42	1.090	21.54
1.003	0.77	1.047	11.66	1.091	21.77
1.004	1.03	1.048	11.9	1.092	21.99
1.005	1.28	1.049	12.14	1.093	22.21
1.006	1.54	1.050	12.37	1.094	22.43
1.007	1.80	1.051	12.61	1.095	22.65
1.008	2.05	1.052	12.85	1.096	22.87
1.009	2.31	1.053	13.08	1.097	23.09
1.010	2.56	1.054	13.32	1.098	23.31
1.011	2.81	1.055	13.55	1.099	23.53
1.012	3.07	1.056	13.79	1.100	23.75
1.013	3.32	1.057	14.02	1.101	23.96
1.014	3.57	1.058	14.26	1.102	24.18
1.015	3.82	1.059	14.49	1.103	24.40
1.016	4.08	1.060	14.72	1.104	24.62
1.017	4.33	1.061	14.96	1.105	24.83
1.018	4.58	1.062	15.19	1.106	25.05
1.019	4.83	1.063	15.42	1.107	25.27
1.020	5.08	1.064	15.65	1.108	25.48
1.021	5.33	1.065	15.88	1.109	25.70
1.022	5.57	1.066	16.11	1.110	25.91
1.023	5.82	1.067	16.34	1.111	26.13
1.024	6.07	1.068	16.57	1.112	26.34
1.025	6.32	1.069	16.8	1.113	26.56
1.026	6.57	1.070	17.03	1.114	26.77
1.027	6.81	1.071	17.26	1.115	26.98
1.028	7.06	1.072	17.49	1.116	27.20
1.029	7.30	1.073	17.72	1.117	27.41
1.030	7.55	1.074	17.95	1.118	27.62
1.031	7.80	1.075	18.18	1.119	27.83
1.032	8.04	1.076	18.4	1.120	28.05
1.033	8.28	1.077	18.63	1.121	28.26
1.034	8.53	1.078	18.86	1.122	28.47
1.035	8.77	1.079	19.08	1.123	28.68
1.036	9.01	1.080	19.31	1.124	28.89
1.037	9.26	1.081	19.53	1.125	29.10
1.038	9.50	1.082	19.76	1.126	29.31
1.039	9.74	1.083	19.98	1.127	29.52
1.040	9.98	1.084	20.21	1.128	29.73
1.041	10.22	1.085	20.43	1.129	29.94
1.042	10.46	1.086	20.65	1.130	30.15
1.043	10.70	1.087	20.88		

More generically, a *total or overall* mass balance can be written:

Rate of mass accumulation = Rate of mass in − Rate of mass out   (2.7)

Mass is always conserved (at least in the absences of nuclear reactions, which are rare in food processing), but *species* are not. For example, how much ethanol enters a fermentor charged with juice versus how much leaves? Ethanol, a species, can be generated. By the same token, sugar can be consumed. In this case, we can state a *species* material balance as:

Rate of species accumulation = Rate of species in − Rate of species out
+ Rate of Species Generation
− Rate of Species Consumption   (2.8)

In this equation, the rates of species in and out are often written as the product of the overall mass flow rate multiplied by a species concentration.

Volume is often used as a unit in processing. For instance, tank volumes and volumetric flow rates are commonly used when describing operations in a beverage plant. Volume is **not** a conserved quantity—mixing 1 L of water and 1 L of ethanol will not yield 2 L of liquid. Adding a salt crystal 1 L in volume to 10 L of water will not result in a solution of 11 L. Nevertheless, when dealing with a system where there are no concentration changes (i.e. pure water, juice without blending or fermentation, etc.), we can use a "volume balance" to solve a problem, so long as we take care to ensure that there are no changes in density.

We divide the analysis of mass and species balances into two cases: **steady state** is when the overall rate of accumulation is zero, while **unsteady state** is when mass or a species accumulates or is lost in a system.

2.2.1 Steady State Balances

Steady state systems are those that do not vary with time. A trivial example might be water flowing through a section of pipe: water flows into the pipe, travels through the pipe, and leaves the pipe. While new material is entering and leaving, the amount of water in the pipe does not change with time. In Equation (2.7), for example, this would make the accumulation term on the left side of the equation equal to zero. In that case, mass flow into the system will equal mass flow out of the system.

Steady state systems are the exception rather than the rule in a winery. Examples might be an RO membrane for water treatment, the operation of a bottling line, or a column distillation for brandy production.

Example 2.1: Steady State Mass Balance at a Bottling Line

A bottling line is producing 60 bottles/min of sparkling wine. Each bottle contains 750 mL of wine with a density of 0.98 kg/L from the tank farm, with an additional 0.01 kg CO_2 per kg wine injected en route to bottling. If each bottle (including packaging) weighs 190 g, what is the mass flow rate of packaged wine leaving the bottling line?

SOLUTION

We can start with an overall mass balance:

$$\text{Rate of mass accumulation} = \text{Rate of mass in} - \text{Rate of mass out}$$

Because this system is at steady state, the rate of mass accumulation is 0 and therefore

$$\text{Rate of mass in} = \text{Rate of mass out}$$

Substituting in for the various mass flows:

$$W + C + B = P$$

where W is the mass flow rate of wine, C is the mass flow rate of carbon dioxide, and B is the mass flow rate of bottles. P is the overall mass flow rate out of the bottling line, which is what we are calculating here. To do this, we need to calculate W, C, and B.

To calculate W, we know 60 bottles of wine per minute are leaving the bottling line. Each bottle has 750 mL of wine, which weighs 0.75 L × 0.98 kg/L = 0.735 kg. For 60 bottles per min, W = 0.735 kg/bottle × 60 bottles/min = 44.1 kg/min. For C, each bottle has 0.01 kg CO_2/kg wine × 0.735 kg wine = 0.00735 kg of CO_2 injected into it. Therefore, C = 0.00735 kg CO_2/bottle × 60 bottles/min = 0.441 kg/min. Finally, the packaging weighs 0.19 kg per bottle. Therefore, B = 0.19 kg/bottle × 60 bottles/min = 11.4 kg/min.

Combining these results:

$$P = W + C + B = 44.1\,\text{kg/min} + 0.441\,\text{kg/min} + 11.4\,\text{kg/min} = 55.94\,\text{kg/min}.$$

Therefore, with the bottling line operating at 60 bottles/min, the mass flow rate leaving the line is **55.94 kg/min**.

Example 2.2: Reverse Osmosis Treatment of High Chloride Groundwater

You are operating a small brewery in Paso Robles during a drought, which increases the concentration of salts in the groundwater. Chloride (Cl^- ions) levels have increased to 500 mg/L, to the point where groundwater is corroding your 304L stainless steel. You install a Reverse Osmosis (RO) membrane system to filter groundwater. The RO can be fed 100 L/min, but only 90% of the water fed to the RO passes through the filter (known as "permeate"). The RO filter removes 95% of chlorides in the product water. The unfiltered water and the rejected chlorides leave the system through a waste line, called "reject" water. The system is operating at steady state.

(a) What is the concentration of chlorides in the permeate?
(b) What is the flow rate of permeate?
(c) What is the mass flow rate of chlorides in the permeate?
(d) What is the concentration of chlorides in the reject?

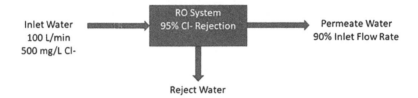

SOLUTION

(a) If 95% of the chlorides in the permeate are removed, 5% must remain. The feed has a chloride level of 500 mg/L, so the permeate has a concentration of 5% × 500 mg/L = **25 mg/L**

(b) If 90% of the water is filtered, and 100 L/min are fed, then 90% × 100 L/min = **90 L/min**

(c) If the permeate flow rate is 90 L/min, and the concentration of chlorides is 25 mg/L, the flow rate of chlorides in the permeate is 90 L/min × 25 mg/L = **2,250 mg/min**

(d) To get the concentration, we need to determine the mass flow rate of chlorides in the reject, in mg/min, and the volume flow rate of reject, in L/min. The feed carries 50,000 mg/min of chlorides (100 L/min × 500 mg/L), 95% of which is rejected. We can calculate the mass flow rate of chlorides in the reject by the difference between the feed and the permeate: 50,000 mg/min – 2,250 mg/min = 47,750 mg/min. We can also find the reject flow rate by difference: 100 L/min (input) – 90 L L/min (permeate) = 10 L/min. To find concentration, we divide mass flow rate of chlorides by the volume flow rate of water: 47,750 mg/min / 10 L/min = **4,775 mg/L** chlorides in the reject water.

2.2.2 Unsteady State Balances

Unsteady state systems evolve with time. Filling a glass of water is an unsteady state process—the amount of water in the glass changes with time. Most winemaking and brewing processes are unsteady state—fermentation is performed batch wise, where a tank of juice transforms into wine over time. Crushing, mashing, brewing, fermentation, additions, and tank emptying/filling are unsteady state processes. Interestingly, distilleries can operate as either steady or unsteady state processes, depending on equipment and scale.

We can analyze unsteady state systems by going back to our mass balance:

$$\text{Rate of accumulation} = \text{Rate in} - \text{Rate out}$$

For unsteady state systems, the left-hand side (the accumulation term) is non-zero, indicating that the system changes with time. We can illustrate the idea of unsteady state mass balances with the following two examples that parallel the examples in the previous section. Example 2.4 uses a total unsteady state mass balance and Example 2.5 includes a species-specific balance.

Example 2.3:   Clarified Juice Surge Tank

Surge tanks are often used when process inlet and outlet flow rates change in time, in order to provide a "buffer" or "surge" volume to keep the process running. You are a winemaker at a large commercial winery, which uses a wine flotation system to clarify incoming crushed and pressed juice. The wine flotation system must be fed a constant 40 gal/min to keep the unit running smoothly. Grape receiving and processing is at a lull, with only 10 gal/min on average of juice coming from the presses. Fortunately, there is a 4,000 gal working volume surge tank between the press and the float, and it is full.

How long until the surge tank is emptied and the floats need to be shut down?

SOLUTION

We start by determining the rate of volume change in the surge tank

$$\text{Rate of accumulation} = \text{Rate in} - \text{Rate out} = 10 \text{ gpm} - 50 \text{ gpm} = -40 \text{ gpm}$$

(Note that "gpm", gallons per minute, is often used in industry instead of "gal/min")

So, the tank is losing a net 40 gal/min. Another way to say this is the net flow rate out of the tank is 40 gall/min

We find the time to empty by dividing the current surge volume by the rate of juice loss:

$$\text{Total Volume}(V) = \text{Flow rate}(Q) \times t(\text{time})$$

$$V = Q \times t$$

$$t = V / Q$$

$$t = V / Q = 5,000 \text{ gal}/40(\text{gal/min}) = 100 \text{ min}$$

At the current rate of grape receiving, the floats can be run for 100 more minutes before they run out of juice feed.

Example 2.4: Inoculation of a Wine Tank

A 150,000 L working volume tank of Pinot Grigio juice is filled with 100,000 L juice, and needs to be inoculated with EC1118. The yeast is hydrated in an inoculum slurry of 20 g/L yeast in juice, and then metered into the fermentation tank at a rate of 10 L/min.

  (a) What is the mass flow rate of yeast into the tank?
  (b) Derive an expression for the concentration of yeast in the tank over time
  (c) If the target biomass concentration is 0.25 g/L, how long should the inoculum slurry pump be run? What volume of yeast slurry will be added?

SOLUTION

  (a) To get a mass flow rate, we multiply the volume flow rate by the concentration of yeast:10 (L/min) × 20 (g yeast/L) = **200 g yeast/min**
  (b) We need to first derive the mass balance of biomass on the fermentor:

$$\text{Rate of accumulation} = \text{Rate in} - \text{Rate out}$$

Since no biomass is leaving the fermentor, we can say that rate out = 0, so:

$$\text{Rate of yeast accumulation} = \text{Rate of yeast in} = 200 \text{ g yeast/min}$$

Concentration = mass/volume. Since the rate of mass addition is steady, the mass of yeast at any given time is the yeast mass flow rate × time:

$$\text{mass of yeast} = t \times 200 \left(g \text{ yeast} / \min \right)$$

$$\text{Concentration} = \text{mass of yeast} / \text{fermentor volume}$$

But, the fermentor volume is changing! We account for this by a balance on the total volume:

$$\text{Volume accumulation} = \text{Volume in} - \text{Volume out} = 10 \, L / \min$$

$$\text{Volume} = \text{Initial volume} + \text{rate} \times \text{time} = 100,000 L + t \times 10 L / \min$$

$$\text{Yeast concentration in tank} = \text{yeast mass} / \text{liquid volume}$$

$$\text{Yeast concentration in tank} = \left(t \times 200 \, g \text{ yeast} / \min \right) / \left(100,000 L + t \times 10 L / \min \right)$$

  (c) We can take the expression from part b, and solve for a yeast concentration in tank = 0.25 g/L:

$$0.25 g / L = \left(t \times 200 \, g \text{ yeast} / \min \right) / \left(100,000 L + t \times 10 L / \min \right)$$

$$t = 126.6 \min$$

The volume of yeast slurry added willbe = 126.6 min × 10 L / min = **1,266 L**

2.3 ENERGY BALANCES

An energy balance is directly analogous to a material balance. When employing an energy balance, we examine energy entering a system and the energy leaving a system:

$$\text{Energy accumulation} = \text{Energy in} - \text{Energy out} \qquad (2.9)$$

Energy balances can be used for any kind of energy—potential energy, kinetic energy, nuclear energy, thermal energy, chemical energy, etc. From a processing standpoint, we typically only care about thermal and chemical energy. For this reason, energy balances in the chemical processing industry (and by extension, in winemaking) are often referred to as *enthalpy balances* or *heat balances*. Energy balances are solved exactly as a mass balance, but instead of tracking pounds, grams, gallons, or liters, we track joules, calories, or BTUs (British Thermal Units) or their associated rates.

One thing that makes energy balances more complex than material balances is measurement. Mass and volume can be measured directly, but there is no instrument that measures "energy" or "heat." Instead, changes in temperature are used to infer changes in enthalpy by invoking "specific heat capacity," the measure of a material's energy change per unit temperature change.

2.3.1 SPECIFIC HEAT CAPACITY

Specific heat capacity is a measurement of a material's ability to "hold" heat and gives a direct proportionality between temperature change and enthalpy change. We define a material's specific heat capacity, C_p, as:

$$C_p = \Delta H / (\Delta T \times m) \qquad (2.10)$$

where ΔH is the change in heat, ΔT is the change in temperature, and m is the mass of the substance.

Water has a "high" specific heat capacity (1 calorie/g × °C, or 1 BTU/lb_m × °F), meaning that it can absorb a large amount of heat relative to the increase in temperature. Steel, for example, has a "low" specific heat (0.12 calorie/g × °C, or 1 BTU/lb_m × °F), meaning that it takes much less heat to raise the temperature of steel.

If we know a substance's specific heat capacity, and can measure the change in temperature, we can calculate the change in heat. This is what allows for the calculation of energy balances.

Example 2.5: Chilling a Bright Beer for Cold Stabilization

A tank with 40,000 gal of bright (yeast free) beer is at 75°F and needs to be chilled to 32°F for cold stabilization and filtering. Assume that the beer has the same density of water (8.3 lb_m/gal) and the specific heat capacity of water (1 BTU/lb_m × °F). How many BTUs of heat need to be removed from the beer?

SOLUTION

In order the solve for the heat removed, we need to rearrange the definition of specific heat capacity to solve for ΔH.

We also have to solve for the mass of beer:

$$m = V \times \rho$$

where V is volume and ρ is density. Then, we can calculate the necessary heat removal:

$$\Delta H = Cp \times \Delta T \times m$$
$$= 1(\text{BTU/lb}_m \times {}^\circ F) \times (75 - 32\ ^\circ F) \times (40,000\ \text{gal} \times 8.3\text{lb}_m/\text{gal})$$
$$= 14,276,000\ \text{BTU}.$$

2.3.2 LATENT HEAT

Changing the temperature of a substance requires adding or remove heat—this heat change is known as "sensible" heat. However, energy is also required to drive phase changes (i.e. boiling or freezing). A pot of boiling water, for example, does not warm up as it continues to be heated—it stays at 100°C. However, what does change is the phase of the water, from liquid to gas. The energy needed to drive this phase change is called "latent" heat. We define the latent heat of a substance as the amount of energy required to enact a phase change on a given amount of material. For example, the latent heat of evaporation of water at 100°C is 2,256.4 kJ/kg, or 40.65 kJ/mol, or 970 BTU/lb$_m$.

We can calculate the energy required to drive a phase change by:

$$\Delta H = \left(\Delta H_{\text{phase change}} \right) \times m \tag{2.11}$$

where "m" is the mass of material undergoing a phase change.

Example 2.6:   Boiling a Pot of Water

How much energy is required to boil 2 gal of liquid water into vapor? Water has a density of 8.3 lb$_m$/gal.

SOLUTION

$$m = V \times \rho = 2\ \text{gal} \times 8.3\ \text{lb}_m/\text{gal} = 16.6\ \text{lb}_m$$

$$\Delta H = \left(\Delta H_{\text{vaporization}} \right) \times m = (970\ \text{BTU/lb}_m) \times (16.6\ \text{lb}_m) = 16,102\ \text{BTU}$$

It is interesting to note in this example that we can calculate the amount of heat necessary to boil of the water. However, it may be even more relevant to consider if we would like to boil this water in 1 min, 1 hr, or 1 day. In any case, the total heat supplied to the water would be the same, but the power needed would be significantly different. Power is energy per time, measured in units like Watts (W, equal to J/s) or Btu/hr, and will be important when discussing heat transfer or sizing utilities in latter chapters.

2.4 SANITARY DESIGN FUNDAMENTALS

The idea of "sanitary design" is common in the food and beverage industry, as well as the pharmaceutical and bioprocess industries. While the level and implementation may vary among industries, the general idea remains the same. In sanitary design, equipment and systems are designed to be "cleanable," so that dirt and potential chemical or microbial contaminants will not be harbored or released by the equipment into the intermediate or final product.

For the wine industry, the idea of sanitary design is important for two main reasons. First, we want to be able to control what organisms grow in the product during fermentation and after fermentation (for stability). Even in the case of "uninoculated" or "natural" fermentations, we want to be able to control the appearance or accumulation of "undesirable" organisms that may enter the fermentation environment. Second, we do not want to introduce other contaminants into the process stream that will affect the ease of downstream processing or final product quality. Examples of this type of contaminant to be avoided would be large clumps of dirt or organisms that break away from the equipment after initial entrapment, paint chips, small pieces of rubber, grease, or rust. This is true of brewery operations as well.

The two main means of accomplishing sanitary design in a winery or brewery setting are by choice of "materials of construction" and how we use these materials—that is, by design of fittings, equipment, and piping configuration. It is possible, though obviously undesirable, to use acceptable materials of construction and end up with an unsanitary piece of equipment because of poor design. Materials of construction and designing for cleanability are the topics that we will focus on for this section.

2.4.1 MATERIALS OF CONSTRUCTION

The characteristics of a "good" material for sanitary design are that they are inherently cleanable, corrosion-resistant, and contain no "leachables." Inherently cleanable materials are ones that have smooth surfaces and are non-porous. Corrosion resistance is important for continued strength of equipment that is in contact with process fluids that may react over time with the materials of construction. It is also undesirable to have corrosion products, like rust on metal, to be in contact with or fall into the product stream. Therefore, material compatibility with your process stream is of major importance. Finally, materials need to be chosen that do not leach into the process stream over time. For example, some plastics may release plasticizers over time. Brass (an alloy—or blend—of mostly copper and zinc) has also been known to leach very small amounts of its component materials into gas or liquids that flow past it. Materials that fit all of these desirable characteristics include glass and stainless steel, as well as various polymer materials such as EPDM (ethylene propylene diene monomer) and PTFE (i.e. teflon, polytetrafluoroethylene). Other types of materials might also be suitable but would depend on the application. If materials get irreversibly stained, they may be too porous for sanitary use. Material like oak would not fit into the definition of a sanitary material, as one of the uses of oak is to leach flavor into a wine or spirit. This does NOT, however, mean that these materials cannot be used for processing. It simply means that special handling or cleaning regimens may have to be incorporated into normal processing, or equipment with these materials may have to be treated as a consumable.

2.4.2 STAINLESS STEEL

As much of the equipment in a winery, brewery, or distillery is fabricated from stainless steel, we will cover this material in greater detail. Stainless steel is basically defined as iron with at least 10.5% chromium added. Oxides of chromium on the surface of the metal are what give the "passive" corrosion-resistant layer. As chromium content increases, corrosion decreases, at least up to about 16% (Figure 2.1). Various types of stainless steel can be found in Figure 2.2. This includes austenitic, ferritic, and martensitic forms of stainless steel, corresponding to the predominant crystal structure of the metal atoms.

Austenitic stainless steel is typically the most desirable for winery or brewery purposes as it is able to be formed at room temperature. Almost all tanks and equipment

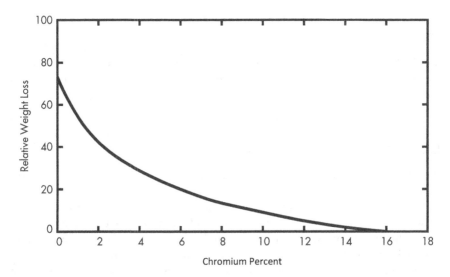

FIGURE 2.1 Corrosion resistance as a function of chromium content. Corrosion resistance increases dramatically for stainless steel with the concentration of chromium in the alloy, at least up to about 16% chromium. A minimum of 10.5% is needed to classify the alloy as stainless steel. (M. Ogawa after Covert and Tuthill 2000).

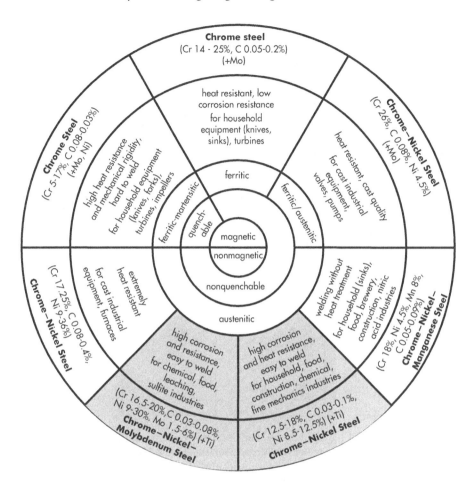

FIGURE 2.2   Types and uses of common types of stainless steel. The austenitic varieties at the bottom of the figure are the ones most typically used in wineries and other sanitary applications such as in the brewing, dairy, biotech, and the microelectronics industries. (M. Ogawa after Lydersen et al. 1994, *Bioprocess Engineering—Systems, Equipment and Facilities*).

for the wine and brewing industries are constructed from 304 or 316 stainless (304L or 316L stainless may also be used). 304 and 304L SS are the most common stainless steels used in the wine industry. The composition of these steels can be found in Figure 2.3. A closer inspection of the composition will show four main components for 304 or 304L stainless. Iron is present for strength. Chromium is present for corrosion-resistance. Nickel allows ductility at room temperature, in addition to corrosion resistance. Increased ductility is achieved because the addition of nickel allows transition from ferritic to austenitic at significantly lower temperatures (blacksmiths, on the other hand, have to do this by heating their steel to extremely high temperatures and then form the steel before it cools back down) as shown in Figure 2.4.

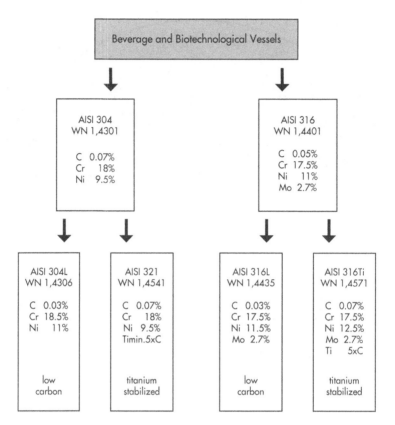

FIGURE 2.3 304 and 316 stainless steel families. These two types of stainless steel and their corresponding low carbon variations (304L and 316L stainless) are commonly used in sanitary industries. 304L is commonly used in wineries. (M. Ogawa after Lydersen et al. 1994, *Bioprocess Engineering—Systems, Equipment and Facilities*).

304L and 316L are typically used where weld strength and corrosion-resistance are critical, as they contain a lower (i.e. L) percentage of carbon. At elevated temperatures during the welding process, formation of chromium carbide can pull chromium out of the structure of the steel, thus allowing a greater probability of later corrosion or weakness. By lowering the carbon level, this problem is lessened. Addition of molybdenum in 316 stainless gives additional pitting resistance.

Once a material is chosen, interior and exterior finishes must also be specified. It is common in the food industry to have both interior and exterior finishes be "2B," which is basically the finish that results from the milling of the stainless steel with no extra polishing. Occasionally, a smoother finish will be applied to the internal surface by grinding or polishing. This type of finish can be specified in "Grit" or "Ra." Using Ra, or arithmetic roughness as measured by an instrument called a profilometer, is generally preferable to using the traditional "grit" measurement. This is because Ra is an actual measurement of surface roughness, while grit is a measurement of the particle density on the polishing surface (similar to using a sandpaper). As the

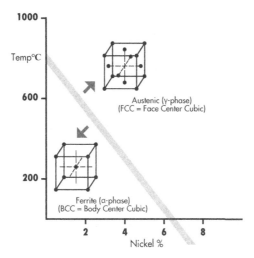

FIGURE 2.4 The effect of nickel content on stainless steel chemistry and formability. As nickel content increases, the transition of the steel from ferritic to austenitic occurs at increasingly lower temperatures. With the nickel content of 304 SS, the metal can be formed at room temperature. (M. Ogawa after Covert and Tuthill 2000).

grinding surface can change with use, a maximum roughness specification is more likely to get a consistent, desirable finish. Ra is calculated from a profilometer trace by breaking the surface roughness scan into a number of small sections and finding the average distance from the mean line, as seen in Figure 2.5.

$$R_a = \frac{\sum_{i=1}^{z} |d_i|}{z} \qquad (2.12)$$

Typical units of Ra are µin (micro-inch) in the US and are sometimes not stated explicitly in literature from manufacturers and fabricators. It is typical to have an Ra of 20–40 µin on winery equipment. While other sanitary industries usually specify maximum Ra's for their stainless steel equipment based on practice or regulation or size of expected microbial contaminant, it is less common in the wine industry to do so. Having said that, an Ra of 30 µin would correspond to microscopic pockets of approximately 0.8 µm, which is slightly smaller than most bacteria in a winery environment, thus assuring a higher level of cleanability.

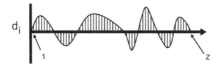

FIGURE 2.5 Trace of deviation from average surface level to determine relative roughness as might be found with a profilometer.

Welding is typically used to fabricate stainless steel equipment, basically for permanently joining two pieces of metal. Welding procedures can be roughly categorized as hand versus machine welded and consumable versus non-consumable electrodes. In the general case, a weld is created by a series of arc strikes from the electrode across a gap (like a lightning strike) that creates enough energy to melt the surrounding metal. When the metal cools, a strong bond between adjoining pieces is formed. Good welds are flush or slightly convex on both sides, uniform, and fill the whole gap between the metal pieces being joined with no mismatch. Characteristics of good and bad welds can be found in Figure 2.6. Most welds in the wine industry are performed using TIG (Tungsten Inert Gas) welding or GTAW (Gas Tungsten Arc Welding), which uses a non-consumable tungsten electrode to create the arc across the small gap to create the energy necessary to melt the two adjoining pieces of metal. While a skilled welder can accomplish desirable welds by hand, an automated process known as orbital or machine welding achieves greater consistency. This type of welding becomes even more desirable given the rarity of examining the inner diameter of welds in the food industry. Orbital TIG welding, especially of pipes, can now be achieved using relatively small and flexible equipment. Tanks used as fermentors are frequently hand welded and the welds are not ground smooth. This can lead to rough spots on the interior of tanks that create cleaning challenges. While finishing the inside of a tank to a higher finish (smaller Ra) will incur extra cost, this may be balanced by the relative ease of cleaning presented by a smooth surface.

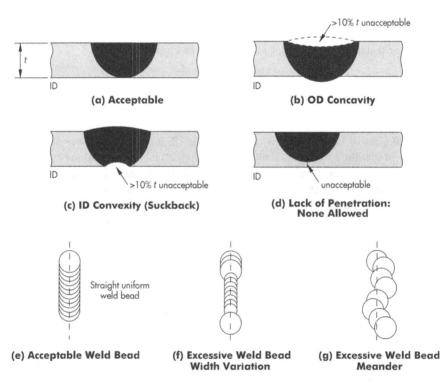

FIGURE 2.6   Characteristics of good and bad welds. (M. Ogawa after BPE-2019 (2019)).

Passivation may be carried out after fabrication. Passivation procedures typically involve contact of the inner surfaces of the equipment with a citric or nitric acid solution. This process is supposed to restore the passive chromium oxide layer to the surface of the metal. The worst-case scenario is that a passivation will thoroughly clean the equipment of any dirt or contaminants left over from fabrication.

Some common problems with stainless steel include heat tint around welds and rouging. Heat tint, dark bands on either side of a weld, is caused by a thicker than normal layer of chromium oxide. It is caused by exposure of the metal to oxygen at an elevated temperature, usually just after a weld if the inert gas purge is not maintained during the cooling process. Drawing the chromium oxide to the surface can weaken the inert nature of the material and promote corrosion. Rouging is essentially rusting of the metal, or a deposit of iron oxide on the surface of the metal. Excessive rouging may be an indication that a more inert material choice is warranted, such as a higher grade of stainless steel. Ultra-pure water systems can sometimes have issues with rouging. It should be remembered that stainless steel is not compatible with chloride (e.g. HCl and bleach), as it may leave the stainless steel susceptible to chloride attack and stress corrosion cracking—especially in areas of the equipment under mechanical stress.

Any choice of material and finish for a piece of equipment will have implications not only the performance of the equipment, but also on the cost. In general, higher performance steels and nickel alloys can be multiple times the cost to procure and fabricate versus 304L stainless.

2.4.3 Design of Fittings and Piping Configurations to Maintain Cleanability

The second major influence on "cleanability" is the design of fittings and piping configurations. It is certainly possible (though highly undesirable) to use inherently cleanable materials to fabricate an uncleanable piece of equipment. Certain designs for valves, fittings, flanges, pumps, etc., are inherently more cleanable than others. The best way to decide whether something is cleanable is to look at or actually touch the equipment and look for crevices that would be difficult to clean. For a small fitting, running your fingers over the interior surface is a good way of deciding on the cleanability of the design. For larger equipment, attention to detail and experience may be the best guide, though focusing on how you will clean the equipment (in addition to how you will use it) is also a good idea. More detail on specific types of fittings is given in a latter section. However, there are some good rules of thumb for deciding whether a fitting or piece of equipment is cleanable. First, threads for connecting two pieces of equipment are never cleanable. It is nearly impossible to keep threads clean. If two pieces of equipment are to be joined in a non-permanent manner (e.g. they are not welded), joining the pieces through a sanitary flange (e.g. Tri-clamp fitting) is the method of choice. Fittings and equipment should be designed to eliminate any places where liquid will pool or accumulate, as these pools will promote microbial growth and make cleaning more difficult.

For piping systems, sloping of lines is important for drainability. Lines that do not drain well will allow growth of organisms. A slope of 0.3–0.5 ft drop per 100 ft drop is usually sufficient. All pipes should slope toward low point drains or drain valves. This

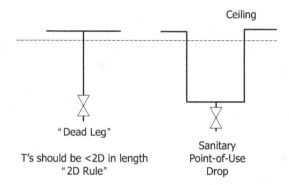

FIGURE 2.7   Schematic of a dead-leg (non-sanitary) drop versus a sanitary u-bend drop.

seems straightforward, but for complicated piping systems or skids (self-contained unit operations), complete draining of the system may be elusive as correct placement of low-point drains may be non-obvious or difficult. Deadlegs that may fill up with stagnant process fluid should also be avoided. Usually these deadlegs are formed as a "T" off of delivery manifold, for instance, delivering water or must or wort to one out of a number of tanks. Use of piping Ts should therefore be minimized and should not be more than two pipe diameters in length whenever possible. Longer deadlegs should come with specific cleaning protocols to address this issue. These longer, problematic deadlegs can be avoided by using point-of-use drops, or u-bends, instead of a straight T. In this way, the bottom of the u-bend can be at the point of delivery and the delivery valve can be a very short T (less than two pipe diameters) or zero static (actually built into the u-bend itself). Therefore, the bottom of the u-bend can be at the point of delivery and the delivery valve can be a very short T (less than two pipe diameters) or zero static (actually built into the u-bend itself). This is illustrated in Figure 2.7.

Similarly, any probes or sensors inserted through the side of a tank like a fermentor need to be mounted at angle downward into the tank so that any leftover moisture will drain off of the probe into the bottom of the tank and out a low point drain.

REFERENCES

Bioprocessing Equipment Standard (BPE-2019). American Society of Mechanical Engineers (ASME), New York, 2019.

Covert, R.A., and Tuthill, A.H. Stainless Steels: An Introduction to their Metallurgy and Corrosion Resistance. *Dairy, Food and Environmental Sanitation*, 20, 506–517, 2000.

Himmelblau, D.M., and Riggs, J.B. *Basic Priniciples and Calculations in Chemical Engineering* (8th Edition). Prentice Hall, Upper Saddle River, NJ, 2012.

Lydersen, B.K., D'Elia, N.A., and Nelson, K.L. (Eds.). *Bioprocess Engineering: Systems, Equipment and Facilities*. John Wiley & Sons, New York, 1994.

PROBLEMS

1. Just after graduating from UC Davis, you find yourself working for the relatively new winery, Puddle Bottom Cellars, in Sonoma down the road from Weeping Mountain. Up until this point, the winery has been paying to custom crush their grapes. Now they are building their own small winery including a

bank of 1,300 gal fermentors that you are asked to help design/specify. Based on the new fermentors at UC Davis, you decide to specify tanks that can be run in a completely automated mode—with a dedicated pumpover pump and sensors for temperature, Brix, volume, and pump flow rate. The bottom of the tank is supposed to slope to the front tank drain in order for the tank to be free draining and to facilitate skin removal out the manway, also at the front of the tank. Unfortunately, you are not given a lot of money to purchase these tanks and the job to fabricate them goes to the low bidder.

(a) When the tanks arrive you start noticing problems. First, the inside surface of the tanks does not look smooth. You borrow a profilometer and get the following data. Calculate the Ra. Is this acceptable for cleanability? Explain your answer.

$$1\,\mu in = 0.0254\,\mu m$$

Segment #	Distance (micro inch)
1	-52.5
2	-15
3	21
4	0
5	78
6	36
7	70.5
8	-7.5
9	-3
10	-81
11	18
12	60
13	-33
14	21
15	31.5
16	-69

Profilometer Trace

(b) After cleaning the tanks for the first time, you notice that there is still standing water in them at the end of cleaning. The drain valve is a diaphragm valve, and at first, you think the pooling of water is because the diaphragm valve has been installed incorrectly. Give an explanation for why someone might think that this is the problem. How would you know whether the drain valves have been installed correctly?

(c) Upon further inspection, you decide that the drain valves have been installed correctly. However, you discover that the bottom of the tank is not well-supported and is actually sagging in the middle (see figure below). How would you suggest fixing the drainage problem to allow better manual or automated cleaning? Give two possibilities and explain your answers, along with any issues that might arise from your solutions. Remember, it is not easy to bend stainless steel plate after it is formed without damaging it.

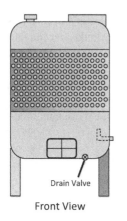

Drain Valve

Front View

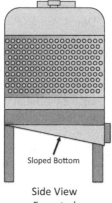

Sloped Bottom

Side View
Expected

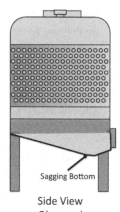

Sagging Bottom

Side View
Observed

2. After completing the design for the Breakfast Winery and going through two harvests there, you decide to travel around the world and get new ideas for winery design. Below are three examples of what you find. For each of them, briefly assess the "cleanability" of the design.

Concrete Egg-Shaped
Fermentor

Concrete Open-Top
Fermentor

Stainless Steel Rectangular
Fermentor

3 Fluid Flow

Fluid flow in wineries, breweries, and distilleries follows the same fundamental physics and engineering analysis as fluid flow anywhere, with the added caveat that the principles of sanitary design must be followed at all times. A notable exception is post-distillation high proof transfer lines. High-proof solutions are considered food from a materials perspective but are not typically at risk of biological contamination. They are, however, a fire/explosion risk and must be engineered accordingly.

3.1 STANDARD VALVES, FITTINGS, AND PUMPS

The presence of skins, seeds, some stem fragments, grain, trub, hop cones, and other suspended solids requires special consideration for the type of valves, fittings, and pumps that are chosen for winery and brewery applications. A further consideration is the ease of cleaning for each of these system components. This includes the materials of construction, finishes, and design of each of these components as described in Section 2.4.

3.1.1 VALVES

The use of valves can range from simply open and shut positions, in which the nature of the flow channel is not of particular importance, to flow control valves in which there is a relationship between the valve stem position and the resulting flow rate. In other words, the two main functions of valves are "blocking" and "controlling." A common feature of some tank designs is the use of flange and clamps fittings that allow the valve to be removed for cleaning (this is especially common in small wineries and breweries). However, some valves will be welded in place, and therefore, will require more stringent attention to cleanability.

3.1.1.1 Diaphragm Valves

These valves (Figure 3.1) are mostly used to control the flow of water or air rather than juice, wort, beer, or wine, though this may have more to do with tradition than for other technical reasons. In other industries such as the pharmaceutical industry, these valves are considered to be the **most** sanitary design. If they are installed at the correct angle (usually around 45° from the upright position), they are free draining around the weir and contain no pockets or crevices for liquid or solids to accumulate. They are distinctive since the base of the pipe in the valve body arches up to about half the pipe diameter and a flexible diaphragm, mounted in the roof of the valve, is lowered down to meet this raised base or weir. This restriction of the cross section can be varied by the height of the valve stem, either manually with a hand wheel or in an automated manner by air pressure controlled by some type of control system/

DOI: 10.1201/9781003097495-3

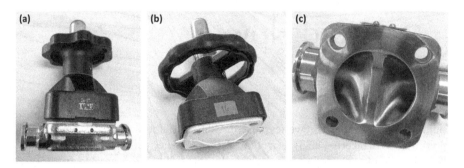

FIGURE 3.1 Sanitary diaphragm valve. A complete valve is shown in (a), with an example of the topworks shown in (b) and the weir shown in (c). Diaphragm vales are by far the most cleanable design and are commonly used in sanitary and even aseptic applications. In order to be cleanable, it must be mounted at the correct angle to allow drainage around the weir.

solenoid bank. Diaphragms are made of several materials including EPDM and EPDM-backed Teflon or Buna (synthetic) rubber. The potential disadvantages of a diaphragm valve are that they require multiple turns for opening or closing (not an issue for an automated valve) and that they require periodic replacement of the diaphragms that will wear out from age and use. However, because of the construction of these valves with a separate base and bonnet, inline maintenance is possible, even when welded into place.

3.1.1.2 Butterfly Valves

Butterfly valves (Figure 3.2) are constructed using a round stainless steel disk the diameter of the surrounding pipe that turns using connections at the top and bottom of the disk (using a handle attached to the top connection). When fully open, the disk is seen to be side-on and simply a thin divider from the top to the bottom of the pipe. When fully closed (using a quarter turn of the handle), the disk is seen face-on, and it completely blocks the pipe. The disadvantages of these valves are that solid fragments can get caught on the disk, even with its thin profile in the open position and plug the pipe. The disk can also catch solids around the seal when put into the closed position, resulting in a leak because of the lack of a complete seal. They are best used

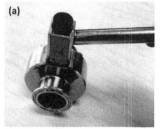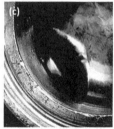

FIGURE 3.2 Butterfly valve. A complete valve is shown in (a), with the disk that stops or controls flow shown in (b). A close-up photo of the where the disk meets the body of the valve is shown in (c). This point of contact is the most difficult part of the valve to clean.

for low solids flows and can be used for controlling the flow rate in a simple way during filtrations or other operations. Because flow will tend to push the disk into the closed position, these valves typically have a mechanical mechanism in which they can lock into several intermediate positions in between fully open and fully closed. Butterfly valves are generally cleanable except for the two places where the disk attaches to seal/valve body (where particles and/or microbes can be trapped). Therefore, they are considered cleanable for the food and beverage industries, but not for pharmaceutical applications. However, special attention needs to be given to how and when these difficult-to-clean spots will be cleaned.

3.1.1.3   Ball Valves

Ball valves (Figure 3.3) can be described as a cube through which the pipe passes from one face to the other. In the center of the cube is a sphere (or ball) through which a hole, the diameter of the pipe, has been created. When this hole is aligned with the pipe, the flow is unobstructed and when it is turned through 90°, the pipe cross section is completely closed and the outer part of the ball is exposed to the process stream or environment. The quarter-turn operation in either direction makes these valves well suited for manual operations where rapid operator action is necessary based on oral or visual cues (e.g. someone seeing or being told to close a valve when a change in color occurs in a process stream). These valves can be used for all lines that will handle mash, musts, and pump-over operations since they are not easily blocked by solids when fully open. An additional feature of ball valves is that they permit the introduction of a mixing impeller through the fully open position even though the tank contains juice or wine. This allows the mixer to be used on several tanks rather than being dedicated to only one (see mixers in a later section). Ball valves are usually low maintenance (the handles are usually the first thing to fail). Ball valves are generally cleanable except for the seat around the ball. This, however, can be problematic, as juice or wine stuck on the outside of the ball can promote microbial growth on the inner parts of the seat and will not be cleaned unless the valve is disassembled, which is not typical at peak operations times. Ball valves are widely used for the food and beverage industries, but not for pharmaceutical applications, because of these cleanability issues. However, special attention needs to be given to how and when the outer part of the ball will be cleaned to reduce issues stemming from their use.

FIGURE 3.3   Ball valve. A complete valve is shown in (a), with a partially open ball shown in (b). When the valve is wide-open (c), the path through the valve is fairly smooth and cleanable. Cleaning the outside of the ball thoroughly is only possibly by disassembling the valve.

3.1.1.4   Gate and Guillotine Valves

Gate valves (Figure 3.4) work by having a wedge-shaped valve stem lowered into a seat when the handwheel is turned. These types of valves are used quite often for garden hoses on houses in the US. The design of the seat in these valves creates a space that cannot easily be drained at any mounting angle. In the wine industry, these types of valves are sometimes used on utility lines (like water), though they are not used extensively for process streams due to issues with cleanability. Guillotine valves, similar to gate valves but generally used for larger applications, are sometimes used for delivery of must to large tanks or to release solids at the outlets of drainers. They are generally equipped with pneumatically or hydraulically operated actuators. However, the recessed seat necessary (see Figure 3.5) for these applications makes it less than ideal from a sanitary design point of view. For newer versions of retort, conical bottom tanks used for red wine fermentations, large guillotine valves are often used at the tank bottom to rapidly release skins by gravity into a waiting press or elevator below. Because of the desirability of this type of fast and safe cap removal, extra cleaning vigilance or attention to design of the guillotine valve may make this a desirable configuration, especially since this action will only be at the end of a fermentation when cleaning is likely to occur at the same time as the closure of the valve.

3.1.1.5   Mix Proof Valves

Mix proof valves (Figure 3.6) are a common type of valve found in sanitary beverage processing, among other applications. These valves use bellows that inflate into a seat when pressurized, closing off flow. Mix proof valves do not come in manual versions, but do come in many automated configurations to allow, for instance, cleaning past a valve when the product stream is not in a pipe. These valves are very popular in breweries for controlling fluid flow to unit operations such as packaging lines or between fermentor manifolds, as well as for complex multi-product manifolds.

(a)

(b)

FIGURE 3.4   Gate valve. A gate valve as pictured in (a) is common on water supplies. The seat for the gate (b) is not free draining in any orientation.

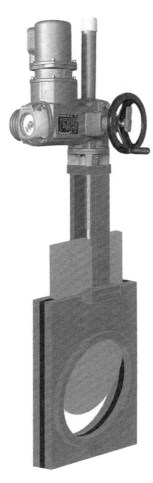

FIGURE 3.5 Guillotine valve. Note the combination of electric motor and manual hand crank on top to actuate the valve. By HerbstrittM: Own work, CC BY-SA 3.0, https://commons.wikimedia.org/w/index.php?curid=1693320.

3.1.1.6 Check Valves

Check valves are used when unidirectional flow is required. That is, check valves only allow flow in one direction. There are several types of check valves available. Some check valves use a swinging gate that only swings in one direction. Others, as in diaphragm pumps discussed below, use a ball in a cage that floats back into the seat when flow is reversed. Still others use a spring-loaded stem as in Figure 3.7. In this type of check valve, flow in the forward direction compresses the spring and allows flow. Flow in the opposite direction forces the stem back into the closed position. Due to their one-way nature, check valves can be difficult to clean, some designs more so than others.

FIGURE 3.6   A mix-proof valve. Photo courtesy of GEA.

3.1.1.7   Automated Valves, Limit Switches, and Failure States

Many valves in the food industry are currently designed for manual operation. In other words, to operate a valve, an operator goes up to a valve and turns a handle. Interestingly, large breweries have embraced automation for some decades. Automation in wineries is progressing rapidly and some wineries have significantly more automation and automated valves than others. This trend is likely to continue. Automated valves are operated through a computer screen or through an automation sequence controlled by a programmable logic controller (PLC) or other types of controllers. The actual opening or closing of the valves is typically initiated by an electrical signal to a solenoid valve that allows compressed air into the valve actuator to change its state. Many times, valves can be specified that are either air-to-open (fail close) or air-to-close (fail open), depending on the application. In the case of automated valves, it is always a good idea for an operator, brewer, or winemaker to be able to tell if the valve is open or closed. This confirmation can take various forms. First, the valve, itself, may have a light or switch on its body that changes position or

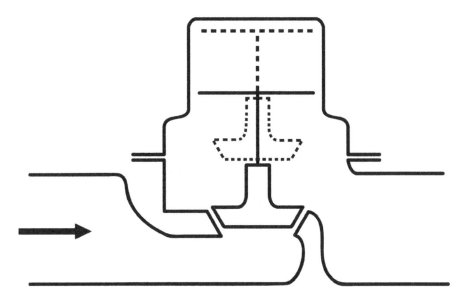

FIGURE 3.7   One type of check valve for maintaining unidirectional flow. (M. Ogawa after Perry's Handbook).

color to indicate the state of the valve (i.e. open/close). Second, a computer operator interface may have a graphic interface that changes color based on the expected state of the valve. Many times, the state shown on an operator interface is the one in which the controller/computer thinks the valve should be. However, what if the valve gets stuck and does not actually move? For this reason, automated valves will many times have a mechanical switch—a limit switch—that actually moves with the valve and gives a mechanical or electrical signal indicating the true state of the valve.

3.1.2   FITTINGS

Most of the pipe fittings that are used in wineries are detachable, and are connected to hoses, fixed pipes, or tanks with clamped fittings (also known as sanitary flanges). Breweries also use sanitary fittings, but breweries (especially larger breweries) are often "hard-piped," where all connections are welded in, avoiding the use of detachable fittings and hoses due to contamination fears. These fittings include bends or elbows, Y's and T's to provide a change in flow direction or the mixing of two streams. Reducers are used to connect pipes or hoses of different diameters. While various styles of flanges/clamps are available, the industry standard is a "tri-clamp" or "tri-clover" fitting. This type of sanitary flange uses an O-ring that is specially formed (i.e. inner and outer diameter thickness) to flatten out when clamped to form a uniformly thick junction with a smooth inner diameter. Tri-clover fittings are extremely common in the food and bioprocess industries (Figure 3.8).

It bears repeating here that threaded fittings should, in general, be avoided, especially where they may be in contact with product, as threads are difficult, if not impossible, to clean thoroughly.

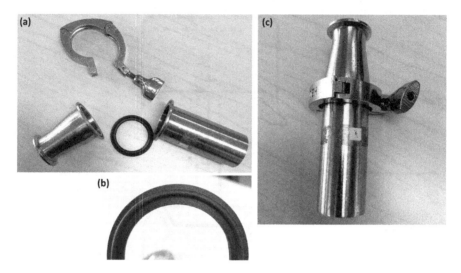

FIGURE 3.8 Sanitary flange. Sanitary flanges are an excellent option when welded connections are not possible. The components of the flange are shown in (a) disassembled, with a close up of the o-ring/gasket in (b). The o-ring/gasket is designed to flatten out and give an even thickness and good seal when assembled and tightened as shown in (c).

3.1.3 PUMPS

Pumps are used to transfer liquids or slurries in wineries and breweries during the course of operations. These fluids can include water, juice, sugar solutions, cleaning solutions, wort, must, wine, beer, and even spent grains or pomace. Pumps that work for one type of liquid, will generally work for all types of liquids. However, not all pumps will work well with suspended solids or slurries, especially when it is important to avoid heavy shearing or maceration of the solids. Therefore, it makes sense that finding a universal pump is not an easy task. This is evidenced by the wide variety of pumps found in winery and brewery settings.

There are two classes of pumps, centrifugal and positive displacement, that work on fundamentally different principles. These will be discussed separately below.

3.1.3.1 Centrifugal Pumps

Centrifugal pumps (Figure 3.9) use the vanes on a rotating impeller to impart energy to the fluid to get it to flow (an analogy would be like hitting the fluid in a bathtub with your hand to push it along). These pumps are designed to take in fluid through the circular pump casing at the center of the impeller, where the fluid is then hit by the rapidly rotating vanes and directed toward the discharge which is tangent to the circular head. Because of their design, centrifugal pumps are not generally self-priming—the operator needs to be able to get the fluid to the pump before it will start pumping. If flow is completely cut off downstream of the outlet (called "dead heading" the pump, for instance by closing a valve), liquid in the pump head will remain there as the impeller continues to turn (very much like a blender!). Dead heading a centrifugal pump will cause the liquid in the pump head to be beat up considerably, but it will not

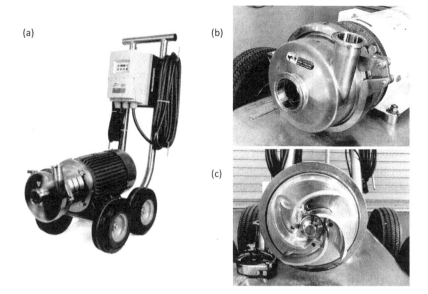

FIGURE 3.9 Centrifugal pump. (a) Example of a centrifugal pump, a TCW Dynahead 216 Centrifugal Pump. Note the attached cart and variable frequency drive, common in winery applications. Photo credit TCW Equipment Company. (b) Detail of the pump head on a centrifugal pump with the inlet at the center and outlet tangent to the pump head. (c) The pump head without the pump casing shows details of the impeller blades, which turn counter-clockwise for this pump.

cause a dangerous elevation in system pressure because there is not a tight tolerance between the tip of the impeller and the inside of the pump casing. Because there is not a need for tight tolerances between metal parts, centrifugal pumps tend to be less expensive than other pumps of similar capacity. If the flow is restricted in any way to the inlet of a centrifugal pump, cavitation can result (to be discussed later) which can be damaging to the pump head and impeller. Cavitation can be a major issue in distilleries, due to the dangers of vaporized ethanol. These pumps generally maintain a smooth flow rate over time, free of oscillations. However, they can only be used in relatively low-pressure applications (which applies to more or less all beverage applications), they create high shear at the impeller tips, and they are not self-priming. Flow rate is also dependent on the pressure downstream of the pump.

There are many manufacturers of centrifugal pumps and designs capable of producing flow rates at a range of mechanical heads. Later in this chapter, we will discuss the process of calculating the required information for pump selection.

3.1.3.2 Positive Displacement Pumps

Positive displacement pumps operate using a completely different mechanism of action from centrifugal pumps. In all positive displacement pumps, a fixed amount of fluid is pulled into the pump with each stroke through creation of a vacuum and then pushed out on the outlet or discharge side (an analogy for these pumps would be sucking up liquid with straw from one side of the bathtub and then spitting it out on the other side). Because they are creating a vacuum at their inlet, positive displacement pumps are self-priming and will draw the liquid to the pump. Due to this mechanism of action, they have a throughput which is directly proportional to their rotational speed (every stroke moves exactly the same amount of fluid). For this reason, the pump flow rate can be varied simply by increasing the rotational speed, assuming the pump is equipped with a variable speed drive. The consistent volume pumped in each rotation also lends itself to metering, such as might be necessary in a winery or brewery for nutrient additions to fermentations or automated preparation of cleaning solutions to name two such uses.

If a positive displacement pump is completely dead headed, pressure will rapidly continue to build. This is because these pumps will continue to pump the same volume with every stroke even if flow downstream has ceased. This can cause severe safety issues that will occur quite quickly. Therefore, it is imperative when using a positive displacement pump to install it with a pressure sensor and/or rupture disk downstream for safety purposes. The pressure sensor can be used to automatically turn off the pump in overpressure situations for automated systems and the rupture disk can safely release pressure when needed at a predetermined pressure and temperature. The advantage of a positive displacement pump is that flow will be constant over time on average, even if this system changes downstream of the pump over time (e.g. a filter that begins to foul). Some positive displacement pumps, by nature of their design, will, however, have pulsatile flow (i.e. a sinusoidal flow rate around a mean).

Unlike centrifugal pumps where there is little variation in style between different manufacturers, positive displacement pumps come in many variations, although all of them use the same mechanism of action. Some of the more commonly used types of positive displacement pumps are described below.

3.1.3.2.1 Rotary Lobe and Circumferential Piston Pumps

A rotary lobe pump is shown in Figure 3.10. It can be seen in the diagram in Figure 3.10 that this pump has two lobes each with three "points" that turn in opposite directions away from the inlet and have a tight tolerance between the end of the lobe and the pump housing. When the lobes turn, the space at the inlet goes from small to large, thus creating a vacuum that draws fluid into the pump. The fluid then moves

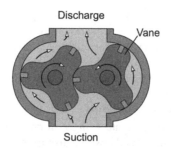

FIGURE 3.10 A rotary lobe pump. The movement of the cavity makes it both low shear and self priming. By Jahobr: Own work, CC0, https://commons.wikimedia.org/w/index.php?curid=41822700.

around the outside of the lobe with the rotation and is forced out on the outlet side of the pump when the space goes from large to small. An equal amount of fluid will be carried through the pump on each rotation. While this can be useful for metering, a closed valve or blocked filter downstream of the pump will cause a rapid increase in pressure because the pump will still be pushing through the same volume of fluid with each stroke. As stated above, this is why pressure gauges and pressure relief devices should always be used downstream of a positive displacement pump. It is better to have a controlled release of pressure than a dangerous, unpredictable one.

A pump that is similar to the rotary lobe pump is the circumferential piston pump. The main difference between these two pumps is the shape of the lobes/vanes. This type of pump is extremely popular in wineries and is sometimes called a "Waukesha" after one of the original suppliers of this type of pump. This type of pump works best with no solids or a small amount of soft, compressible solids.

3.1.3.2.2 Diaphragm Pumps

Diaphragm pumps (Figure 3.11) have two circular diaphragms, usually constructed of a sturdy synthetic rubber, that oscillate back and forth within a sealed housing and generally operate on compressed air instead of electricity like most other pumps. The two diaphragms are attached to one piston whose direction is also controlled by the compressed air and an attached mechanical mechanism. As one diaphragm is being drawn back, a vacuum is created that draws fluid in through a check valve. As the piston pushes the diaphragm back, the fluid is pushed through a second check valve to the pump discharge. Because the piston is attached to both diaphragms, as the pump is drawing fluid in to one side of the pump it is simultaneously pushing fluid out of the other side of the pump with all flow controlled by check valves. These

(a)   (b)   (c)

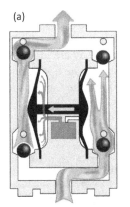  

FIGURE 3.11 Diaphragm pump. (a) Schematic showing how ball check valves allow one side to pull liquid and prime, while liquid on the other side is expelled downstream. Both cavities have the central diaphragm as one wall, which reciprocates to drive fluid. By There's No Time – Own work, CC BY-SA 4.0, https://commons.wikimedia.org/w/index. php?curid=48661796. (b) Diaphragm pump mounted on a cart. (c) Cutaway model diaphragm pump shows the ball check valves and diaphragm, along with the mechanism that converts the air pressure to the back and forth motion of the piston.

pumps have speed control by regulation of the compressed air flow that provides their power. Because they require the compressed air source, they work well for uses where they do not have to move around the winery or brewery frequently. For wineries, they can be found as dedicated, fixed-in-place pumpover pumps, as an example. Solids only interfere with the pumping action when they disrupt the check valves. This type of positive displacement pump can only build up pressure to that of the compressed air source used to drive it, thus supplying a built-in safeguard against system over-pressurization.

3.1.3.2.3 Progressive Cavity Pumps

A progressive cavity pump (Figure 3.12) uses a stainless steel rotating rotor inside a stationary helical stator, usually constructed of a softer, elastomeric material, to provide tight tolerance between the two surfaces. After the fluid is drawn in by an increasing gap between the rotor and stator at the inlet, there is a helical volume which travels from the inlet to the outlet as the rotor rotates. The fluid is released at the discharge side of the pump as the volume of the gap decreases. This pumping action results in very little shear stress on the fluid being pumped, and progressive cavity pumps are considered among the gentlest of pumps. Non-abrasive solids such as those found in wort or must are easily and gently pumped by this type of pump, making them well suited as a must pump attached to a crusher–destemmer. In fact, in other food industries, this type of pump is used to pump nearly solid material with very little free moisture (e.g. fruit purees and ground meat). If abrasive solids are used with this pump, the rotor and/or stator can be damaged, which will degrade the tight tolerances between them and reduce pump performance (especially during priming).

3.1.3.2.4 Flexible Impeller Pumps

Flexible impeller pumps (Figure 3.13) work well for smaller volumetric flow rates. They use a synthetic rubber impeller with thin fingers that run from the center of the impeller to the housing. While the fingers are all the same length, they are mounted in a housing that is off-round on one side. This acentric surface bends the fingers and reduces the volume between them. As the fingers pass the pump inlet, they spread out

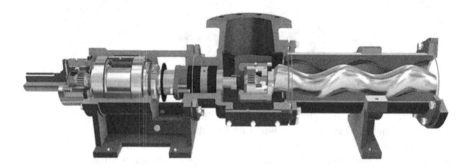

FIGURE 3.12   Progressive cavity pump. This pump is shown in cutaway. The helical rotor can be seen inside the stator on the right side of the image. Courtesy of Summit Pump, Inc.

(a)

(b)

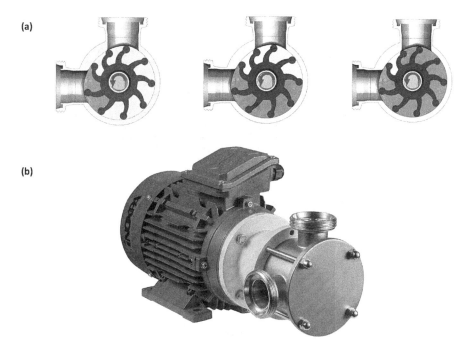

FIGURE 3.13 Flexible impeller pump. (a) Schematic and (b) example of a flexible impeller pump. Courtesy of INOXPA.

creating a larger space that creates a vacuum. As the fingers reach the outlet, they begin to deform again to push the fluid out. While some pumps can be run dry without consequence, flexible impellers will rapidly degrade in performance if run dry. Friction will increase tremendously which will wear down the fingers until they no longer seal with the side of the pump housing. Without the tight tolerance between the impeller and the housing, this type of pump loses its ability to be self-priming. Solids or particulates will also interfere with seal between the fingers and the housing, so these pumps are better used for fluids low in solids.

3.1.3.2.5 Piston Pumps

Piston pumps (Figure 3.14) use the action of a piston to draw in fluid on one side (as the piston moves up and creates a vacuum) and then subsequently pushes the fluid out the outlet as the piston moves back down. The fluid flow is maintained in the right direction using check valves, as with the diaphragm pump. To minimize the pulsatile flow that would result from this action, piston pumps generally have two pistons operating on the same motor—one piston is moving up and drawing in fluid, while the other is moving down and pushing the fluid out. To further dampen the pulsatile nature of flow in these pumps, they are often equipped with reservoirs at the inlet and outlet called pulsation dampeners. Some of the designs of piston pumps can be quite large or heavy so they are often used fixed in place, for instance as a must pump on a crush pad.

FIGURE 3.14   A piston pump. Courtesy of TCW Equipment Company.

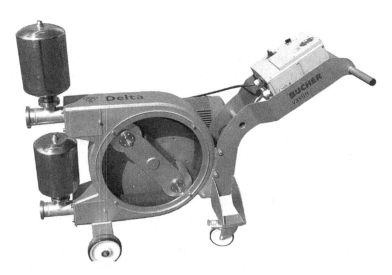

FIGURE 3.15   A peristaltic pump. Note how the rollers rotate and impinge against the tubing. This impingement forces a bolus of fluid to move, thereby ensuring flow. No mechanical parts directly touch the fluid in the tubing, making theses pumps highly sanitary. Courtesy of Bucher-Vaslin.

3.1.3.2.6   Peristaltic Pumps

Peristaltic pumps (Figure 3.15) work by compressing a section of flexible tubing with rotating rollers inside a pump head. As the roller moves along the pump head, the vacuum needed to draw in fluid is created as well as the force to push the fluid out on the other side. The fluid being pumped only touches the tubing and not the pump. This has made smaller versions of this type of pump popular in other industries (e.g. the biopharmaceutical industry) for pumping sterile or corrosive fluids. When the tubing wears out, it can easily be replaced. Larger, washdown versions of this type of pump were introduced into manufacturing environments in the late 1980's. This type of pump has recently become popular for wineries as must and general pumps. This pump is also often used in large breweries for yeast culture transfers, due to its aseptic nature (since the fluid is only touching the tubing which can be easily sanitized). Because a flexible tube is used, this type of pump is reserved for low-pressure applications and tubing wear or deformation needs to be monitored to avoid significantly reduced flow rates or tubing failure.

3.1.4   MIXERS

A Guth mixer (Figure 3.16) is a mixer that fits into a full tank through a ball valve. The collapsed impeller on a shaft is attached to the outside of the ball valve using a tri-clamp fitting. The valve is then opened, and the shaft is pushed through. When the shaft is all of the way through the valve, the impeller automatically opens fully and can be operated through an external motor. When mixing is completed, the mixer can be withdrawn and used on other tanks. Tank mixing is a good idea after additions such as inoculum, nutrients, or other processing agents, though sterility concerns make this application uncommon in the brewing industry.

Inline, or static, mixers (Figure 3.17), where multiple liquid streams are introduced into a baffled pipe which induces turbulence, are increasingly popular. In smaller facilities, an inline mixer allows the blending of two tanks into a third blending tank

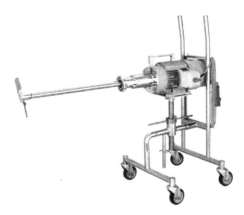

FIGURE 3.16   A Guth mixer. This unit is designed to fit through a ball valve on the side of a tank. Courtesy of Scott Labs.

FIGURE 3.17 A static mixer. Multiple disparate streams are introduced into turbulent, intimate contact to facilitate rapid inline mixing without moving parts. Courtesy of the Koflo Corporation.

simultaneously with the transfer. In larger facilities, inline mixing can be used to produce blends directly in line from the tank farm to a packaging facility.

3.1.5 Handling of High Proof Spirits

The equipment and concepts discussed above are applicable to the movement of any fluid, be it wine, beer, or distilled spirits. However, **pumping high proof ethanol using a normal winery or brewery pump could be hugely dangerous (explosion hazard) and should be avoided**.

While winery and distillery pumps are extremely similar, there are critical operational and safety distinctions between the two. High proof ethanol has a flash point of 12–16°C (53–61°F), depending on the concentration, and as such is hazardous at below-room temperature. Distilleries are typically rated by the National Fire Protective Agency as Class 1 Division 1; and spirits blending/bottling plants are typically Class 1 Division 2. As such, all liquid handling equipment, especially pumps, sensors, and valves are typically "intrinsically safe" and explosion proof, where the electrical portions of the equipment are sealed away from access to any combustible gasses (i.e. ethanol). Winery and brewery pumps are almost never intrinsically safe/ explosion proof, as non-distilled alcohol is not a combustion hazard. As such, there is a substantial hazard present in using winery process piping and pumps for high proof ethanol movements. Any such movements of high-proof ethanol should only be made using appropriately rated pumps in facilities designed for this type of work.

3.2 SIZING PIPES/HOSES AND PUMPS

Sizing pipes and pumps is necessary in many processing situations. These include the design of new facilities or facility expansions, as well as in the ordering of new equipment such as hoses or pumps for existing facilities. For existing equipment, these calculations can be used to find expected flow rates and pressure characteristics.

3.2.1 SIZING PIPES

Sizing pipes is really the process of deciding on the inner diameter of the pipe or hose to be used for a given fluid transfer at a desired flow rate. Therefore, the question is, if you know how fast you want to make a transfer (e.g. tank-to-tank or tank-to-barrel), how do you choose the correct diameter conduit. In general, to find an acceptable diameter, we choose the size to give turbulent flow with a "reasonable" pressure drop.

Turbulent flow is flow in which the streamlines (lines which particles follow) are not straight. This ultimately means that the apparent velocity is constant across the diameter of the pipe. This provides a sweeping or cleaning action that minimizes material adhering to the walls of the pipe. The opposite of turbulent flow from a fluid dynamics point of view is laminar flow. In laminar flow, the streamlines are straight and parallel to the sides of the pipe. Unlike in turbulent flow, the velocity profile across the pipe in laminar profile is parabolic with the maximum velocity in the center of the pipe and zero at the pipe wall. The zero velocity at the pipe walls is often called the no-slip boundary condition in the engineering literature. In our case, the important point is that the fluid will be stagnant at the wall. As we have noted previously, stagnant fluid is never compatible with sanitary design. This is the reason for targeting turbulent flow.

In order to check for turbulent flow, the fluid velocity, v, in the pipe is calculated first. Velocity is equal to the volumetric flow rate, Q, divided by the cross-sectional area of the pipe:

$$v = \frac{Q}{\frac{\pi D_i^2}{4}} \tag{3.1}$$

Here, the flow rate is chosen based on how long a transfer or operation is desired to take. The cross-sectional area of a pipe with a diameter, D_i, is equal to $\pi D_i^2/4$. Diameter is used instead of radius because, in practice, no one ever speaks of using a $1''$ radius hose—it is always a $2''$ diameter hose. A velocity of 3–10 ft/s is generally sufficient to give turbulent flow. However, flow can still be turbulent for velocities lower than this range. The best method for assessing whether flow is turbulent is by calculating the Reynolds number, Re, for the fluid:

$$Re = \frac{\rho v D_i}{\mu} \tag{3.2}$$

where ρ is the fluid density, v is velocity, D_i is the inner diameter of the pipe, and μ is the fluid viscosity. The Re is a dimensionless number that has been correlated experimentally with identification of turbulent versus laminar flow. A Re >> 2,300 usually corresponds to turbulent flow. Note: When using dimensionless numbers, make sure that all dimensions cancel! If your units do not all cancel out, your number is not dimensionless and the resulting number will not be meaningful. Dimensionless numbers are dimensionless. While it sounds repetitive, the largest problem that we have seen with these calculations is that a student's Re is not dimensionless. Don't be that person!

As velocity increases, the pressure drop due to friction also increases. A velocity that is too high can, therefore, cause erosion and heating, in addition to increasing the size of a pump necessary for the transfer. For this reason, pressure drop must also be calculated when choosing a pipe diameter to be sure that it is in a "reasonable" range. Pressure drop in a pipe due to friction can be calculated using the following relationship:

$$\Delta P = 2f_f \frac{L}{D_i} \frac{\rho v^2}{g_c} \tag{3.3}$$

where f_f is the Fanning friction factor and L is the length of pipe. The constant g_c is present to convert lb_m to lb_f as discussed earlier. This constant is equal to 1 and dimensionless in metric units. The Fanning friction factor has been found experimentally to be a function of the Re and the dimensionless roughness of the pipe, ε/D_i. A graph of this relationship can be found in Figure 3.18. It should be noted that a Darcy friction factor also exists in the literature. The Darcy friction factor is four times the Fanning friction factor. Know which one you are using! A "reasonable" pressure drop is one that is less than 2 psi/100 ft of pipe length. Average pressure drops are on the order of 0.5 psi/100 ft. Therefore, at this stage, 100 ft is typically substituted for L.

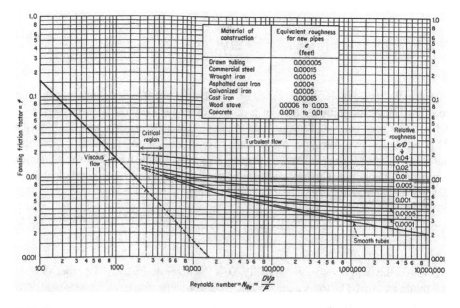

FIGURE 3.18 Fanning friction factor as a function of Reynold's number. From Peters and Timmerhaus.

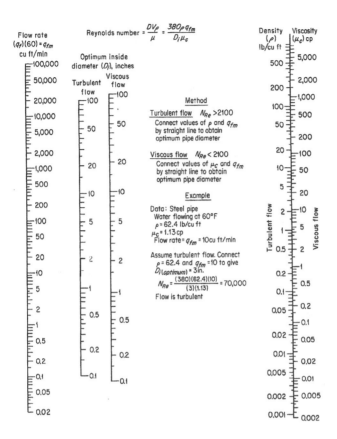

FIGURE 3.19 Nomograph to estimate pipe size with reasonable pressure drop in turbulent flow. From Peters and Timmerhaus.

In order to find the pipe diameter that gives turbulent flow with a reasonable pressure drop, you can choose D_i and then calculate v, Re, and ΔP, repeating this procedure until all conditions are satisfied. Alternatively, you can use a nomograph like the one attached from Peters and Timmerhaus (Figure 3.19), though double-checking your choice is always a good idea. To use this nomograph, one simply needs the desired volumetric flow rate and density of the fluid, both in the correct units. When these two points are connected with a line or straight-edge, the optimal pipe diameter can be found on the scale for turbulent flow.

3.2.2 Sizing Pumps

In order to size pumps, the desired flow rate and total energy that you are pumping against must be known. In order to calculate this Total Mechanical Head (H_m), an energy balance is performed on the system. The following equation for the energy balance for an incompressible fluid, called the Bernoulli equation, can be used:

$$H_m = (Z_2 - Z_1) + (P_2 - P_1) + \frac{v_2^2 - v_1^2}{2g} + H_f \qquad (3.4)$$

where the terms on the right-hand side of the equation are called, respectively, the static head, the pressure head, the velocity head, and the friction head.

To calculate each of these terms, one must first choose the two points in the system over which the energy balance will be applied. This corresponds to the point you are pumping from (Point 1) and the point you are pumping to (Point 2). Sometimes multiple energy balance points will be possible. However, the worst-case scenario for pumping (i.e. the one that will require the greatest H_m) is the best choice to assure pumping at an acceptable rate throughout processing.

Once the balance points are chosen, then each of the component heads are calculated. The static head is the net height differential against which you are pumping. This head is independent of the fluid path and simply dependent on the height difference between the end of the fluid path and the beginning of the fluid path. The pressure head is the net pressure differential between the target and the origin. For example, if you are pumping from a tank at atmospheric pressure to the top of a tank being held at 5 psig, then the pressure head would be 5 psig. The velocity head represents the amount of energy that needs to be added to the fluid to overcome inertia. For winery situations, the velocity head is often negligible. The friction head or losses due to friction can be calculated knowing the characteristics of the pipe, fittings, and equipment in the transfer path. Friction in each part of the fluid path must be considered. For straight pipe, the equation above for pressure loss (using the Fanning friction factor) can be used. Various correlations have been developed for finding "equivalent lengths" for given fittings. One convenient way of determining equivalent length is the nomograph for common fittings (Figure 3.20). For equipment not included on this nomograph such as valves and cartridge filters, vendors will often be able to supply this information in the form of actual data or correlations. Once all friction losses have been calculated, then H_m, can be determined. With the desired flow rate and total mechanical head, it is then possible to go to a pump curve or characteristic to find the required size for the pump. While pump curve styles vary by vendor, much of the information found on them is the same. For centrifugal pumps, choice of impeller diameter and motor can be made from the curve, in addition to efficiency (of converting electrical to mechanical energy) and the required net positive suction head (NPSH$_R$) that will be discussed later (Figure 3.21). For a positive displacement pump, rotational speed (rpm) and motor size can be chosen for a given flow rate and head. Figure 3.22 illustrates such a pump curve.

3.2.3 NET POSITIVE SUCTION HEAD (NPSH) CALCULATION AND CAVITATION

NPSH is the pressure available at the inlet or suction side of a pump. Every pump has a characteristic NPSH required for efficient pumping. This NPSH$_R$ can usually be found on a pump curve or from the pump manufacturer. If the available NPSH, NPSH$_A$, is less than the NPSH$_R$, then cavitation may occur. Cavitation is the process

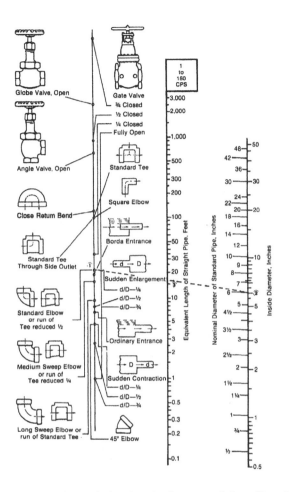

FIGURE 3.20 Nomograph for equivalent length of common fittings. From Lydersen et al.

of forming vapor bubbles at the low-pressure pump inlet that later implode at the high-pressure discharge side of the pump head. These implosions can damage the pump head and/or seals and should be avoided. Cavitation frequently sounds like rocks or gravel bouncing around in the pump head. To be sure that the available NPSH is high enough it can be calculated in a fashion similar to the calculation of the total mechanical head, H_m, but only the section of equipment upstream of the pump is used as follows:

$$NPSH_A = H_a \pm \Delta Z - H_{fs} - H_v - H_{vap} \qquad (3.5)$$

where H_a is atmospheric pressure at a given altitude, ΔZ is the height difference between the starting point and the pump (+ if pump is lower, − if pump is higher), H_{fs}

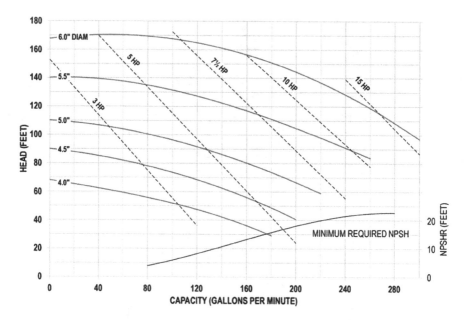

FIGURE 3.21 Pump curve for a TCW Dynahead 216 Centrifugal Pump, the same pump shown in Figure 3.9b. Courtesy of TCW Equipment.

is the friction head upstream of the pump, H_v is the velocity head (which will normally be negligible), and H_{vap} is the vapor pressure of the fluid being transferred at the given temperature.

If you are pumping water, cavitation is not usually an issue. However, there are several factors that could make it more of an issue. For instance, placing numerous fittings just upstream of the pump may cause cavitation by increasing the friction head component, thus decreasing NPSH. Another example would be if you have a lot of dissolved gas (e.g. carbon dioxide) in your fluid, as you might with an actively fermenting beer or wine. The dissolved gas will increase the vapor pressure of the fluid, thus decreasing available NPSH. Similarly, high ethanol or high temperature fluids will also have a higher vapor pressure.

In distillery applications, specially designed low-NPSH pumps are typically selected, especially when transferring distillate directly from the still. Available NPSH drops with increasing temperature, as the liquid is closer to boiling, and distillate is typically removed from a still at or near the boiling point. As such, extra care must go into the design of distillery liquid transfer systems, as cavitation is a particular safety hazard with combustible materials.

3.2.4 DENSITY AND VISCOSITY EFFECTS

Fluid density affects the pump sizing calculations in two ways. First, in the calculation of the individual heads, the conversion factor of 2.31 ft H_2O/psi needs to be divided by the specific gravity of the fluid. Second, motor power for a centrifugal

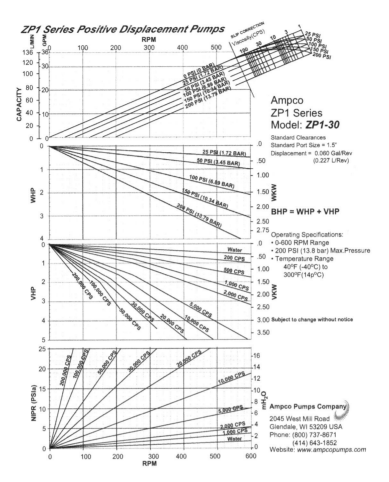

FIGURE 3.22 Pump curve for an Ampco ZP1-30 positive displacement pump. Courtesy of TCW Equipment.

pump needs to be multiplied by the specific gravity after finding it on the pump curve, as higher density fluids will require more power to pump at the same rate as lower density fluids.

Centrifugal pumps are particularly affected by viscosity. If a fluid viscosity is significantly different than that of water, correction charts exist for the pump efficiency, flow rate, and total mechanical head. All must be corrected prior to sizing the pump using the pump curve. For positive displacement pumps, the corrections needed are not as uniform and pump vendors must be consulted. Most fluids pumped in a winery or brewery setting will not present a problem in this regard. However, a high sugar dosage for sparkling wine or a high solids must, as examples, should be approached with caution, as their viscosity or apparent viscosity may warrant corrections.

Example 3.1: Pipe and Pump Sizing For the Golden Gate Winery

THE GOLDEN GATE WINERY CONCEPT (JONAS MUELLER)

The most significant detail regarding "Golden Gate Winery" is its location: just 15 min from downtown San Francisco, in the heart of the Marin Headlands. Since capital and regulations are not limiting factors in this project, we can focus on distinction to make our concept work.

The recreational area of the Marin Headlands makes a perfect spot because of its proximity to the city, its untouched nature, and its unused potential to produce high-quality wines. The whole production of 100,000 cases will be sold directly to the customer at the tasting room.

The top-wines are a Late Harvest Riesling and a single-vineyard Nebbiolo, both from vineyards surrounding the winery. The former is only sold in 375 mL bottles, the latter in regular 750 mL bottles. The golden color of Riesling and the relation to fog of Nebbiolo are used in the marketing concept. The main production, however, comes from grapes produced or purchased in nearby Sonoma. The only two other wines are a dry Riesling and a Pinot noir. The reds are fermented in open wood fermentors, pressed with basket presses, and aged in small oak barrels. The whites are fermented and aged in stainless steel and pressed with a normal bladder press. There are no doctrines being followed, but the maxim should always be to do as little as possible to the wine and, if a treatment becomes necessary, it should be as gentle as possible.

The labels are kept very simple for the base red and white wine. However, the "Golden Dew" and the "Red Dew" show pictures of the bridge with the Marin Headlands in the back.

PROBLEM

In the Golden Gate Winery, each pressing of Riesling juice will give 2,600 gal that will be transferred into a 10,000 gal tank. Three press lots will be added to each tank to give a total volume of 7,800 gal. The details of the system are shown in the diagram below.

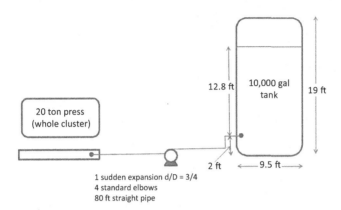

Each press lot will take 2 hr. The density of the juice (ρ) is 70 lb_m/ft^3 (22–24 Brix) and the viscosity (μ) is 2 cp. As the assistant winemaker at Golden Gate Winery, you are given the task to size the pipe and choose a positive displacement pump capable of this transfer.

SOLUTION

First, we size the pipe:

$$Q = \frac{2,600 \text{ gal}}{120 \text{ min}} = 21.7 \frac{\text{gal}}{\text{min}} \times \frac{\text{ft}^3}{7.48 \text{ gal}} = 2.9 \frac{\text{ft}^3}{\text{min}}$$

where gal/min is many times written "gpm."

Using the nomograph with the density given in the problem, we find that the D_i should be approximately 1.5″ (between 1.5″ and 2″). We will choose 1.5″ inner diameter and first calculate the velocity:

$$v = \frac{Q}{A} = \frac{4Q}{\pi D_i^2} = \frac{4 \times 2.9 \frac{\text{ft}^3}{\text{min}}}{\pi \left(\frac{1.5}{12} \text{ ft}\right)^2} \times \frac{\text{min}}{60 \text{ s}} = 3.9 \frac{\text{ft}}{\text{s}}$$

Then, with the velocity, we can calculate the Reynolds number to decide if this diameter will give turbulent flow:

$$Re = \frac{\rho v D_i}{\mu} = \frac{\left(70\,\frac{lb_m}{ft^3}\right)\left(3.9\,\frac{ft}{s}\right)\left(\frac{1.5}{12}\,ft\right)}{2\,cp \times 0.000672\,\frac{lb_m}{ft\,s}}{cp} = 25,391$$

This Re definitely indicates turbulent flow and indicates a fanning friction factor of 0.006 for "smooth tube." Next, we check to see if the pressure drop is "reasonable":

$$\Delta P = 2f_f\,\frac{\rho v^2}{g_c}\,\frac{L}{D_i} = \frac{2(0.006)\left(70\,\frac{lb_m}{ft^3}\right)\left(3.9\,\frac{ft}{s}\right)^2(100\,ft)}{\left(32.2\,\frac{lb_m ft}{lb_f s^2}\right)\left(\frac{1.5}{12}\,ft\right)\left(144\,\frac{in^2}{ft^2}\right)}$$

$$\Delta P = \frac{2.2\,psi}{100\,ft}$$

This is too high (greater than 2 psi/100 ft), so we need to increase the diameter to 2" and repeat the calculations:

$$v = \frac{Q}{A} = \frac{4Q}{\pi D_i^2} = \frac{4 \times 2.9\,\frac{ft^3}{min}}{\pi\left(\frac{2}{12}\,ft\right)^2} \times \frac{min}{60\,s} = 2.2\,\frac{ft}{s}$$

$$Re = \frac{\rho v D_i}{\mu} = \frac{\left(70\,\frac{lb_m}{ft^3}\right)\left(2.2\,\frac{ft}{s}\right)\left(\frac{2}{12}\,ft\right)}{2\,cp \times 0.000672\,\frac{lb_m}{ft\,s}}{cp} = 19,097$$

which is still turbulent. For this Re, the fanning friction factor is 0.0062 for "smooth tube." Next, we check to see if the pressure drop for this diameter is "reasonable":

$$\Delta P = 2f_f\,\frac{\rho v^2}{g_c}\,\frac{L}{D_i} = \frac{2(0.0062)\left(70\,\frac{lb_m}{ft^3}\right)\left(2.2\,\frac{ft}{s}\right)^2(100\,ft)}{\left(32.2\,\frac{lb_m ft}{lb_f s^2}\right)\left(\frac{2}{12}\,ft\right)\left(144\,\frac{in^2}{ft^2}\right)}$$

$$\Delta P = \frac{0.54\,\text{psi}}{100\,\text{ft}}$$

which is reasonable. Therefore, 2″ is an appropriate pipe size.

Next, we need to size the centrifugal pump. To do this, we set up the total mechanical energy balance.

$$H_m = H_z + H_p + H_v + H_f$$

$$= (Z_2 - Z_1) + (P_2 - P_1) + \left(\frac{v_2^2 - v_1^2}{2g} \right) + H_f$$

$$H_m = (2\,\text{ft}\,H_2O - 0\,\text{ft}\,H_2O) + (12.8\,\text{ft}\,H_2O - 0\,\text{ft}\,H_2O) + 0\,\text{ft}\,H_2O + H_f$$

Calculating the friction head:

$$H_f = \left(80\,\text{ft} + 1\,\text{expander} \times \frac{1.25\,\text{ft}}{\text{expander}} + 4\,\text{elbows} \times \frac{5.5\,\text{ft}}{\text{elbow}} \right)$$

$$\times \frac{0.54\,\text{psi}}{100\,\text{ft}} \times 2.31 \frac{\text{ft}\,H_2O}{\text{psi}}$$

$$H_f = 1.29\,\text{ft}\,H_2O$$

Therefore, substituting the friction head into the above equation for total mechanical head:

$$H_m = 16\,\text{ft}\,H_2O = 6.9\,\text{psi}$$

$$Q = 21.7\,\text{gpm}$$

With these two values, we can go to the pump curve below and find that we need the pump to operate at about 375 rpm with a 0.5 hp motor (this comes from adding the 0.1 hp and 0.4 hp together).

It is important to note here that the static head and pressure head can change over the course of the fluid transfer. In this case, the static head does not change over time, but the pressure head increases from the beginning of the pumping, when the receiving tank is empty, to the end of the pumping when there is a significant pressure head at the inlet to the tank because of the height of liquid over this feed valve. In a case such as this, where the static head and/or pressure head are changing, one should choose the time that is the worst case scenario (highest

H_m) for sizing the pump as we did here. This will assure that the fluid transfer will be completed at or before the desired time. There are situations, such as pumpovers in a tank, in which the H_m will not vary over time—therefore, it is important for you to think through this in detail prior to starting your calculations.

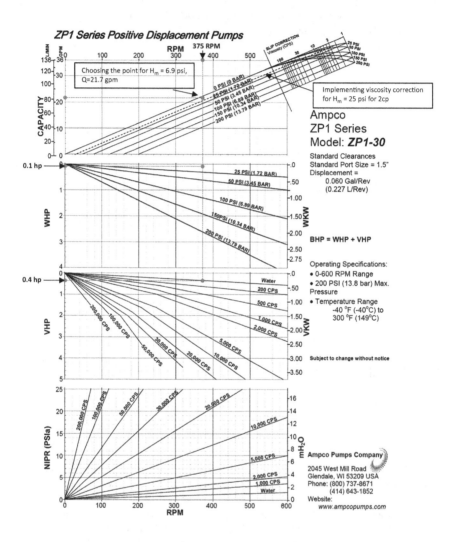

3.3   FLUID FLOW DUE TO GRAVITY

Gravity flow is sometimes desirable in a winery setting, though is quite rare in breweries. In large distilleries, gravity flow may be employed to remove product from high up on a distillation column. In wineries, it may provide the gentlest means of transferring juice, wine, or must (though we have found that no sensory impact can

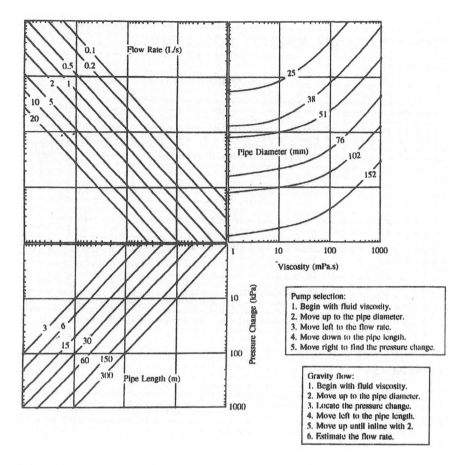

FIGURE 3.23 Nomograph for pipe sizing with gravity flow. From Boulton et al.

be detected after pumping *wine*). In addition, it does not require the capital outlay of a pump. Calculations for gravity flow can be performed in much the same way as for pump flow. However, in the case of gravity flow, the total mechanical head to be supplied by a pump is zero. The driving force for flow in this case is the static head or height differential between the start and end of the fluid flow. Assuming a negligible velocity head, the static head (height difference) will simply be equal to the friction head (using Bernoulli's equation as in the previous section). Using this kind of approach, fluid velocity could be found for a given pipe or hose size or, on the other hand, hose length could be chosen to give a particular velocity or flow rate. Given that one cannot explicitly solve for these parameters (e.g. the friction factor is a function of the Re which is a function of velocity and pipe diameter), an iterative solution approach would be necessary. Alternatively, nomographs such as the one shown in Figure 3.23 from Boulton et al. can be used for this type of calculation. The use of this type of nomograph is illustrated in Example 3.2.

Example 3.2:   Gravity Flow for the Golden Gate Winery

PROBLEM

If, instead of the pump chosen in Example 3.1, we decide to use gravity flow to move the juice from the press pan to the tank, can we move the juice fast enough? We will use a 2″ (50 mm) pipe for the transfer and the drop will be 2 m from the press pan to the top of the tank, which will now have to be on a level below the press. This corresponds to 19 kPa head. Overall, there are 5 m of pipe between the press pan and tank top.

SOLUTION

We can solve this problem using the nomograph in Figure 3.23. We start by drawing a vertical line from the viscosity of 2 cp (= 2 mPa s) up to the 2 in (= 51 mm) pipe diameter curve. From this intersection, we draw a horizontal line across to the flow rates. We then start a second line at a pressure drop/driving force of 19 kPa. We move horizontally over to the length of 5 m. From this point, we move vertically until we intersect with our original horizontal line. This gives us the flow rate, which is approximately 8 L/s or 125 gpm, which is significantly higher than the flow rate leaving the press, so the gravity flow is more than sufficient.

REFERENCES

Lydersen, B.K., D'Elia, N.A., and Nelson, K.L. (Eds.). *Bioprocess Engineering: Systems, Equipment and Facilities.* John Wiley & Sons, New York, 1994.

Peters, M., Timmerhaus, K., and West, R. *Plant Design and Economics for Chemical Engineers.* McGraw-Hill Education, New York, 2002.

PROBLEMS

1. You have decided to redesign your crush pad to increase capacity for your highly successful proprietary white wine blend, *Venology*. In the new design, a press with a 6 ton capacity (approximately 1,500 gal) is to be used. The juice will be pumped from the press pan using a centrifugal pump followed by a tube-in-tube chiller to rapidly cool the juice down to 55°F prior to settling, racking, and inoculation. Settling will take place in a new 2,000 gal tank. The setup is shown in the diagram below.

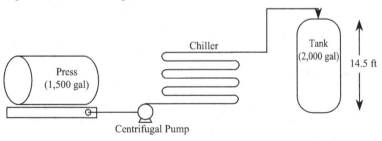

Between the press and the tank, there is 220 ft of hose and pipe with eight standard elbows and six close u-bends as part of the chiller system.

(a) Assuming that the first 1,000 gal of juice will be come out of the press in the first half-hour of pressing, calculate the size of the pipe and the size of the pump using a pump curve that you find online (vendors such as SPX flow or Alfa Laval typically post curves online).

The juice (at approximately 24°Brix) will have a density of 70 lb_m/ft^3 (SG = 1.12) and a viscosity of 2 centipoise. The Fanning friction factor for this system is $f_f = 0.007$. NOTE: It is reasonable for these values of density and viscosity to assume that the pump curve will not be significantly modified.

(b) Qualitatively (i.e. do not perform calculations), how would the answer differ if you were to fill the tank from the bottom?

2. Your brewery is built into the side of a hill with a cellar/cave underneath for barrel finishing of your signature imperial stout. It is your usual practice to blend beers from barrel in a blending tank that is also in the cellar. Then, the blended beer is pumped up to the canning line filler on the floor above via a filter bank containing a 1 μm cartridge filter and a 0.45 μm cartridge filter (having a 20 psi and 30 psi pressure drop, respectively). A centrifugal pump, also located in the cellar, is used for the transfer. The general setup is shown below.

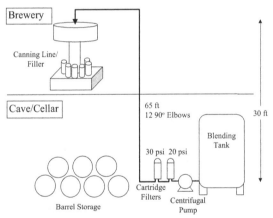

(a) The blending tank holds 10,000 L (= 2,646 gal = 354 ft^3). You have a high-speed canning line that will allow you to finish bottling this beer in 2 hr. What size pipe should you use? Using a pump curve at the same website as in Problem (1), choose the necessary impeller and motor size. Assume a beer density of 62.4 lb_m/ft^3, a viscosity of 2 cp (= 0.001344 lb_m/s ft), and smooth pipe (same as "smooth tube").

(b) One day, the brewmaster decides that you should move the centrifugal pump and cartridge filters up to the canning line area (instead of in the cellar). The blending tank will remain in the cellar. Is this a good idea (i.e. will it work)? Explain.

(c) What pump would you choose (rpm and motor size) for this application if you were choose a rotary lobe pump.

3. After years of performing manual pumpovers on your red fermentations, you finally find the capital to purchase fixed-in-place pumps for all of your large red fermentors. This will allow you to automate pumpovers, thereby freeing up valuable winery staff time for other critical tasks. As shown in the figure below, the total tank volume is 12,500 gal with a working volume of 10,000 gal (= 1,337 ft^3). As an extraction strategy, you decide to pump over one-half of the working volume twice per day (i.e. 5,000 gal two times per day). Each one of these pumpovers should take 60 min. The density of the must can be assumed to be 63 lb$_m$/ft^3, and the viscosity is 2 cp.

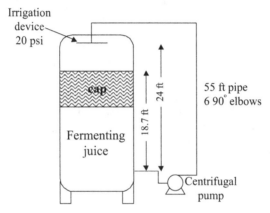

10,000 gal = 1,337 ft^3 working volume

(a) Size the pipe for the pumpover system to assure turbulent flow and a reasonable pressure drop. Assume that the pipe finish of Ra = 50 μin can be considered "smooth tube." (NOTE: 0.000672 lb$_m$/s ft = 1 cp)

(b) Size the centrifugal pump for this system using a pump and pump curve that you select. Assume that the irrigation device has a pressure drop due to friction of 20 psi.

(c) After getting used to the automated pumpovers, you decide to add an inline heat exchanger in the pumpover line after the pump. The heat exchanger manufacturer tells you that this equipment will have a 20 psi pressure drop. Will your choice of pump change? If so, to what will it change?

4. You are doing an internship at Casa Ingeniero in Chile. The master distiller has decided to barrel age the distillery's pisco (a type of brandy) in used Bourbon American oak barrels in the cellar. After aging for 24 months in barrel, you now need to pump the pisco from the barrels into a blending tank on the floor above (see diagram below). This transfer will be accomplished using hoses through a hole in the floor. In order to complete the transfer for all of the barrels in a reasonable time, you decide that you must be able to completely empty a 225 L (= 59.5 gal = 7.96 ft^3) barrel in 10 min.

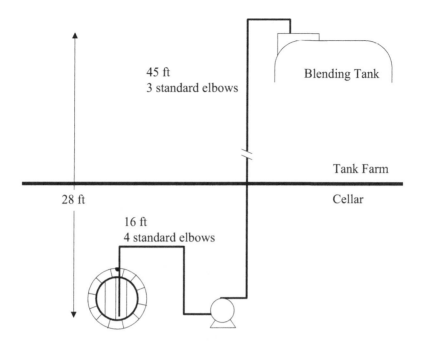

(a) You decide to use a positive displacement pump for the transfer. Why would it be difficult to use a centrifugal pump for this purpose?
(b) What size (i.e. diameter) hose would be best for this transfer? You can assume that the hoses used have a "smooth" surface. Be sure to verify that you have turbulent flow and a "reasonable" pressure drop. The pisco has a density of 58.6 lb_m/ft^3 and a viscosity of 0.9 centipoise (1 cp = 0.000672 lb_m/ft s).
(c) What is the total mechanical head that you will be pumping against in this system? Use the information on the diagram above for this calculation.
(d) Are there any special considerations you must keep in mind for this pump application?

5. During the design of the Wickson–Hall Winery, you start thinking about the design of the bottling line. For your Sauvignon blanc wine, you decide to bottle directly from the 10,000 gal tank that you have used to prepare the final blend. To do this, you will be using a centrifugal pump and pumping through a series of three cartridge filters (1 μm, 0.45 μm, and 0.22 μm pore sizes) to the filling machine as seen in the figure below. The pressure drops across these filters are 5 psi, 10 psi, and 20 psi, respectively. You estimate that there will be about 50 ft of straight pipe between the tank and the filler, five standard elbows, and one close return U-bend right before the filler. Because of the slope of the land, the top of your filler is 5 ft below the height of the tank racking valve.

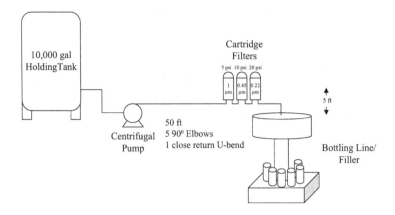

(a) The filler can fill 120 bottles/min (each with 750 mL = 0.20 gal = 0.027 ft³). Assuming a wine density of 61.8 lb$_m$/ft³ and a viscosity of 1.55 centipoise (1 cp = 0.000672 lb$_m$/ft s), what size (i.e. diameter) should the pipe be between the storage tank and the filler? Be sure to verify that you have turbulent flow and a "reasonable" pressure drop. Assume that the 304 L pipe used is "smooth tube."

(b) Using the information above and a centrifugal pump and pump curve that you select, choose the necessary impeller and motor size for the pump.

(c) After the bottling line is built, you begin to notice that you are getting significant oxygen pick up in the filler. To reduce this, you decide to pressurize the filler reservoir with nitrogen gas, but you now need to know whether the pump you chose in part (b) will still be able to maintain the desired flow rate. Upon opening the pump head, you discover that the original pump vendor had actually sent you a pump with one impeller size larger than you had specified. How much pressure can you apply in the filler and still get the desired flow rate from the pump? Show your work.

6. You have finished your fermentation for a lot of red ale in a stainless steel fermentor in your tank room. You decide to use gravity flow to fill barrels for long-term storage and aging. Your barrel room is in the cellar below the tank room 15 ft below the bottom of the tank.

(a) Using a 1.5 in hose with two 90° elbow fittings, how long will it take to fill each barrel? Assume a viscosity of the ale of 1.6 mPa S and use the nomograph provided

(b) How will these times change for a 2 in line?

(c) How would your answer in (b) change if your hose were actually 30 ft long instead of 15 ft (but keeping the same overall height difference of 15 ft)?

7. You are crushing 2,000 gal lots of Pinot noir. To reduce shear from pumping, you decide to design a gravity-feed winery. Therefore, the destemmer/crusher is in a room above fermentation tanks. There is a 30 ft straight drop into first tank. The other two storage tanks have an additional 30 ft run of pipe between them and two additional elbows.

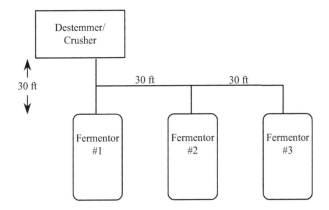

(a) Using a 1 in line, how long will it take to fill each of the tanks? Assume an apparent viscosity of the must of 10 mPa S and use the nomograph from Figure 3.23.

(b) How will these times change for a 2 in line?

(c) What size lines (i.e. inner diameter) do you think would be best for this process? Give your reasoning based on the results for (a) and (b).

8. After leaving Davis, you get a job as Assistant Winemaker at J Block Winery and are in charge of the crush crew pressing off your Cabernet (4,000 gal of must). Instead of having your crew climb into tanks to shovel out skins, you decide to sluice the skins over to the press. To do this, you will first hook up a hose to the drain valve of Tank 1 and use a centrifugal pump to pump 2,000 gal (= 267 ft^3) of skin-free wine over to the racking valve of Tank 2, leaving the skins in Tank 1.

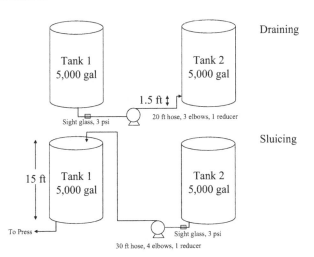

(a) If you want the draining to take 45 min, what size hose should you use? Be sure to check that you have turbulent flow and a "reasonable" pressure drop. Assume that the hose can be considered as "smooth tube."

$$\rho_{wine} = 61.8 \ lb_m/ft^3$$

$$\mu = 1.5 \ cp$$

$$1 \ cp = 0.000672 \ lb_m/ft \ s$$

(b) Using the information in the "Draining" figure above, size the centrifugal pump (impeller size and motor size) using a pump and pump curve of your choice.

(c) After draining the skin-free wine from Tank 1, you then attach a new set of hoses (with the same diameter) and sluice the skins in Tank 1 over to the press using the wine in Tank 2. Using the information in the "Sluicing" figure above, calculate whether you can use the same pump for this transfer.

9. Your first job out of UC Davis is at Mezcla Vineyards. The winemaker at Mezcla is very cautious about making sure that juice for their barrel-fermented Chardonnay is fermenting and well-mixed before putting it into barrels. To do this, she inoculates the juice in a tank and then mixes twice per day for three days before emptying the tank into barrel. The winery used to have a Guth mixer to accomplish this mixing, but it broke just before you joined the winery. The winemaker has come up with a new idea for mixing. She has you connect hoses to the drain valve of the tank, and from there to a rotary lobe pump. From the pump, you connect a pressure sensor and a rupture disk assembly, and then finally you connect this assembly to an L-shaped racking arm positioned in the racking valve port. The outlet is pointed straight up as seen in the figure below. Now, when you turn on the pump, you get a jet that comes out the racking arm and "rolls" or mixes the tank.

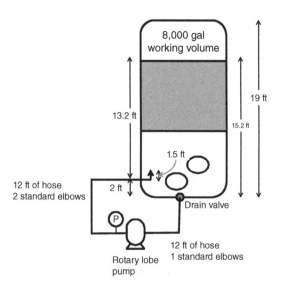

(a) Why is it important to have a pressure sensor and rupture disk in line after the pump in this setup?

(b) If you would like to pump through 4,000 gal (= 535 ft³) in 50 min to achieve the mixing, what hose size should you choose? You can assume that the initial juice has a density of 68 lb_m/ft^3 and a viscosity of 2.5 cp and that the hose can be considered "smooth tube." Be sure to check that you have turbulent flow and a "reasonable" pressure drop with this diameter.

$$1 \text{ cp} = 0.000672 \text{ lb}_m/\text{ft s}$$

(c) Given the setup in the diagram above, size a rotary lobe pump for this process (i.e. choose an rpm and hp). In addition to the hose lengths and fitting shown in the diagram, we have found out from the vendor of the pressure sensor/rupture disk that this piece of equipment has a pressure drop of 2.5 psi at the expected flow rates. The racking arm is a total of 3 ft long with one elbow.

(d) If a cellar worker has not been careful with the setup and the tri-clamp fitting next to the pressure sensor comes apart in between mixings, how quickly would the 6,000 gal (22,680 L) above the racking arm drain out by gravity? Assume the bottom of the tank is 3.5 ft (= 1.07 m) off the ground.

$$1 \text{ m H}_2\text{O} = 9.5 \text{ kPa}$$

$$1 \text{ in} = 25 \text{ mm}$$

$$1 \text{ m} = 3.28 \text{ ft}$$

10. When it opens, The Blue Note Winery will be one of the most sustainably designed wineries in Sonoma County. As part of that sustainable design, this winery will have an automated cleaning system that will lower water usage and chemical usage significantly. The system (called a clean-in-place or CIP system), shown in simplified form below, makes all of the cleaning solutions in a centralized location in the winery and then pumps them out (through a heat exchanger) to each piece of equipment to be cleaned. The diagram below includes just the cleaning solution tank and the large fermentor that is farthest away from the cleaning system. The cleaning system is located on the ground floor, which is the floor below the second or tank floor.

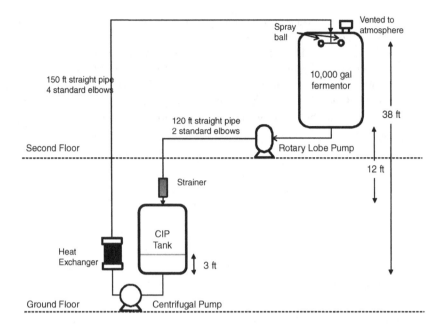

(a) Through experience, people have found that the flow rate needed through this system for good cleaning for this size fermentor is 75 gpm. If the cleaning fluids are water or close to water in properties, what size pipe should we use? Show that this size pipe gives turbulent flow and a reasonable pressure drop. Assume that all pipes can be considered "smooth tubes."

$$\mu = 1 \text{ cp} = 0.000672 \text{ lb}_m / \text{ft s}$$

$$\rho = 62.4 \text{ lb}_m / \text{ft}^3$$

(b) If there will be a 6 psi pressure drop through the heat exchanger and a 25 psi pressure drop through the spray ball device used for cleaning, size the centrifugal pump (impeller size and motor size) used to distribute the cleaning solution from the CIP Tank to the fermentor. All other distances, fittings, and heights are shown on the figure above. Note: Because you are pumping the cleaning solutions in a closed loop, the level of the cleaning agent in the CIP tank will not change—it will remain at 3 ft for the main part of the cleaning cycle.

(c) In order to get good cleaning in the fermentor, you have to drain the cleaning solutions out at the same or slightly faster rate than you are adding it. In this way, you will not get any pooling of liquid in the bottom of the fermentor (which leads to poor cleaning). Generally, the CIP return pumps are positive displacement pumps. Assuming the same flow rate as you used in the problem above, size the circumferential piston pump (i.e. rpm and hp) used to return the fluid to the CIP Tank. There is one inline strainer just prior to the CIP tank to take out particulates. It has a pressure drop of 8 psi.

11. **[Integrative Problem]** In thinking about the crush area of the Arboreal Winery in more detail, you decide that you would like to add all nitrogen (in the form of DAP), SO_2, and acid to your Riesling juice at the press pan. These three chemicals will be stored in concentrated form in 55 gal drums across the street from the crush pad, and each will have its own dedicated pump and line to the press pan that travels underneath the road in a utility tunnel (see figure). To make the additions, the operator will specify the working volume of the fermentor and the desired addition, and the appropriate pumps will turn on for the correct time to dispense this amount of chemical through the top of the drum to the press pan.

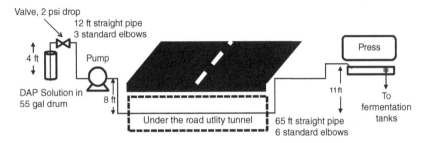

(a) The addition system for the DAP solution is shown above. The concentrated solution is made up at 20 g DAP/L. Since the working volume of the largest Riesling fermentors is 2,400 gal (= 9,072 L), 12 gal of DAP solution will be added to raise the concentration by 0.1 g DAP/L in the fermentor. To get good mixing, you decide that all of this DAP should be added at the press pan during a 5 min period near the beginning of pressing. This will allow the nitrogen to mix in as the juice enters the tank. Given the flow rate necessary to do this addition, size the pipe for this transfer. Be sure to check for turbulent flow and a reasonable pressure drop. You can assume "smooth tubes."

$$\mu = 1 \ cp = 0.000672 \ lb_m/ft\,s$$

$$\rho = 64 \ lb_m/ft^3$$

$$1 \ ft^3 = 7.48 \ gal$$

(b) Given the fluid path shown in the figure, size the pump. Use an appropriate pump for this application. Explain your answer. Note that the DAP solution comes from the top of the drum.

(c) In your travels through the Barossa Valley in Australia, you visit one high volume producer who is dosing oak chips into fermentors at (or just after) the press pan from a hopper. You like this idea and decide to investigate further, just in case you decide to venture into white wines other than Riesling (or you decide to add some oak character to your Riesling). Chip Wood, a salesperson from an oak alternatives company, tells you that his oak chips give exactly the same extraction kinetics and profile as if you aged your wine in a barrel. In addition, he tells you that his product is even cheaper than other oak chips because they do not season their wood out-side—they dry it in a kiln instead—prior to chipping the wood and then toasting it. Name one advantage of using oak chips for your processing over barrels. Also, give three reasons why you may want to seek out an oak alternative company with a better salesperson and/or product after listening to what Chip has to say.

12. After getting the concrete tanks up and running, the consulting winemaker for Pedunculate Oaks Winery decides that he would like to perform some of his red fermentations in barrels. As you are the one that will be responsible for getting the skins out of the barrels afterward, you suggest the alternative approach of using small stainless steel fermentors like we have in the Teaching and Research Winery at UC Davis, followed by barrel aging. Your consulting winemaker reluctantly agrees and tells you to design a prototype fermentor. You do this from memory of your experience at Davis. Below is a diagram of the system. Your pumpover device is designed to take fermenting juice from the bottom of the tank and then spray it over the cap. The spray device has a pressure drop of 5 psi.

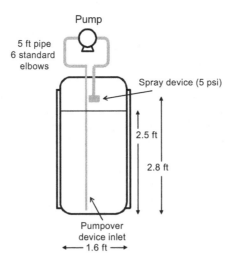

(a) Assuming that you would like to pumpover a maximum of one working vol-ume in 40 min, what size pipe should you use for the pumpover device? Confirm that you have turbulent flow and a reasonable pressure drop. You can assume "smooth tubes."

$$\mu = 1 \text{ cp} = 0.000672 \text{ lb}_m/\text{ft s}$$

$$\rho = 70 \text{ lb}_m/\text{ft}^3$$

(b) Given this pipe size and the schematic above, calculate the total mechanical head, H_m, for the pumpover pump, and choose an appropriate pump for this application

(c) In addition to designing the fermentor, you also need to figure out how you will press these small fermentors. Again, you turn to the system you saw at UC Davis. In this press, the fermentor vessel becomes the basket for a specially designed basket press. The press head actually stays stationary and the fermentor is raised by the press. Juice comes out through slots in the press head and then is siphoned by gravity into a drum or barrel below (after the flow is primed by a cellar worker). Since the head doesn't move on this press, it is always 2.5 ft above the drum. The siphoning is accomplished using a 10 ft length of 1 in diameter tubing. What flow rate do you expect to come out of the tubing. Assume that you can ignore any bends in the tubing and that

$$1 \text{ m} = 3.28 \text{ ft}$$

$$1 \text{ m H}_2\text{O} = 9.5 \text{ kPa}$$

$$1 \text{ inch} = 25 \text{ mm}$$

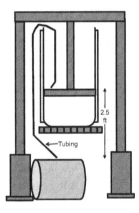

13. The new Jess Jackson Sustainable Winery Building (JSWB) being designed and built at UC Davis will make the Teaching and Research Winery self-sufficient in water and energy. One of the unit operations that this new building will house will take rainwater collected off the roof of the RMI building complex and filter it (using reverse osmosis or RO) so that it is pure enough to use for cleaning the winery equipment. There will be an 8,000 gal RO Hold Tank located in the JSWB. When cleaning in the winery uses up water, water will be transferred from the RO Hold Tank to a Cleaning Water Hold Tank in the winery that will hold 2,000 gal of water and supply the clean-in-place (CIP) system. These two tanks are connected by underground pipes that travel under the access road as shown in the diagram below.

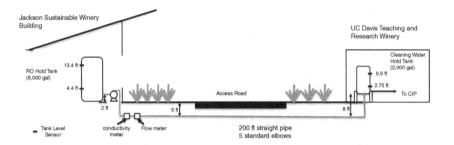

(a) If the water is to be transferred from the RO Hold Tank to the Cleaning Water Hold tank at a rate of 1,200 gal (160.4 ft³) in about 40 min, what size pipe should be buried under the road? Confirm that you will have turbulent flow and a "reasonable" pressure drop. Assume that the pipe can be considered "smooth tubes."

$$\mu = 1\ \text{cp} = 0.000672\ \text{lb}_m/\text{ft s}$$

$$\rho = 62.4\ \text{lb}_m/\text{ft}^3$$

(b) The controller of this system is set up to turn on the RO system and start filling the RO hold tank when the level gets down to 4.4 ft (2,000 gal) of water and to turn off the RO system when the hold tank reaches a level of 13.4 ft (6,000 gal). Similarly, when the Cleaning Water Hold Tank is down to a level of 2.75 ft (500 gal) of water, the transfer pump in the diagram will turn on and transfer water from the RO Hold tank. Water transfer will stop when the Cleaning Water Hold Tank reaches 9.9 ft (1,800 gal) of water. Both a flow meter and a conductivity meter are inline, each with a 1 psi pressure drop (according to manufacturer's information). Using the diagram above and an appropriate pump curve, size the pump for this application.

$$\rho = 62.4\ \text{lb}_m/\text{ft}^3$$

14. **[Integrative Problem]** After the first vintage, the Godello (white) wine being produced at Bodegas Impetu has become extremely popular. As the head winemaker, you decide to double production. The vineyard and tank space prove to not be a problem. However, the barrel room will not be large enough to store this extra capacity. Since you feel that the Godello requires oak contact to improve its palate, you need to find a solution to this barrel storage problem. Luckily, your winery is run by a "group of passionate, artistic, and scientifically driven University of California-Davis alumni," and one of them remembers visiting a brewery in California that used a hops "torpedo" to flavor their beer. You decide to use this idea to give your wine oak contact instead of using barrels. To set this up, you draw wine from the fermentor racking valve and pump it through a stainless steel cylinder packed with oak cubes and then back

to the fermentor drain valve until it has extracted the right amount of oak character. After the extraction, the wine will be stored in the fermentor. This is shown in the diagram below.

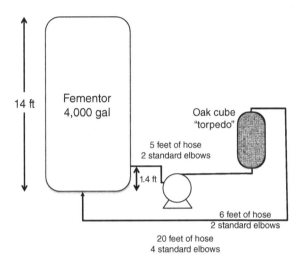

(a) From preliminary experiments, you think that five passes of the wine over the oak cube "torpedo" should be enough to get sufficient extraction. You would like to accomplish this in less than 4 hr. What hose size should you use for this operation? Confirm that you will have turbulent flow and a "reasonable" pressure drop. Assume that the hose can be considered "smooth tubes" and that the 4,000 gal (= 535 ft³) tank is 95% full.

$$\mu = 2 \text{ cp}$$

$$1 \text{ cp} = 0.000672 \text{ lb}_m/\text{ft s}$$

$$\rho = 64 \text{ lb}_m/\text{ft}^3$$

$$1 \text{ ft}^3 = 7.48 \text{ gal}$$

(b) You purchase this size hose for this job, but on the day that you are going to start the oak extraction, one of the harvest interns takes the 6 ft length of hose and uses it for something else in the winery. You are left with having a 2 in hose for this 6 ft length. Assume that the fermentor is 95% full with wine and the oak cube torpedo has a pressure drop of 30 psi. If you had originally specified a C216 centrifugal pump, 3,500 rpm with a 4.5 in impeller and a 5 hp motor, would this pump still work with the substitute hose?

(c) With all of the equipment set, you now turn to choosing the best oak cubes to put in the "torpedo." You would like the extraction to be fast and uniform and to enhance the tropical notes in your Godello with some coconut aroma. Woody, the salesman from E. Robinson Oak, has a good selection of cubes. He tells you about a few of the lines that he sells. One of the lines is made from *Quercus robur* from the Limousin forest in France. These 2.5 cm cubes are cut from staves that have been toasted on one side using an open flame. Another line is made from *Q. alba* and is sold in 1 cm cubes that have been toasted (after cutting) using a new device that monitors and controls temperature in the toasting oven. The final line is also *Q. alba* in 1 cm cubes, toasted with the same new device as the second line. However, Woody tells you that this line would be even better because it comes from wood that has been seasoned in a yard in Missouri for 50% longer. It costs 50% more, but Woody says that it is worth it because of the extra flavor you will extract. Which line would you choose? Briefly explain your answer.

15. **[Integrative Problem]** After completing your work with the fermentors, the UC Davis winemaker decides to have you work on the design of the solar powered hot water generation for the facility. As shown schematically below, these generators from the company, Cogenra, will be mounted on the flat part of the roof of the Jess Jackson Sustainable Winery Building. Initially, there will only be one unit that consists of a parabolic glass array that focuses light on the water pipe that runs down the center of this array. The stream of cool water coming from the rain water storage tank starts at 68°F and ends up heated to 120°F, after which the heated water is stored in a 500 gal, insulated, hot water storage tank and can be used for cleaning or temperature control. The source of the water is from rain stored in a series of 40,000 gal tanks.

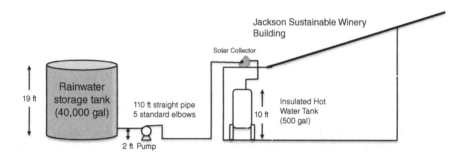

(a) if you calculate that you will need to fill the 400 gal of working volume during the 5.5 hr that your sun exposure is maximized, what is the flow rate of cooling water that is needed from the rainwater storage tank to the insulated hot water hold tank?

(b) What size pipe should be used for this transfer of water to the photovoltaic panel and then on to the storage tank? Check to make sure you have turbulent flow and a reasonable pressure drop.

$$\mu = 1\,cp$$

$$1\,cp = 0.000672\,lb_m/ft\,s$$

$$\rho = 62.4\,lb_m/ft^3$$

$$1\,ft^3 = 7.48\,gal$$

(c) Given the measurements and the fittings in the diagram above, calculate the total mechanical head necessary for the pump. The current plan is that the rainwater storage tanks will always have at least 5 ft of water in them at any time. Select a pump for this application.

(d) If water will enter the stainless steel tube in front of the solar collector at 68°F from the neighboring rainwater storage tank and leave the collector at 120°F, how long do the collector and pipe (in front of the collector) have to be to make this work. Assume the outside of the pipe gets to a uniform 140°F across its entire length when exposed to the sun. Assume the overall heat transfer coefficient for this system is 120 Btu/hr ft² °F.

$$c_{p,water} = 1.0\,Btu/lb_m\,°F$$

$$\rho_{water} = 62.4\,lb_m/ft^3$$

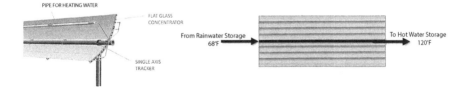

16. Your Estate Chardonnay at Puddle Bottom Cellars is so popular that you decide you need to make another white wine. You decide to make a fruit-forward style Sauvignon blanc and find fruit in another vineyard in Sonoma about 1 hr to the northeast. Because this is not your vineyard, you do not have as much control over the timing of harvest. The fruit therefore arrives at your winery at a temperature of 80°F. Because you plan to ferment the Sauvignon blanc at 55°F, you decide to send the wine directly from the press pan through a rough filter with a 10 psi pressure drop due to friction to a large must chiller (a series of jacketed pipe and u-bends) and then into the fermentor through the racking valve of the tank, one level below.

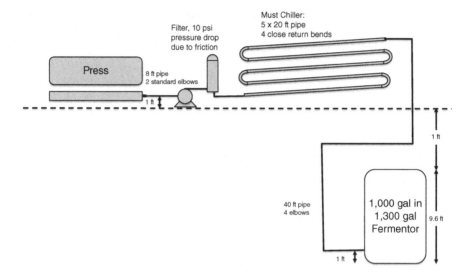

(a) The winery is using a Bucher XPlus 62 membrane press that can fit 8 tons of destemmed grapes. You expect just over 1,000 gal of juice from this fruit with about 2/3 of it (667 gal = 89.2 ft³) in the first hour of the press cycle. What size pipe should you use to give turbulent flow and a reasonable pressure drop? Show all work. Assume "smooth tubes."

$$\mu = 2 \text{ cp}$$

$$1 \text{ cp} = 0.000672 \text{ lb}_m/\text{ft s}$$

$$\rho = 68 \text{ lb}_m/\text{ft}^3$$

$$1 \text{ ft}^3 = 7.48 \text{ gal}$$

(b) Given the diagram above, size a pump for this application.

17. Recently, the Department of Viticulture and Enology installed a rainwater capture system in the Robert Mondavi Institute complex. It captures all of the rainwater that hits the roof of the North, South, and Sensory buildings and transfers this water through a rough filter to rainwater storage tanks behind the Jackson Sustainable Winery Building. In the coming year, equipment will be installed to filter this water down to potable (i.e. drinkable) levels using a reverse osmosis system, and it will subsequently be used for cleaning the winery. The system has two main parts. One part collects the rainwater off the roof in a gutter and delivers it by gravity to sump (underground tank) in the middle of the courtyard. The second part transfers the water from the sump using a sump pump to the water storage tanks behind the Jackson Building. This is shown schematically below.

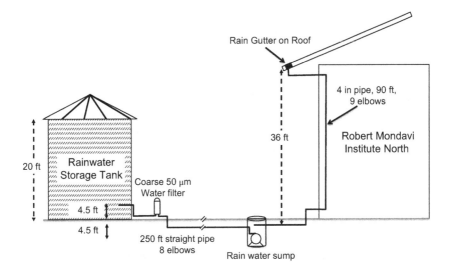

(a) Calculate the maximum flow rate of rainwater possible by gravity flow for this 4 in (= 102 mm) pipe using the information in the diagram above. If we expect to collect 10,000 gal (= 37,800 L) of rainwater for each inch of rain to fall on the roof of the North building alone, what is the maximum rain fall that we can get in an hour and not overload the gutters of this building? The viscosity of rainwater is 1 cp (= 1 mPa s).

$$9.5 \, kPa = 1 \, m \, H_2O$$

$$1 \, m = 3.28 \, ft$$

(b) Once the water flows by gravity into the sump, it must be pumped out to the water storage tanks behind the Jackson Building. If we have 0.1 in of rain per hour on the complex and the water flow rate off the North Building represents 40% of the rain being collected off all three buildings (and all directed to the same sump), what size should the pipe be that brings all of this water from the rainwater sump pump to the storage tanks? Be sure to check to make sure you have turbulent flow and a "reasonable" pressure drop. Assume "smooth tubes."

$$\mu = 1 \, cp$$

$$1 \, cp = 0.000672 \, lb_m/ft \, s$$

$$\rho = 62.4 \, lb_m/ft^3$$

$$1 \, ft^3 = 7.48 \, gal$$

(c) A Flygt sump pump has been chosen for this transfer. Will this pump be sufficient? Show your work. The coarse filter prior to the storage tank has an 8 psi pressure drop, and the storage tank can be filled up to the 20 ft height.

18. At the 2014 Winery Olympics recently held at the Teaching and Research Winery in Davis, CA, Team USA put in an impressive performance (not gloating—just stating the facts). One of their strongest events was the siphoning competition in which teams were required to fill five 750 mL bottles with water from a stainless steel reservoir using a 3/8 in (= 9.5 mm) inner diameter plastic tube and a manual siphon. The stainless steel reservoir was originally sitting on a table 2.5 ft (= 0.76 m) off the ground and the bottles were placed on the ground.

(a) How fast should a team be able to fill all five bottles? Assume 5 s in between each of the bottles and a siphon hose length of 3 m. Use the attached nomograph and mark your lines on this sheet.

$$\mu = 1.0 \text{ cp} = 1.0 \text{ mPa s}$$

$$9.5 \text{ kPa} = 1 \text{ m H}_2\text{O}$$

(b) Observers noticed that, unlike the other teams, Team USA lifted the stainless steel reservoir 1 m above the table (which is already 0.76 m off the ground). How much faster did this allow them to fill the bottles? Assume that the same 3 m siphon hose was used. Was this the reason for their impressive performance or was it just the skill of Team USA's siphoner?

4 Fruit Receiving and Processing

Fruit receiving is a critical first step in wine production. While this section is grape oriented, the same concepts can be applied to apples (or other fruit) for cider. Fruit receiving is also a necessary first step in brandy production, as the wine must be fermented prior to distillation.

4.1 HARVEST DECISIONS

One of the main decisions that need to be made (other than the timing of harvest) is whether to hand pick or machine harvest the fruit. Hand picking is generally associated with high quality wine production in some wine-producing regions around the world. It results in less fruit damage and reduces skin contact between skin and juice (that inevitably forms with older machine harvesters) during transport. Hand picking, using a trained crew, allows sorting during picking and again after delivery to the winery. On the other hand, machine harvesting is less labor-intensive. In addition, machine harvesting allows more control over harvest timing because it precludes a reliance on the availability of picking crews, especially when weather patterns force many wineries to pick at nearly the same time. Machine harvesting sometimes leaves mold-infected fruit on the vine as well, thus automating part of the sorting process. It is interesting to note that studies performed at UC Davis by the Oberholster lab (Hendrickson et al., 2016) have shown little difference between wines made from machine-harvested and hand-harvested fruit.

Another decision that needs to be made, especially when designing a crush pad, is the size of the picking bin that will be used. Commercially, 30 lb picking bins and ½ ton bins are most common for small to medium-sized wineries while gondolas that each hold 6–7 tons (with four of these bins per tandem trailer) are common for large wineries. This choice especially affects the fruit staging and unloading area and initial conveyance systems, as unloading gondolas requires mechanical hoists to lift one side of the hinged bins. Use of ½ ton bins precludes the need for mechanical hoists but requires crush pad space for bin storage and forklift maneuvering around the receiving hopper.

The timing of picking greatly affects fruit temperature. Picking at night or early morning can result in fruit that is 30–40°F cooler than fruit picked in the afternoon. As we will see later in the text when we discuss refrigeration, must chilling is one of, if the not the largest, draws for refrigeration systems in a winery. By picking while fruit is cool, we can save energy and purchase a smaller refrigeration system.

Finally, picking fruit at locations remote to your winery presents several extra challenges over estate fruit. These include increased skin contact during transport, increased oxidation, and fewer opportunities to return skins and stems to the

DOI: 10.1201/9781003097495-4

vineyard. Therefore, sometimes it will make sense, when the grapes are remote, to crush and/or press remotely and transport the resulting juice or must back to the winery.

4.2   RECEIVING AND SORTING

4.2.1   ONSITE WINERY GRAPE SAMPLING AND JUICE ANALYSIS

Sampling grapes on delivery to the winery can help to facilitate later stages of processing. It can also be the basis of rejection of grape deliveries if the analysis is done in near real time. For this reason, sampling of the grapes and rapid analysis as the grapes arrive and are weighed is becoming more common. In larger wineries, it is becoming common to use robotic samplers in gondola bins that transfer a liquid sample of juice to an analyzer that measures sugar concentration, pH, and titratable acidity within a few minutes (Figure 4.1). If more information is desired, the use of clinical chemical analyzers has been implemented in winery settings. These analyzers can measure any analyte using assays based a spectrophotometer. Liquid handling robots measure reagents and handle mixing and measurement. Tests using this type of equipment include initial nitrogen (organic and inorganic) and malic acid, among others. This information can then be used to plan for necessary additions and adjustments prior to the must reaching the press or tank.

4.2.2   RECEIVING HOPPERS AND CONVEYOR SYSTEMS

4.2.2.1   Helical Screws and Moving Belts

When grapes arrive at the winery in bins, they are typically dumped into hoppers that feed the first unit operation, which could be a sorting table, crusher/destemmer, or press. This means that hoppers must be equipped with a means of moving the fruit. There are two common means of moving the fruit. The most common method currently is to use either a horizontal or inclined helical screw (as seen in Figure 4.2).

(a)   (b)

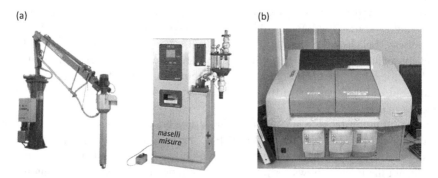

FIGURE 4.1   (a) An automated grape sampler, used to determine the characteristics of harvested grapes, including a robotic sampling arm and liquid analyzer. (b) An automated laboratory analysis unit, which can be used to rapidly screen for Brix, nitrogen, organic acids, and multiple other parameters. Photo credit for (a) Maselli Misure S.P.A.

FIGURE 4.2 Grape receiving using a traditional hopper and helical screws. Grapes are received in ½ ton bins and dumped into the hopper with the helical screw. The fruit moves up the inclined helical screw to a crusher and then to a press.

This type of equipment generally works well although some berry breakage is expected due to the shear force generated at the edge of the screw. Perhaps a gentler means of transporting grapes is by use of conveyor belts. A belt system (as seen in Figure 4.3) can transport whole clusters or machine-harvested fruit (assuming a juice recovery system or sloped belt edges are in place). It is frequently used in conjunction with direct-to-press or whole-cluster pressing.

FIGURE 4.3 Grape receiving using a conveyor system. Grapes are received in ½ ton bins or truck gondolas and dumped into a hopper with two conveyor belts that move toward the center (a). Fruit is conveyed to the presses along movable boom (b) for whole-cluster pressing (belt is not shown on boom) into membrane presses (c).

4.2.2.2 Vibrating or Tilting Hoppers

Hoppers are often loaded using a specially modified forklift, especially in the case of half ton picking bins. Trying to feather the forklift to slowly empty the bins onto a conveyor or elevator can be time consuming and difficult. One alternative to this is the use of a vibrating hopper. A vibrating hopper (Figure 4.4) can often fit an entire bin of grapes and then uses a vibrating mechanism with a slight incline to gradually move the fruit into the next unit operation (e.g. a sorting table). Use of these hoppers requires the forklift operator to lift the fruit far above the ground in order to reach over the top of the hopper. This can be awkward and make it difficult to see the operation from the seat of the forklift. An alternative to this is to use a tipping hopper. This type of hopper (Figure 4.5) starts in a low position, making it easy to empty fruit into. It then pivots to dump fruit through a narrower opening into the next unit operation. Tipping fruit collection bins that can be driven through vineyard to collect hand- or machine-harvested fruit are also now available. This type of equipment can preclude the need for a forklift in the grape processing process.

4.2.2.3 Elevators

When fruit needs to be elevated to reach the top of a unit operation, another option is to use a grape elevator. Elevators contain specialized belts with ridges or dividers to help propel fruit uphill (Figure 4.6). They typically have a built-in juice collection system to collect juice from broken or damaged fruit. The angle of the incline can usually be adjusted and also the height of the base.

4.2.3 MANUAL AND AUTOMATED SORTING

4.2.3.1 Manual Sorting

As a means of quality control, many wineries (and especially smaller ones) insert a sorting unit operation at this point in the process (i.e. prior to destemming and/or crushing). The main goal of this step is to remove material other than grape (MOG), along with unripe or moldy fruit that will affect the quality of the final wine. The two technologies often used for sorting are moving conveyor belts (Figure 4.7) or

(a)   (b)

FIGURE 4.4 Vibratory hoppers for grape receiving. (a) Side view of a vibratory hopper, (b) grapes are dumped into the hopper via forklift and then distributed by vibration onto a sorting table.

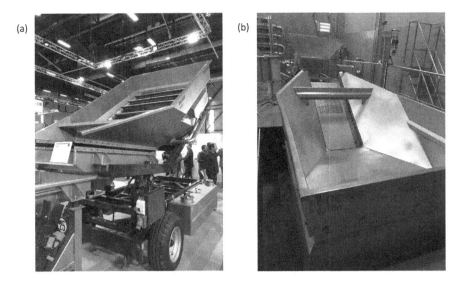

FIGURE 4.5 Tipping hoppers. (a) This hopper can be pulled through the vineyard while being loaded with hand or machine harvest fruit. (b) This hopper receives fruit from larger bins. Courtesy of Bucher-Vaslin.

FIGURE 4.6 Grape elevator. This type of elevator is used to move grapes up an incline using a conveyor belt with ribs. Courtesy of Bucher-Vaslin.

(a) 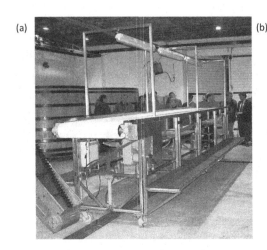 (b)

FIGURE 4.7   Sorting tables used for quality control. In (a), the sorting table can be lined with people on both sides. In (b), the table is maned by 1–2 workers. In both cases, the tables are used to identify and remove MOG (material other than grape), or moldy/unripe grapes.

vibrating tables (Figure 4.8). With conveyor belt systems, people stand on either one or two sides of a moving conveyor belt, manually picking out the material to be removed. Capacity is increased by having a longer table, more people, and a faster conveyor belt. Vibrating tables work in a similar way, though in this case, table vibration spreads out the fruit and moves it along the gradually inclined surface, thus making it easier for people to pick out the undesirable fruit and MOG.

Some of the sorting of MOG has become automated, allowing higher throughput and lower personnel costs. Automation has come through the use of magnetic tables

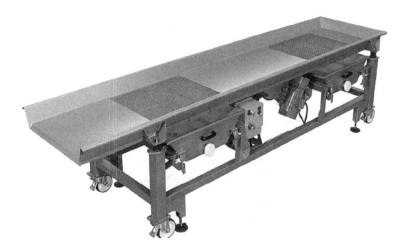

FIGURE 4.8   Vibrating sorting table. Grapes travel down the table using a slight incline and vibration. Free-run juice can be collected separately through the slots in the table. Courtesy of Bucher-Vaslin.

for removal of metal MOG, vibrating tables with holes that allow berries to fall through while larger MOG remains on top (typically with destemmed fruit), or with a series of air blow steps that are able to blow out the light MOG (e.g. leaves) while not affecting the path of the berries.

4.2.3.2 Automated Optical Sorting

Trials with equipment that differentiates berry color (e.g. raisined or moldy fruit), size, and shape, and rejects everything but ripe, unmoldy fruit have proven relatively successful in terms of the ability to automate this process (Figure 4.9). In this case, it is necessary to first destem the fruit, but to bypass the crusher, prior to sorting. Real time image analysis is performed on the berries traveling along a conveyor belt. The operator must use a calibration load of grapes and the sorter software to identify either the undesirable or desirable fruit, depending on the manufacturer of the equipment. An array of air blow nozzles is activated upon recognition of an undesirable object and undesirable berries or MOG are blown out of the process path using compressed air. These devices can be quite effective at eliminating undesirable berries and MOG at rates of up to 10 tons/hr. On the other hand, wines from sorted and unsorted Chardonnay berries did not differ significantly in most aroma and flavor descriptors (Falconer et al., 2006). Other studies have corroborated these results for other types of berries/wine (e.g. Hendrickson et al., 2016).

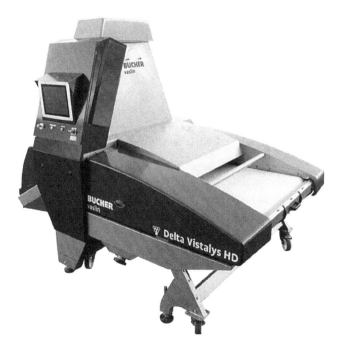

FIGURE 4.9 Optical sorter. Optical sorters can use size, shape, and color to differentiate between acceptable grapes and unwanted material. They take an image of the destemmed fruit on the conveyor belt and use the image to decide what to retain or discard. This unit can process up to 10 tons/hr. Courtesy of Bucher-Vaslin.

4.2.3.3 Automated Density Sorting

Recently, equipment has been developed to sort berries based on density (Figures 4.10 and 4.11). While multiple companies manufacture this type of equipment, the principle is the same. Destemmed intact berries are unloaded into a bath of known density. The density is chosen to be lower than the density of the desired grapes and higher than that of unripe grapes and MOG (e.g. stems and leaves). A conveyor brings ripe berries from the bottom of the bath to be further processed, while the unripe berries and lighter material float to the top of the bath and are collected in an overflow. Overflow liquid is separated from the undesirable solids and recycled to the bath. For these units, care must be taken to keep the bath clear of unwanted microbes that may affect downstream processing.

FIGURE 4.10   Sorting grapes based on density. Grapes are dropped into a liquid of a given density, so that ripe berries will sink while unripe berries and MOG will float. Courtesy of Bucher-Vaslin.

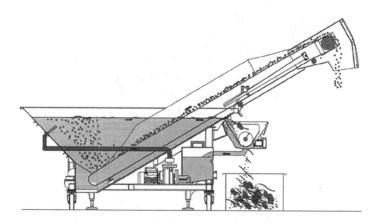

FIGURE 4.11   Schematic of sorting by density. Courtesy of Bucher-Vaslin.

4.3 CRUSHING AND DESTEMMING

It is most typical to remove the grape berries from the stems (destemming) and then crush the berries (crushing) to begin the release of juice prior to pressing (for whites) or fermentation (for reds). The equipment for accomplishing these tasks is often combined into one piece of equipment. While these operations are typical, there are variations to their use, and it is not uncommon for winemakers to skip one or both of these steps. Reasons for skipping these steps include:

- The (white) fruit will be whole-cluster pressed ("direct-to-press") to minimize skin contact time and phenolic extraction
- The fruit is to be used to produce a carbonic maceration wine style with some or all intact berries.
- The fruit is in bad condition from mold or is purposely *Botrytis*-infected.
- The fruit is partially dried prior to processing to increase sugar concentration.
- The winemaker would like to have some stem contact or ferment with whole clusters

Choosing one of these alternative approaches will have consequences, not only on aroma or flavor, but also on the speed of processing grapes as they enter the winery and on the capacity of the subsequent steps, such as presses and fermentors.

4.3.1 TRADITIONAL CRUSHERS

Crushers come in two classifications, destemmer–crushers and crusher–destemmers, depending on which operation happens first. Destemmer–crushers (Figure 4.12) are considerably more common. In this case, clusters are loaded into the destemmer–crusher and an auger will move them into a slowly rotating horizontal cage. Within that cage, there are rapidly rotating paddles or fingers, usually in a helical pattern, that knock the berries off of the stems. The berries then fall through the perforated rotating cage, and the stems continue to be propelled out the end of the destemmer–crusher into a waiting bin or conveyor for disposal or composting. The perforated cage rotates to keep the holes from blinding as the grapes accumulate post-destemming. After falling through the rotating cage, the grapes pass through interlocking gears or crusher rollers, where they are broken open and begin to release their juice. They are then collected by the must pump and transferred to the next unit operation. Because the stems do not pass through the crusher, it is thought that the resulting must will have less "stemminess," though the time of contact in the crusher of a crusher–destemmer is truly minimal. An example of a common destemmer crusher is shown in Figure 4.13. Typically, the gap between crusher gears is adjustable to control the degree of berry breakage. Crusher gears are often times removable as well, if whole berries are desired.

A crusher–destemmer crushes the whole cluster in the crusher gears and then destems the broken berries from the stems. It is not clear that there is a significant sensory difference between the two approaches.

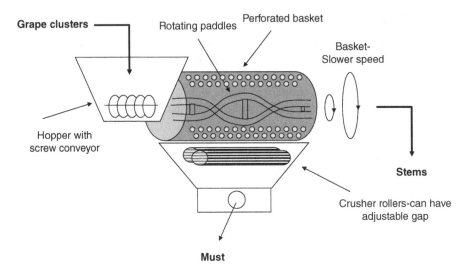

FIGURE 4.12 Schematic of a destemmer/crusher. After fruit is loaded into the hopper, it moves slowly by screw into the large perforated basket that rotates slowly to avoid blinding holes by grape berries. The rotating paddles knock the fruit off the stems. Berries fall through the holes in the basket, and then pass through the crusher gears to release juice. Crushed berries are then transferred out via must pump. Throughput is determined by perforated basket diameter an paddle speed.

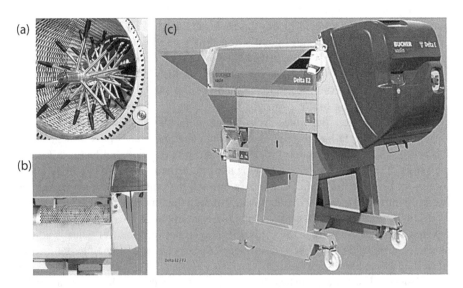

FIGURE 4.13 Images of a destemmer unit. (a) Paddles inside the perforated basked. (b) External view of the perforated basket. (c) Actual destemmer unit, with sides closed for operation. Courtesy of Bucher-Vaslin.

4.3.2 Traditional Crusher Capacity

Traditional crushers are available with capacities of 1–5 tons/hr to as much as 50–100 tons/hr or more. There are two determinants to the capacity. First, the diameter of the rotating cage is important, with increasing diameter corresponding to increasing throughput. Second, the speed at which the helical paddles or fingers turn is also directly proportional to the crusher throughput. While the speed of the paddles is often controllable on commercial models, choosing a rotational speed that is too high will cause partial clusters (stems with some of the berries) to be ejected out the end of the destemmer, and yield will be lost. If the speed is too low, throughput is lost though destemming will be thorough. It is common for one motor to drive the rotating cage, helical paddles or fingers, and both crusher rollers. The change in speed and direction of these four rotating parts is accomplished with a series of gears and chains or belts leading from the main motor shaft.

4.3.3 Non-Traditional Destemmers

Alternative designs for destemming grapes have been implemented that work on a different mechanism. One of these involves motorized fingers that pull the berries off of the stems as they travel up a perforated incline conveyor belt (Figure 4.14). Berries fall through the holes and are collected at a chute at the bottom of the incline. Stems continue to rise on the conveyor and are collected in a bin after the stems fall over the

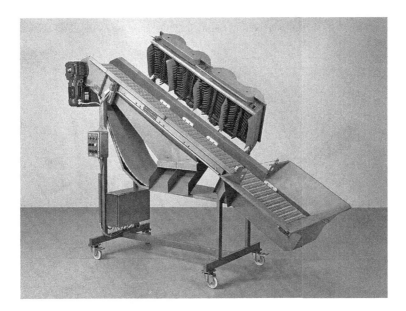

FIGURE 4.14 Euroselect destemmer. This unit works by having rotating fingers pull the berries off the stems as they move along an inclined perforated conveyor. Berries fall through the holes in the conveyor and are collected at the bottom of the incline. Courtesy of Scharfenberger GmbH & Co. KG / Germany.

top of the incline. Another type of destemmer works by feeding the grape clusters through oscillating perforated cylinders and then separates berries from stems using a rolling grate. These destemmers (Figure 4.15), that have capacities up to 8–10 tons/hr (e.g. the Oscillys by Bucher Vaslin), are claimed to reduce MOG and shot berries/raisins. Related to this design is one that acts like a mechanical harvester and knocks berries off the stems (e.g. one from Pellenc) (Figure 4.16).

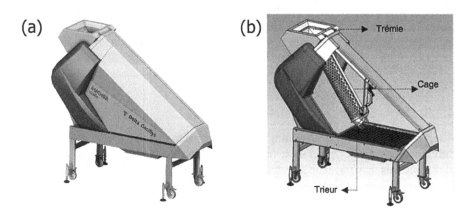

FIGURE 4.15 Oscillating destemmer. This unit works by feeding fruit through an oscillating basket. Fruit falls through the bottom rolling grating, while stems travel off the end of the rollers.

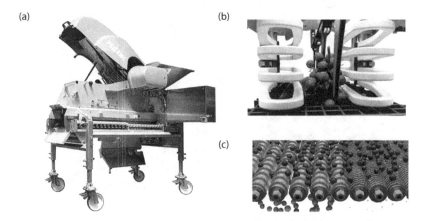

FIGURE 4.16 A Pellenc destemmer. This machine (a) uses fingers (b) to shake fruit off of the stems, similar to a mechanical fruit harvester. Berries, stems, and other solids are then sorted mechanically with a series of rollers. (c) Courtesy of Pellenc.

4.3.4 Sizing Destemmers and Crushers

A destemmer–crusher is best sized by understanding the peak and average expected daily loads of fruit during harvest. As in an example at the end of this chapter, this estimate can be made using information on total expected tonnage of each variety to be harvested and historical harvest dates for a given geographic area. With this information, along with maximum length of workdays (depending on personnel availability), a choice of destemmer can be made easily from manufacturer's throughput information. In larger wineries where there may be separation between processing of red and white grape varieties, different solutions may be needed for each case.

4.3.5 Crusher Location and Stem Removal

The destemmer–crusher location needs to be chosen to facilitate the delivery of grapes to the feed hopper and the removal of stems from the outlet of the rotating cage. This generally means that the destemmer—crusher is located outdoors, close to the fermentors or presses. The feeding process can occur using an elevator, though occasionally destemmers are located below grade so that grapes can simply fall into the feed hopper from a conveyer belt or large receiving hopper above. If the destemmer is at grade, stems can be removed in bins by forklift or pallet jack for a small to medium operation. Larger operations, or when destemmers are below grade, will necessitate the use of conveyers or elevators for stems to a higher throughput system for stem removal. Stems can be removed whole and spread in the vineyard or chopped first with a specialized cutting apparatus.

4.4 DRAINERS

A drainer is a piece of equipment designed to facilitate the separation of skins from free run juice. While this is generally accomplished in a press, or at least while loading a press, there are times when adding this extra step can be desirable. In the case where some skin contact is desired prior to pressing, for instance, use of a drainer can be useful and more efficient than other options. This would include cases where the contact would be quite limited, such as rosé wines from red grapes. It would also include the idea of extended skin contact for white wine production which tends to go in and out of favor. For these reasons, it is important to discuss the idea here. In addition, the general design of these drainers that dates back to the 1960's, has more recently led to a similar interesting design for red wine fermentors that will be discussed in Chapter 6.

A basic drainer (Figure 4.17) is an upright, cylindrical tank, many times with a conical retort bottom (i.e. pointed to the side toward a press). The must is loaded into the drainer and juice will travel through side-mounted screens to be collected as "free run." As the juice continues to drain by gravity, skins will build up around the screen and act as a depth filter (to be explained in more detail in a latter chapter) and actually clarify the solids out of the juice to the point where additional settling and racking are not necessary prior to inoculation and the beginning of fermentation. When all of the free-run juice has drained, a large guillotine-type valve is opened at the bottom of the

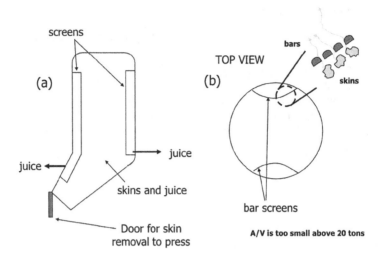

FIGURE 4.17 Schematic of a drainer. (a) Side view of a drainer showing the location of the screens and door for removal. (b) Top view of the drainer with an inset illustrating retention of skins by a section of bar screens.

drainer and the solids are emptied by gravity into a press mounted below. To make this set up more efficient, multiple drainers are clustered to feed a single press (Figure 4.18). An additional benefit of drainer use is that it allows for a smaller-capacity press, as free run is diverted from the press and only the de-juiced solids will be present in the press. Above a capacity of about 20 tons of grapes, drainers become impractical as the ratio of screen surface area to must volume will decrease with increasing size, thus increasing the draining time significantly.

FIGURE 4.18 An example of four jacketed drainers mounted over a press.

In addition to increased skin contact, another potential issue with drainers is a lack of temperature control. While efforts have been made to equip drainers with cooling jackets, the heat transfer possible through this arrangement is very limited as there is minimal mixing in a drainer.

4.5 PRESSES

Presses represent a unit operation in which the main purpose is to separate juice or wine from skins. This includes free run, as well as juice or wine still associated with the skins that requires energy in the form of mechanical pressure to facilitate the release. This unit operation can be batch or continuous, though batch presses that are repeatedly loaded and unloaded, are more common—especially at small to medium size wineries. As pressure is increased, either over time in a batch press or over the length of a continuous press, the characteristics of the juice or wine change. This fact can serve as both a complication and an opportunity for the winemaker. These differences are a function of both the press design and the operating parameters chosen by the winemaker. The various types of common presses are described in the next sections.

4.5.1 BATCH PRESSES

When specifying and operating a batch press, it is important to consider all parts of the cycle, as they all take time and will affect throughput. These steps include filling, applying pressure, and releasing pressure (sometimes repeatedly), and then emptying the skins and other solids. Cleaning in between lots of grapes can also add significantly to cycle times. The filling time is determined by the design and size of the press and the means for filling it, which could include a must pump from a destemmer–crusher (e.g. white juice) or tank (e.g. red wine) or from a conveyer belt or elevator. In some cases where grapes are pressed whole cluster (direct-to-press), large presses are filled by tipping grapes from bins directly into the press through a hopper using a forklift. The press cycle is made up of one or more steps in which the pressure is increased to some value, held, and then released. For many batch presses, the equipment has the ability to mix or tumble the solids in between pressings to redistribute solids and make dejuicing more efficient. The pressure on presses is gradually increased following a schedule to achieve the desired juice quality and yield. Some winemakers will separate "free run" juice from juice released at progressively higher pressures, as phenolic extraction will increase as conditions are made more stringent in the press. A complete press cycle is likely to take 3–4 hr.

Batch presses have evolved over the years and many types are still used. They range from simple basket presses used by home and small-scale winemakers to bladder and membrane presses in which compressed air is used to press the skins against stainless steel screens.

4.5.1.1 Basket Presses

Large hydraulic basket presses (Figure 4.19) have enjoyed a resurgence of popularity in recent years. This is most likely for three reasons. First, the technology has improved including: (1) the baskets are now larger and constructed out of stainless steel, (2) the baskets can be removed and repositioned by forklift, and (3) the control

(a)   (b)

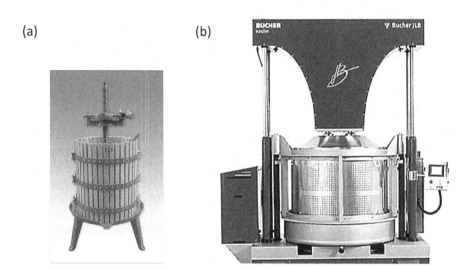

FIGURE 4.19 Basket presses. (a) A traditional basket press with wood slats and a manual capstan (courtesy of Greco Store S.A.S.) and (b) a commercial scale hydraulic basket press with a stainless steel forkliftable basket (Courtesy of Bucher-Vaslin).

of pressure is more precise. Second, there is a perception that basket presses are more gentle than other types of presses. This perception may, in fact, come from the reality that this type of press is far less efficient than other types in dejuicing skins as discussed further below. Third, wineries have found that there is marketing appeal to using basket presses.

The efficiency of dejuicing has to do with the design of these presses and the fact that the pressure is applied to the skins perpendicular to the flow of the juice. As the head is lowered on the basket of skins (and other solids), juice or wine flows radially from the center of the press cake out through slots or perforations in the basket. As the head continues to lower, the channels through which the juice or wine is flowing become smaller and smaller until they are shut off completely by the compression of the cake. Because this typically happens before all liquid is removed from the center of the cake, it is common to end up skins that are still quite wet in the middle of the cake, while the outer skins are quite dry as shown in Figure 4.20. Because the yield cannot reach that of the other types of presses described below, even by raising the head stirring the skins and re-pressing, less phenolic/skin extraction is likely to occur. While other types of presses can achieve skins that are nearly "paper dry," this does not occur with a basket press. Shear forces do form along the side of the basket as the moving head is moving tangentially to the basket (as opposed to perpendicular to the basket).

4.5.1.2 Moving Head Press

A moving head press is essentially a horizontal basket press. In this type of press, grapes are loaded into the cylindrical tank with a screen on its side. Either one moving head or two moving heads (at opposite ends of the cylinder) are then activated to

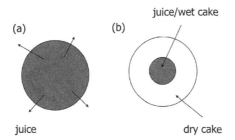

FIGURE 4.20 Schematic of the top view of a basket press (a) before pressing and (b) after pressing. Notice that the outer part of the pressed cake acts as a filter for the juice from the center, eventually blinding and leaving the center of the cake wet with lost liquid.

press the grapes as the volume decreases. The juice will drain out of the screen. The heads are often connected by chains that act to break up the press cake as the heads recede and the chains become taught. This action acts to improve the yield of subsequent pressings. Here again, shear forces do form as the heads move along the screen and the pressing action is perpendicular to the liquid flow.

4.5.1.3 Membrane and Bladder Presses

The most common type of press being used today in commercial wineries is the membrane press. Membrane presses are composed of a horizontal stainless steel cylindrical tank with a sturdy membrane mounted across the center of this cylinder lengthwise. To load the press, the membrane is drawn back using a vacuum to expose the entire cylindrical volume. Presses are either loaded through one or two sliding doors on the side of the cylinder or through an axial feed on a swivel mount on the end of the cylinder. Either way, loading is periodically paused to close the press and rotate it to evenly distribute the loaded solids. After loading, the press is sealed and the on-board air compressor is used to inflate the membrane to squeeze the solids again the opposing side of the cylinder so that juice will flow into perforated juice channels mounted on this wall. The juice or wine is then collected in the press pan below the press, from which it is pumped back to a tank for further processing (see Figure 4.21 for a schematic and examples of a Bucher press). Willmes, a different press manufacturer, has employed a slightly different technology, incorporating two membranes with central juice channels (Figure 4.22).

There are many alternative pressing strategies employed, ranging from successive stages of increasing pressure, say 0.5, then 1, then 2 atm, with tumbling for cake breaking between each pressure, to two or three cycles of the maximum of 2 atm. Some membrane presses are now being equipped with flow meters that can be used to control the membrane pressure to achieve a constant flow rate, though this feature is not yet being widely used in the industry. Another additional feature that can be specified is having a reservoir of inert gas attached to the press to minimize oxygen contact when the membrane is deflated in between pressings.

Della Toffola manufactures a bladder press that has a central inflatable bladder that presses evenly against all sides of the surrounding cylinder (see Figure 4.23). In one Australian study, it was shown that this press gave a higher yield with less

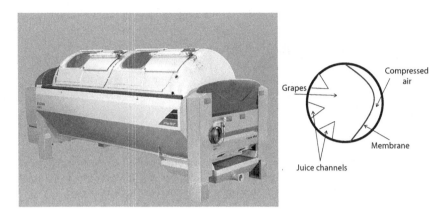

FIGURE 4.21   A membrane press. Courtesy of Bucher-Vaslin.

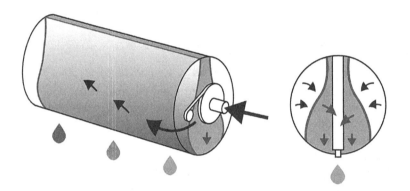

FIGURE 4.22   Membrane press with central juice channels (M. Ogawa).

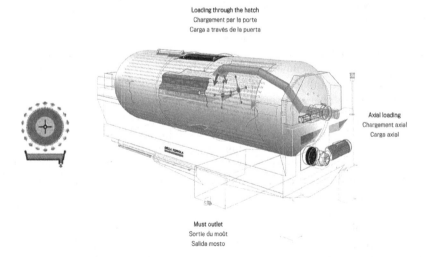

FIGURE 4.23   Central bladder press. The central bladder inflates to press an annulus of grapes. Courtesy Della-Toffola USA.

phenolic extraction. This could be related to the fact that all pressure is perpendicular to the press surface, thus reducing shear forces (and potentially skin tearing).

4.5.2 PRESS CAPACITY FOR GRAPES AT VARIOUS STAGES OF PROCESSING

Batch presses are available in capacities from 4 tons to over 100 tons. As can be seen in the attached table from Bucher (Figure 4.24), press capacity changes with the type of must being processed. For example, for a given press size, the press can hold macerated skins (e.g. from a red must) from 1.5 to 2 times as much original grape weight than for crushed, destemmed (but unmacerated) white grapes. For whole cluster pressing, the press capacity is approximately 1/3 of that for crushed, destemmed grapes. Membrane presses will have a practical minimum capacity as juice channels need to be covered with grapes to apply pressure. Practically, this means that they will likely not function well under 20% of their stated capacity. In sizing presses, peak daily loads during harvest should be considered. Press cycles can take up to 3 hr or longer when loading, emptying, and cleaning are also considered. This, together with planned work schedules, can help determine the capacity and number of presses needed.

4.5.3 CONTINUOUS PRESSES

While batch presses go through cycles of filling, pressing, and emptying, continuous presses, as the name implies are continuously doing all three steps. This is accomplished using a large auger or helical screw to convey grapes against a restriction that

Dimensions and technical data

XPert IT	Whole grape* (kg)	Destemmed grape* (kg)	Macerated grape* (kg))	Rated power (kW)	Length (mm)	Width (mm)	Height (mm)	Empty weight (kg)	Trough capacity (L)
100	6000	16000 to 20000	30000	11,6	5756	2200	2580	4000	460
115	6900	18400 to 23000	34500	14,1	6255	2200	2520	4150	460
150	9000	24000 to 30000	45000	14,1	6313	2455	2720	4550	460
250	15000	40000 to 50000	75000	19	7143	2925	3350	6800	850
320	19200	51200 to 64000	96000	20,7	7726	3100	3430	8100	850
450	27000	72000 to 90000	135000	34	9070	3340	3420	12000	1300

* For information only, those data can change depending on grape variety, ripeness and filling conditions. Grapes weight used before destemming or maceration.

FIGURE 4.24 Capacity data for Bucher-Vaslin presses. Note the difference in capacity between fruit and must for the same unit. Over fermentation, skins release remaining juice, thereby taking up less volume. Therefore, whole clusters take up more volume than destemmed and crushed fruit which takes up more volume than macerated skins, when considering the same initial weight of fruit. Courtesy of Bucher-Vaslin.

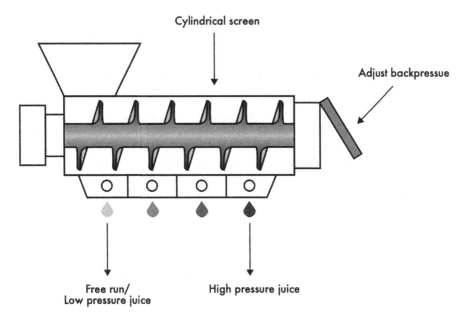

FIGURE 4.25   Schematic of a continuous press (M. Ogawa).

builds pressure (Figure 4.25). Grapes enter through a receiving hopper and continuously move along with the screw at higher and higher pressure until they emerge from the restriction at the end of the screw completely pressed and free of most juice. This restriction is typically adjustable and is made up a hinged door that can be opened or closed to supply more pressure at the end of the screw. Along the length of the screw, juice is collected at multiple points through screens below the screw. These collection points correspond to increasing pressures as the collection points get closer to the exit restriction. These fractions can then be combined or kept separate, depending on the style of intended wine. Continuous operation and the large size of these units (both the helical screw and the cylinder in which it is contained) means that they are capable of significantly higher throughput than most membrane presses. Throughputs of 50–100 tons/hr are possible, so these units are typically only found in large wineries.

While screw presses can increase throughput due to their continuous nature, they do have their disadvantages as well. First, the nature of their pressing mechanism can cause greater damage to grape skins and seeds than other types of presses, thus increasing phenolic extraction above that found with other presses. Other chemical constituents of the grapes, such as minerals and structural carbohydrates, may also be over-extracted using this equipment, which may alter downstream juice preparation or wine clarification. Second, the solids content of the resulting juice will likely be higher than from batch presses. This can lead to the necessity of additional steps for juice clarification prior to inoculation, as high juice solids content has been associated with lower resulting wine quality.

4.6 DISSOLVED AIR FLOTATION

An emerging technology for large-scale wineries is the use of Dissolved Air Flotation (DAF) for juice clarification. DAF is a technology initially developed for wastewater solids removal but has adapted surprisingly well to the winery environment. Figure 4.26 illustrates the process flow diagram for a DAF system.

In DAF, either fresh or recycled juice is dosed with a flocculant (a material which makes solids clump together) such as bentonite, and then fed into a closed pressure vessel, known as the Air Drum. The juice is then dosed with high-pressure air. The high-pressure operation of this tank ensures that air is dissolved at higher concentrations than would be possible at 1 atm. The pressurized juice is then pumped into an open-top tank. This sudden drop in pressure (to atmospheric pressure) causes air to come out of solution. This air forms bubbles on the surface of solids. These bubbles cause the solids to float to the top of the tank, where they can be skimmed off, as seen in Figure 4.27.

There are two major process variables available to control the level of solids removal. First, flocculant dosing rates can be increased or decreased, with higher flocculant rates promoting solids removal. Note that there are limiting returns on flocculant dosing, and beyond a certain addition rate, there will be no improvement in overall solids removal. Second, the residence time of juice in the open-air tank can be adjusted, with high residence times (slower feed and draw rates) resulting in superior clarification, at the expense of reduced throughput.

DAF is well suited to juice handling, as juice oxygenation does not necessarily have a negative impact. DAF is not employed post fermentation, where oxygenation can lead to major wine faults.

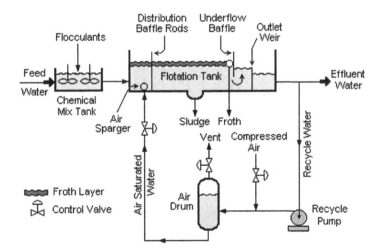

FIGURE 4.26 Process flow diagram of a dissolved air flotation system. By Mbeychok, CC BY-SA 3.0, https://commons.wikimedia.org/w/index.php?curid=2619433

FIGURE 4.27 Open top of a winery dissolved air flotation unit. Note the rotating paddles, which suck up and remove the solids that have floated to the top surface. Clarified juice is removed beneath this level, with a weir to ensure no top-floating solids are entrained.

4.7   SIZING GRAPE RECEIVING AND CRUSH EQUIPMENT

Example 4.1:   Sizing Grape Processing Equipment for the Weeping Mountain Winery

THE WEEPING MOUNTAIN WINERY CONCEPT (TIANA REED)

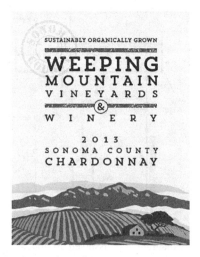

The Weeping Mountain Vineyards and Winery (WMVW) is a moderately sized winery located on Sonoma Mountain in the newly established AVA, Petaluma Gap. WMVW strives to achieve and maintain a level of sustainable and organic

viticulture practices and wine production in order to preserve the breathtaking grounds from which the grapes are grown. All grapes are harvested from Estate blocks located on 200 acres of land. Sonoma Mountain, previously coined "Weeping Mountain" by native ancestors, produces well-balanced grapes for the production of well-balanced wines due to the long growing season as a result of its foggy cool climate. This, in turn, allows for sustainable and organic agriculture practices. The Chardonnay will be fermented in stainless steel, with half of it aged in stainless steel and the other half aged in French Oak for 9 months. The Pinot noir will be fermented in stainless steel and aged in French Oak for 9 months. The Primitivo will be fermented in stainless steel and aged in stainless steel for 9 months. Luma Fusion will combine Merlot, Grenache and Primitivo. All three will be fermented separately in stainless steel. The Merlot will be aged in stainless steel for 9 months, the Grenache will be aged in stainless steel for 9 months, and the Primitivo will be aged in stainless steel for 6 months, and the American Oak for 3 months.

Front labels display the majestic "Weeping" Sonoma Mountains, vineyards, and morning fog found in the relaxed wine country farm town of Petaluma, while the back label describes the history of the land, engaging the particularly environmentally aware consumer demographic between 25 and 60 years. This brand also invites those with an appreciation for nature, history, and community. Additionally, a "petting zoo" with baby doll sheep (and other small farm animals) used to organically maintain the vineyards allow for a family friendly environment.

PROBLEM

The challenge in sizing equipment for this winery is that the Chardonnay and Pinot noir deliveries will overlap significantly, making most of the grapes predicted to be delivered within a very short period of time with just a few other types of grapes delivered later. Given the necessary tons of each of five types of grapes, size the equipment for grape processing for the Weeping Mountain Winery.

SOLUTION

In order to size equipment, we first need to predict a typical delivery schedule for grapes for this winery. To do this, we start with a table of expected grape tonnage for each variety and the expected mean delivery date.

Grape	Tonnage (Tons)	Mean Delivery Date	Standard Deviation (Days)*
Chardonnay	360	Day 9	3
Pinot noir	360	Day 12	3
Primitivo	121.5	Day 32	3
Merlot	45	Day 22	2
Grenache	13.5	Day 30	2

* Delivery range will be broader for a greater number of vineyards.

Mean delivery dates for grapes can be estimated using one of two sources: (1) the book "Wine Grape Varieties of California" or (2) McIntyre et al., AJEV, 33(2): 80–85, 1982. In the former source, grape varieties are categorized from early to late varieties. Then, typical harvest dates are given for each of these categories. We can use these data to plot the expected grape delivery pattern for the Weeping Mountain as can be seen in this plot.

Predicted Weeping Mountain Grape Delivery Schedule

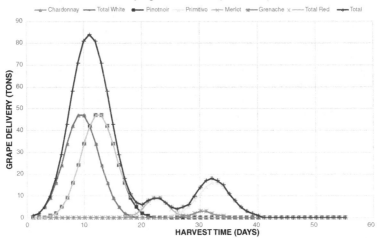

From this chart, we can start to make decisions about the unit operations in the grape receiving area of this winery.

For sorting the grapes, we could use no sorting, manual sorting, or an optical sorter. As there does not seem to be a special need for a more expensive optical sorter in this case, we will likely choose a manual vibrating sorting table. This will require multiple cellar workers to remove MOG.

As the Chardonnay will go direct-to-press without destemming, our peak day for destemming will be at the peak of Pinot noir where we expect 47 tons/day of fruit. If we would like to keep the destemming to 8–9 hr, where possible, two possibilities would be a Bucher Delta E2 which is rated at 12–20 tons/hr or an Oscillys 200 rated at 8–10 tons/hr. While either of these will work for the job, we chose an Oscillys 200 to reduce MOG, shot berries, and raisins as this was important for this particular application. The Oscillys is currently about twice as expensive as the Delta E2, however, so if capital outlay is important, we may make a different decision.

For presses, it is likely that the whole cluster pressing of Chardonnay will determine the size of the press. For the Chardonnay, the largest lots are 16 tons and the largest delivery day for Chardonnay will likely be 47 tons. The largest pressing day for red wines will be around day 22 or 23 when we will be pressing the peak of Pinot noir—also about 47 tons of fruit from three fermentors. Using this information, we can look at a table of possible choices, in this case from Bucher Vaslin.

Press	Whole Cluster (Tons)	Destemmed White (Tons)	Fermented Red (Tons)
XPlus 50	3	8–10	15
XPlus 62	3.7	9.9–12.4	18.6
XPlus 80	4.8	12.8–16	24
XPert 150	9	24–30	45
XPert 320	19.2	51–64	96

Given these data, a few possible scenarios were considered. For Option 1, we could have a XPert 320 for whites and a XPlus 62 for reds. This would allow us to press whites in three loads on our peak day and to press reds in three loads on that peak day. At approximately 3 hr/cycle, we could achieve this in a 9-hr period which would be reasonable during harvest. Option 2 would be two XPert 150 presses. These would be run in parallel for Chardonnay and one would be used for reds. This would be on the edge of working for some of the smaller lots as we would like to operate them at greater than 20% of their stated capacity. Option 3 would be to purchase two XPlus 62 presses. These would be run in parallel to whole cluster press about 8 tons of white grapes, and individual presses would work well for reds. If we change our mind and decide to destem whites, this option would work really well. In the end, we decided on Option 3 with the thought that we would make a case-by-case call on whether to press whole cluster or destem depending on style requirements and expected delivery on any given day.

The layout of the crush area would be as shown in this diagram, with parallel presses fed from a single belted delivery hopper, vibrating manual sorting table, and destemmer for whites—all situated centrally to the tanks. For reds, the presses could be moved into the winery or the must transferred out of the winery.

Tank Level

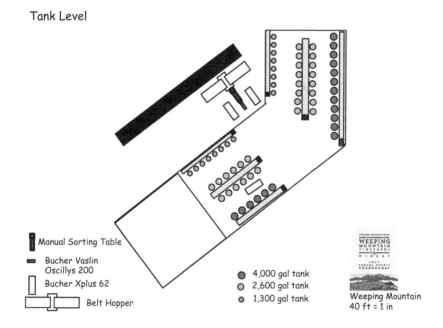

Manual Sorting Table

Bucher Vaslin Oscillys 200

Bucher Xplus 62

Belt Hopper

● 4,000 gal tank
○ 2,600 gal tank
◉ 1,300 gal tank

WEEPING MOUNTAIN VINEYARDS
WINERY
1813
SONOMA COUNTY
CHARDONNAY

Weeping Mountain
40 ft = 1 in

REFERENCES

Boulton, R.B., Singleton, V.L., Bisson, L.F., and Kunkee, R.E. *Principles and Practices of Winemaking*. Chapman and Hall, New York, 1996.

Christensen, L.P., Dokoozlian, N., Walker, M.A., and Wolpert, J. (editors). *Wine Grape Varieties in California*. University of California Agriculture and Natural Resources, 2003.

Falconer, R., Liebich, B., and Hart, A. Automated Color of Hand-Harvested Chardonnay. *American Journal of Enology and Viticulture*, **57**, 491–496, 2006.

Hendrickson, D.A., Lerno, L.A., Hjelmeland, A.K., Ebeler, S.E., Heymann, H., Hopfer, H., Block, K.L., Brenneman, C.A., and Oberholster, A. Impact of mechanical harvesting and optical berry sorting on grape and wine composition. *American Journal of Enology and Viticulture*, **67**, 385–397, 2016.

McIntyre, G.N., Lider, L.A., and Ferrari, N.L. The chronological classification of grapevine phenology. *American Journal of Enology and Viticulture*, **33**, 80–85, 1982.

PROBLEMS

1. **[Integrative Problem]** After several years at Puddle Bottom Cellars, you get a chance to be head winemaker at a medium-sized winery in the Central Valley of California called Desert Oaks. The idea is to make a high-quality red wine from relatively inexpensive fruit. You are satisfied with the aroma of the Cabernet grapes you are receiving, but you would like to get more tannin extraction to improve the mouthfeel.

   (a) You are currently using a membrane press to remove the skins. What other type of press could you use that might increase phenolic extraction? Explain your answer by describing how that press works compared to the membrane press.

   (b) When you install this new press, the press is actually elevated so that gravity flow can be used to fill barrels directly below from the press pan/juice collection tank. If you want to fill two 225 L barrels in 5 min through a 60 ft long, 2 in (51 mm) diameter hose, how high does the press pan have to be above the barrel?

5 Brewery Upstream Processing

Brewing is a critical first step in beer production, as well in the production of whiskeys and grain-based vodkas.

5.1 BREWERY UPSTREAM PROCESS FLOW

Fermentation processes are typically separated into two categories: "upstream," or raw materials preparation through fermentation, and "downstream," what happens to the material post-fermentation. Beer production is extremely similar to wine production in fermentation and downstream processing, but quite distinct in pre-fermentation processing. Wine raw materials (i.e. fruit) contain readily fermentable sugars (i.e. fructose and glucose), which need only be expressed from the fruit for fermentation. Beer, on the other hand, requires first the extraction of carbohydrates, followed by the enzymatic conversion of polysaccharides to fermentable sugars. This fermentation media is referred to as a "mash," akin to juice or must in winemaking.

It is important to note that harvested grain must go through several processing steps by a maltster prior to being ready for brewing. The *malting* process involves steeping the harvested grain to allow it to germinate, which produces the enzymes needed to convert the starches to fermentable sugars, followed by kilning to arrest germination. In this chapter, we assume that the brewer has sourced prepared cereal grains from a maltster, and so will focus exclusively on the brewing process.

The brewing process typically follows the process flow, as shown in Figure 5.1: *milling*, where grain particle size is reduced to enhance mass transfer and extraction; *mashing*, where starches are extracted from grain and then enzymatically converted to fermentable sugars; *lautering/solids removal*, where the mash is filtered to remove grain particles; and *brewing/kettling/boiling*, where the mash is boiled in the presence of hops to (i) extract hop acids and oils, (ii) concentrate that mash, and (iii) sterilize the mash. Trub must be removed post boiling. Finally, beer is typically *aerated* prior to fermentation to ensure sufficient dissolved oxygen (DO) for an effective fermentation.

5.2 MILLING

A mill is a unit operation which reduces particle size by mechanical attrition. For example, the standard rotating blade coffee grinder is a mill, more specifically a *burr mill*. Numerous types of mills exist, but the most common employed by the brewing industry is the *roller mill*.

Roller mills operate by feeding solids between counter-rotating cylinders spaced a fixed distance apart. Particles are forced through the gap, and any particles larger than the gap are crushed. This gap is typically adjustable, which allows for control of

DOI: 10.1201/9781003097495-5

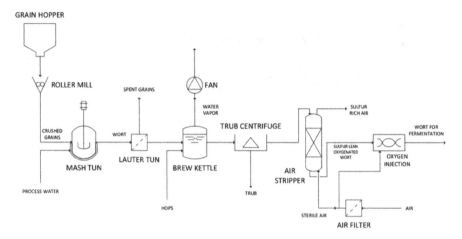

FIGURE 5.1 Brewing upstream (pre-fermentation, or "hot side") process flow diagram.

the product particle size. Figure 5.2 shows a schematic of a *two-roller mill*, while Figure 5.3 shows an actual two-roller mill. This is the simplest time of roller mill, where only one set of rollers is present.

Two-roller mills are common in smaller breweries, as they are mechanically very simple and typically require lower capital expenditure. Multi-roller mills permit finer control of particle size and can also be used to separate the grain husk from the seed without damaging the husk, an important consideration for lautering, where it acts as a filter aid.

While milling can control particle size, there is a question of what size particle should be the goal of this unit operation. This question can be answered by

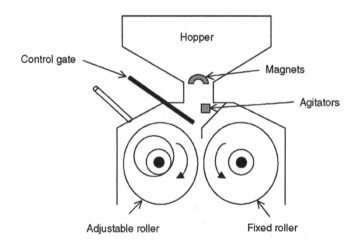

FIGURE 5.2 Schematic of a two-roller mill. Source: Government of Saskatchewan https://www.researchgate.net/figure/Flour-roller-mill-Source-C-2008-Government-of-Saskatchewan_fig7_287312732

FIGURE 5.3   A two-roller mill. The distance between rollers can be set to dictate how fine the grind size is. Courtesy of The Vintage Shop. https://www.morebeer.com/products/monster-mill-mm2-grain-body.html?gclid=EAIaIQobChMI96aw4PG96AIVlONkCh0UxAXNEAQY CCABEgKY-vD_BwE

examining the role of particle size in starch extraction. Extraction is a mass transfer driven process, which follows Fick's Second Law:

$$\frac{dc_i}{dt} = D_i \nabla^2 c_i \tag{5.1}$$

which states that the rate of concentration change of species "i" by diffusion $(\frac{dc_i}{dt})$ is a function of the diffusion coefficient (D_i) and the concentration gradient $(\nabla^2 c_i)$. We can envision a grain particle as a sphere with radius "R" and imagine that starch is being extracted into water in a well-mixed system, such that there is no diffusion barrier from the outer edge of the grain particle into solution. With some manipulation, we can derive that the fraction of starch extracted as a function of time, F(t), is:

$$F(t) = \frac{6}{\pi^2} \sum_{n=1}^{\infty} \frac{1}{n^2} e^{-D_{starch} n^2 \pi^2 t / R^2} \tag{5.2}$$

Practically speaking, this means that the amount of starch extracted increases with time and the starch's diffusion coefficient, and decreases with increasing particle radius. A brewer cannot meaningfully change the diffusion coefficient of starch in grain, though they likely wish to minimize processing time. As such, creating smaller particles (decreasing R) is the most straightforward way of aiding starch extraction. This means milling.

It is important to note that, while grain particle size reduction is critical to starch extraction during brewing, excessive milling can lead to its own host of problems. Mash viscosity is a strong function of particle size, with mash eventually becoming an unfilterable paste. As such, it is critical to be able to quantify particle sizes leaving the mill in order to determine the desired mill gap spacing.

Grain does not leave the mills in a single, discrete particle size. Instead, crushed grain will have a range of particles, with some being rather large (>3 mm) and some

FIGURE 5.4 A sieve stack tester, used for determining the particle size distribution. Particles fall down the mesh of each pan, until they reach a pan with a smaller grate size than the particle itself. The stack is usually placed on either a vibratory table or mounted with an impact hammer to promote sorting. Courtesy of Forestry Suppliers. https://www.forestry-suppliers.com/product_pages/products.php?mi=74891&itemnum=53636&redir=Y

being quite fine (<0.1 mm). This *particle size distribution* can be readily measured by *sieving*, a simple analytical technique where particles are fed into a stack of mesh sieves (Figure 5.4), which is then agitated to promote settling. Coarser sieves are on top, and finer sieves down below. After a fixed amount of agitation time, the sieves are separated, and the mass of particles retained in each sieve are weighed. If a particle is trapped above a sieve, it must be larger than that sieve but smaller than the sieve above it. In this way, the particle size distribution of the ground malt leaving the mills can be quantified.

5.3   MASHING

After milling, grain is transferred to a mashing vessel, often referred to as a *mash tun* or *mash tank* (Figure 5.5). The mash tun, as a piece of equipment, serves two purposes. First, the mash must be heated to promote the enzymatic conversion of starches

FIGURE 5.5 Interior of a mash tun. Temperature is typically controlled via a steam coil under the liquid level, with agitation to ensure homogeneity and good mass and heat transfer. By Ikiwaner (talk): Own work, CC BY-SA 3.0, https://commons.wikimedia.org/w/index.php?curid=2199333; https://www.undiscoveredscotland.co.uk/usfeatures/maltwhisky/mashtun.html

to fermentable sugars. Second, the mash must be agitated to ensure the proper leaching of starches, proteins, and other components from the grain particles. The agitation also ensures even heating throughout the process.

Agitation is typically achieved by use of impellers, which are submerged into the liquid and spin to provide mixing. Mixing power is a function of fluid properties (denser, more viscous fluids are harder to mix), impeller design, and rotation speed (i.e. rps, rotations per second). The power required to turn an impeller can be estimated from Figure 5.6, where the Power Number (N_p), a dimensionless number, is defined:

$$N_p = \frac{Pg_c}{\rho N^3 D^5} \tag{5.3}$$

where P is the power input required, g_c is the mass to force conversion constant (i.e. 32.2 (lb_m·ft)/(lb_f·s^2)), ρ is density, is rps (in 1/s), and D is impeller diameter; and where the impeller Reynold's Number (Re) is defined:

$$Re = \frac{D^2 N \rho}{\mu} \tag{5.4}$$

where μ is the dynamic viscosity of the fluid. It should be noted that this version of the Re is defined differently than the one for fluid flow in a pipe from Chapter 3, but has the same general form, as the velocity used previously was a linear velocity and the relevant velocity here is an angular velocity proportional to N × D. In any of these cases, it is important to note that these relationships only hold when dimensionless numbers are truly dimensionless. Be sure to use units for each variable that make this true!

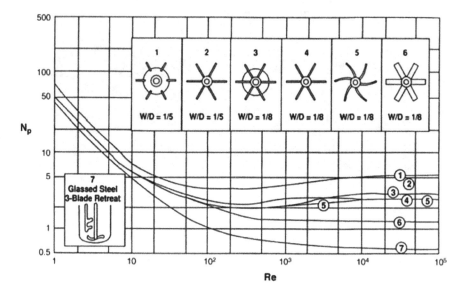

FIGURE 5.6 Power Number versus impeller Reynold's Number for various impeller geometries, as calculated by D.S. Dickens. "W" is the height of the impeller blade. https://onlinelibrary.wiley.com/doi/pdf/10.1002/0471238961.1309240908051318.a01.pub3.

Example 5.1:   Power Required to Stir 5,000 gal of Mash

You are designing a mash tun that will need to process up to 500 gal of mash. The mash is expected to have the following properties:

$$\mu = 13 \times 10^{-4} \, lb_m / (s\,ft)$$

$$\rho = 67 \, lb_m / ft^3$$

You plan on a 10 ft diameter impeller, with flat blades and an impeller breadth to diameter ratio of 1/5 (i.e. Impeller #2 in Figure 5.4). You plan on a maximum stirring rate of 100 RPM. Determine the power required for such a mixer.

First, calculate the impeller Reynolds number:

$$Re = \frac{D^2 N \rho}{\mu} = \frac{\left(10ft\right)^2 \left(\dfrac{100}{60s}\right) 67 \, lb_m / ft^3}{13 \times *10^{-4} \, lb_m / (s\,ft)} = 8.6 \times 10^6$$

From Figure 5.4, curve #2 levels off at Np ~ 5 for Re > 104, so the Np = 5 for a single impeller. From this, we can calculate

$$N_p = \frac{Pg_c}{\rho N^3 D^5} \rightarrow P = \frac{N_p \rho N^3 D^5}{g_c} = \frac{5\left(67\,\text{lb}/\text{ft}^3\right)\left(\dfrac{100}{60\text{s}}\right)^3 \left(10\text{ft}\right)^3}{32.2\left(\text{lb·ft}\right)/\left(\text{lbf·s}^2\right)} = 48{,}165\,\frac{\text{ft}\times\text{lbf}}{\text{s}} = 87\,\text{hp}$$

This impeller will require 87 hp of work to turn it.

It is important to note that in Figure 5.5, different impeller performance is due to the differing amounts of fluid flow resistance and shear at the impeller. Note that the curve for configuration 6, a pitched blade impeller, is substantially lower than for configuration 4, a flat blade impeller of the same geometry. While the impact of impeller design and placement in mash tuns has not received substantial scrutiny, it is important to realize that different designs will have different power requirements and varying amounts of shear.

The mashing vessel serves a second critical function: heating. While recipes vary vastly between breweries, a mash must always be heated to >65°C. This is almost always achieved through heat exchange between saturated steam, utilizing steam heating coils inside the volume of the mash tun.

This kind of heat transfer can be analyzed using the enthalpy balance approaches outlined in Chapter 2.

Note that steam feed rate can be controlled via a feedback controller, with the caveat that this often leads to temperature undershoot and overshoot in non-optimized control loops. The authors have had good success with feed forward systems, which incorporate calculations like those in Example 5.2. Control systems are discussed in more detail in Chapter 12.

Example 5.2: Wort Mash-In Heating

You are processing 10,000 L of mash which needs to be heated from 25°C to 70°C. The mash has a specific heat of 3.4 kJ/(kg × °C) and a density of 1.2 kg/L. Your process steam is saturated steam at 121°C, which has a latent heat of condensation of 2,200 kJ/kg.

(a) *How much heat will be required to warm the mash?*
(b) *If the temperature change takes place in 30 min, what is the power requirement?*
(c) *What saturated steam flow rate will be required to satisfy this power requirement?*

SOLUTIONS

(a) $Q = m \times Cp \times \Delta T$

$m = V \times \rho = 10,000 \text{L} \times 1.2 \text{kg} / \text{L} = 12,000 \text{kg}$

$\Delta T = 70°C - 25°C = 45°C$

$Q = 12,000 \text{kg} \times 3.4 \text{kJ} / (\text{kg}°C) \times 45°C = 1,836,000 \text{kJ}$

(b) Power $= P = Q / \Delta t = 1,836,000 \text{kJ} / (30 \text{min} \times 60 \text{s} / \text{min}) = 1,020 \text{kW}$

(c) $Q = m_{steam} \times \Delta H_{evap}$

$P = dQ / dt = (dm_{steam} / dt) \times \Delta H_{evap}$

$dm_{steam} / dt = P / \Delta H_{evap} = (1,020 \text{kJ} / \text{s}) / (2,200 \text{kJ} / \text{kg}) = 0.464 \text{kJ steam} / \text{s}$

Saarni et al. (2020) determined relationships for "Total Extract Yield" (kg total extracted into mash/kg grain added), "Fermentable Extract Yield" (kg fermentable sugars extracted/kg grain added), and "Fermentable Extract" (kg fermentable sugars/kg total extract) as functions of temperature ("T," in °C), the grist ratio (GR, kg water/kg grain), pH (pH units).

Y_{TE} (Total Extract Yield)

$$Y_{TE}(\%GW) = -262.72 + 4.67(T) + 81.53(pH) - 14.72(GR)$$
$$+ 0.20(T \cdot GR) - 0.04(T^2) - 7.85(pH^2) \tag{5.5}$$

Y_{FE} (Fermentable Extract Yield)

$$Y_{FE}(\%GW) = -1,367.34 + 27.14(T) + 200.40(pH) - 23.95(GR)$$
$$- 0.38(T \cdot GR) - 0.21(T^2) - 18.57(pH^2) \tag{5.6}$$

FE (Fermentable Extract)

$$FE(\%TE) = -1,411.83 + 29.44(T) + 177.10(pH) + 56.07(GR)$$
$$- 0.66(T \cdot GR) - 0.22(T^2) - 16.16(pH^2) - 1.88(GR^2) \tag{5.7}$$

These relationships were derived for 100% barley mashes (i.e. no adjunct additions) in a 90 min mashing period.

Occasionally, temperature swings may be incorporated in the mashing recipe, necessitating a cooling step prior to mash out. Cooling water is often added directly to the mash (as opposed to through a heat exchanger) to rapidly drop the mash temperature; mash tuns with built in chillers are exceedingly rare. Again, an enthalpy balance, this time to calculate the average temperature, can be useful. As such, mash recipes are designed with the addition of extra water in mind.

Finally, the mash must undergo mash out. This final temperature increase serves two purposes. First enzymes are deactivated, ensuring that the brewer can control the extent of conversion from starches to fermentable sugars post mashing. Second, mash viscosity decreases with temperature. A low viscosity mash is critical to ensure flowability during the lautering process.

5.4 LAUTERING/SOLIDS REMOVAL

Lautering is another example of a unit operation that is idiosyncratic to brewing. A *lauter tun* is essentially a depth filter (see Chapter 7) which uses grain hulls and solids as a filter aid. Figure 5.7 shows the internals of a lauter tun. A lauter tun is comprised of three major components: a false bottom, a rake, and an irrigation system.

A "false bottom" is a flat bed of stainless steel sheet metal with numerous slits cut into it. These slits are typically on the order of 1 mm wide. This false bottom sits several inches above a catch pan, which collects liquid that percolates through the slits. These slits are, of themselves, insufficient to properly filter the mash. However, as the mash is transferred to the lauter tun, solids accrete on top of the false bottom, building its own filter bed. Liquid that has percolated through the false bottom is

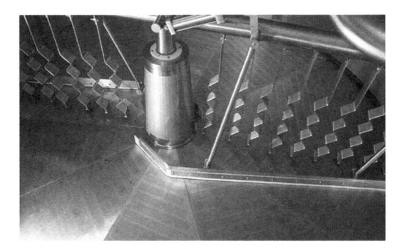

FIGURE 5.7 A lauter run. The false bottom holes are visible, as is the rake. When the filter bed formed by grain hulls begins to plug, the rake will turn slowly to help open up new flow channels. Care must be taken to balance high flow rate and good solids filtration. By Clément Bucco-Lechat: Own work, CC BY-SA 3.0, https://commons.wikimedia.org/w/index.php?curid=35949300; https://www.bevindustry.com/articles/90987-tips-for-improving-todays-brewhouse

circulated over the top of the false bottom and grain bed, which continues to build up the grain bed and filter the media.

A major process issue with the lauter tun is the buildup of filtration resistance. As explained in Chapter 7, filtration resistance increases as the filter fouls. In the case of a lauter tun, the false bottom serves as the particle filter, and the grain bed serves as the filter cake. Eventually, gravity flow is insufficient to drive wort through the grain bed. To reduce this resistance, a "rake" is used to break up the grain bed and permit the wort to flow again.

Finally, insufficiently clarified liquid from the catch pan below the false bottom must be recirculated. This involves a controller and a feedback loop, where the liquid being drawn from the catch pan is continuously monitored for turbidity. Once the liquid is sufficiently clarified, the recirculation stops, and the wort is allowed to carry forward in the process. The buildup of filter resistance, as well as the loss of a standing liquid head, means that some amount of wort will be trapped due to the filter bed's resistance. Additionally, the grain solids will be soaked with wort. To increase yields, it is common to irrigate water over the bed to push residual wort through the filter bed and extract sugars dissolved in the grain solids. High-end breweries will typically have density meters on the lauter tun effluent. Therefore, this irrigation can be cut off when the dissolved sugars fall below some critical threshold.

Lauter tuns are, frankly, a primitive processing technology. The lack of driving force other than gravity flow makes lautering a slow process. The tradeoff between filtration resistance and filtrate clarification has the same issue; while rakes can break up the bed and increase filtration flow rates, they simultaneously degrade filtration performance. The main advantages of lauter tuns are (i) relatively low operating costs, as low-pressure liquid pumping and the power required to turn the rake are the main operating expenses, and (ii) they are traditional, and the brewing industry is comfortable with them. While crossflow filters and decanter centrifuges (Chapter 7) are starting to be used for solids separation in the brewing process, lauter tuns will remain common for the foreseeable future.

5.5 BOILING/KETTLING

After solids removal, the clarified wort must be boiled. This typically takes place in a *brew kettle*. This serves three purposes: sterility, extraction, and concentration.

Most critically, boiling serves to totally sterilize the wort prior to fermentation. So long as sanitary design practices are observed downstream of the brew kettle, microbial spoilage can largely be avoided. This step is not necessary in wine, as wine environments (>10% v/v ethanol, 2 < pH < 4) are quite inhospitable to spoilage microorganisms, while beer (<6 % v/v ethanol, 4 < pH < 6) is relatively welcoming to spoilage organisms. See Chapter 11 for details on microbial deactivation kinetics.

Boiling is also critical for solids extraction. This typically means hop cones, though hop pellets, citrus peels, and other spices are often included. High temperatures and the roiling mixing during boiling promote rapid and complete extraction. Brew kettles are usually designed with an accessible manway for direct addition of solids during boiling, though high-end brew kettles may include automated solids handling systems for hops and other flavorants.

Finally, boiling can be used to reduce the volume of the wort, thereby concentrating it. It is not uncommon to reduce volume of the wort by over 10% over the course of boiling. To promote this, a fan or blower is typically installed at the vapor outlet of the brew kettle, which serves to reduce the pressure of the brew kettle, thereby reducing the boiling point of the liquid in the brew kettle, promoting evaporation. At a 2 psi vacuum (equivalent to a brew kettle pressure of 12.7 psia), the boiling point of water will have dropped from 100°C to 96°C.

5.6 AERATION

The last common step prior to beer fermentation is aeration. While all alcoholic fermentations are anaerobic, initial dissolved oxygen concentrations are critical to healthy fermentations. Oxygen is a prerequisite to sterol synthesis is cells, which is required to maintain cell membrane fluidity, especially in the increasingly high ethanol concentrations faced in wine and beer fermentations. It also promotes higher yeast cell concentrations during the growth phase of the fermentation. While dissolved oxygen capacity is a strong function of temperature, it typically hovers around 7–8 mg/L.

Aeration can be coupled with sulfur compound removal via *air stripping*, which is achieved via an *air stripper* (Figure 5.8). An air stripper is a unit operation which enables substantial mass transfer between a liquid (in this case, wort) and a gas

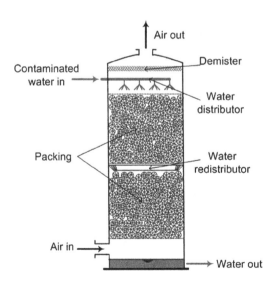

FIGURE 5.8 Schematic of an air stripper, with air ascending and liquid descending. Note the packing present to enhance mass transfer. This packing can also be in the form of trays, as seen in Chapter 8. When stripping wort, sulfur-rich and oxygen-lean wort enters from the top, and sulfur-lean and oxygen-rich wort exits from the bottom. By (Wikimedia Commons) KudzuVine assumed (based on copyright claims): No machine-readable source provided. Own work assumed (based on copyright claims), FAL, https://commons.wikimedia.org/w/index.php?curid=1940870; https://en.wikipedia.org/wiki/Air_stripping

stream (in this case, air) by passing the streams through each other in a countercurrent fashion, with air ascending and wort descending, with a contact surface present to enhance mass transfer. The air stream serves to strip sulfur compounds from the worth and carry them away to either atmosphere or a Volatile Organic Compound (VOC) capture system.

Air stripping serves to also introduce air into the wort. If an air stripper does not introduce sufficient oxygen, or if an air stripper is not used, *air injection* can also be used. This utilizes direct injection of pressurized air into a flowing wort stream. This system often incorporates a feedback controller, where DO is measured downstream of the air injection, and air injection rate is ramped up or down to ensure saturation.

Critically, any air injected into the wort post-boil must be sterile filtered to ensure the microbial integrity of the wort. This is achieved in a manner exactly analogous with wine sterile filtering, via the application of 0.22 μm normal-flow membrane filters.

REFERENCES

Dickey, D.S. Mixing and Blending. In *Kirk-Othmer Encyclopedia of Chemical Technology*. 6th Edition, John Wiley & Sons, 2000.
Saarni, A., K.V. Miller, and D.E. Block. A multi-parameter, predictive model of starch hydrolysis in barley beer mashes. *Beverages*, 6, 60, 2020.

6 Fermentor Design and Heat Transfer

6.1 GENERAL FERMENTOR FUNCTIONALITY AND DESIGN

Fermentors (spelled with an "o," as "fermenter" refers to the organism or person doing the fermenting) are used for fermenting a grape juice, must, or wort through a primary or secondary fermentation. For wine fermentors, it is generally best to design for use with red and white wines, though there can be specific needs associated with one or the other, based on the level of solids or temperature, among other factors. Beer fermentor design is somewhat different, as almost all beer fermentors are designed to operate at (i) elevated pressures to produce a carbonated product, and (ii) are typically designed with conical bottoms to permit yeast harvesting and reuse. Regardless of their use, all fermentors need to be capable of filling, mixing, fermenting, emptying, and cleaning. In this section, we will discuss the design of a generic fermentor, in addition to factors that need to be considered for any more specific type of fermentor. These include materials of construction, geometry, seismic stability, venting (or not) of carbon dioxide, and cleaning. Subsequent sections will address issues specific to fermentors that will produce white wines, red wines, carbonated beers, and carbonated wines.

6.1.1 DESIGN OF A GENERIC WINE FERMENTOR

A schematic of a generic wine fermentor is shown in Figure 6.1. Most wine fermentors are cylindrical in nature. Filling the tank is commonly accomplished through a drain valve at the bottom of the fermentor or through a manway at the top of the tank. A racking arm is another outlet that exits the tank about 10% of the way up the height of the tank. The internal part of the racking arm can be rotated to remove liquid at various heights above the layer of tank sediment that forms during processing. An additional sample valve is used for taking liquid samples for Brix and other chemical analyses. Side manways allow access to the inside of the tank, and perhaps more importantly, to remove skins and seeds in red wine fermentations. A properly sized vent allows the carbon dioxide generated during fermentation to vent into the atmosphere. It is very common to have a way to control the temperature of the must during fermentation, and this generally takes the form of a heat transfer jacket around the outside of the tank, a temperature sensor that senses must temperature, and a controller that opens or closes a valve that allows cooling fluid (or possibly heating fluid) to flow through the jacket when needed.

DOI: 10.1201/9781003097495-6

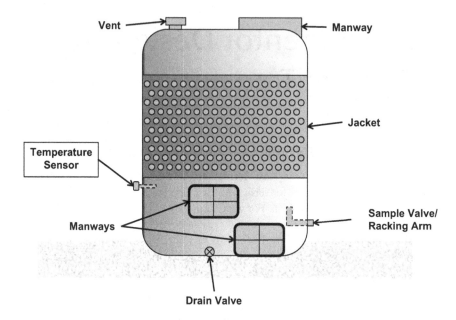

FIGURE 6.1 Schematic of a jacketed fermentor.

6.1.2 MATERIALS OF CONSTRUCTION

Wine fermentors are made of a wide variety of materials. However, the most common material is stainless steel. As discussed in Chapter 2, 304 or 304L stainless steel would be the most likely grade for fermentors using a mill finish without further polishing. Polishing, especially of welds, can increase ease of cleaning and sanitization. This is likely the most cleanable of materials used for fabrication of fermentors. Other materials include plastic (e.g. plastic picking bins), wood (e.g. oak upright tanks), and concrete (in various shapes and sizes). Use of these materials comes with concerns of cleanability. While these concerns may not be enough to preclude use of these materials, fermentors fabricated in this way need special attention during cleaning and sanitization. In addition, use of materials that do not transfer heat efficiently may require additional heat transfer fins or surfaces to be added to the inside of the fermentor. This increases system complexity, reduces working volume, and makes cleaning more difficult.

6.1.3 GEOMETRY

Many different geometries are used for fermentors, so much so that it seems that geometry may not be critical given the centuries of experience with winemaking. However, geometry will indirectly affect key factors in fermentor performance. For instance, tank geometry will affect temperature control. Because the heat will be generated proportional to the volume of the fermentation and it will be removed proportional to the surface area of the fermentor, the ability to remove heat from a fermentation will scale with A/V (surface area to working volume ratio) which is proportional to $1/D_t$. That is, larger tanks will be more difficult to cool than smaller tanks. Geometry will also be a factor in how easily a tank can be cleaned. Horizontal tanks can be challenging to clean as there is not a single spray location in these tanks that will be guaranteed to provide complete coverage during cleaning. Tall, skinny tanks can be challenging to clean from the bottom but may be easily cleaned from above. Aspect ratios of tanks will also help determine the footprint necessary for the tanks, as well as the necessary ceiling height. These factors should be given consideration prior to design and purchase.

6.1.4 SEISMIC STABILITY

Many wine producing regions are seismically active. It is important to know the kind of seismic region in which your winery is built and any associated guidelines or regulations for tank construction and/or anchoring. From years of experience with stainless steel tanks in earthquakes, it is clear that some designs are preferable (Figure 6.2). Legs welded to the bottom of a tank tend to lead to failure at the bottom of the tank with legs penetrating up through stainless skin. Legs that are welded to the side of tanks or secured in lugs attached to the side of tanks have weathered earthquakes better. Tanks resting on concrete platforms can be stable, though this seems to be somewhat dependent on how the tank is secured to the platform. Completely rigid fastening of the tank to the platform can lead to tearing of welding seams at the knuckle of the tank (where the wall of the tank meets the bottom).

6.1.5 VENTING CARBON DIOXIDE

Based on the stoichiometry of alcoholic fermentation, the volume of carbon dioxide evolved during a wine fermentation is approximately 60 times the volume of the fermenting juice or must. Therefore, a wine fermenting at the rate of 2 Brix/day will evolve five working volumes of carbon dioxide daily on average. For this reason, the tank needs to be vented to allow the gas to escape so pressure does not build up. As red wines will usually have higher peak fermentation rates, the vent should be designed or specified with these fermentations in mind. It is also important for vents to allow ingress of gas. This way, when tanks are being emptied, air can flow into the

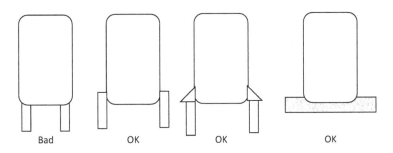

FIGURE 6.2 Tank support options for seismic stability. From experience in seismically active winery production zones, some supports are more stable than others. Even a tank supported on a concrete slab can fail in an earthquake if it is fastened incorrectly.

tank to replace the lost liquid volume and a vacuum will not form that can collapse or damage the tank. This can be a major, expensive problem, and therefore many times, a top-mounted manway is opened prior to transfer, just in case the vent has inadvertently been clogged during processing.

6.1.6 CLEANING

There are a variety of cleaning methods used in the wine industry from hand scrubbing to the use of high-pressure wash units and fixed in place spray devices. The "dirt" left in the fermentors likely depends on the type of wine produced and the processing performed, in addition to the time between cleanings. For instance, a tank used for long-term storage of a wine with high levels of tartrate will likely have higher levels of soiling than a tank used for an overnight settling operation. There is often more dirt in a ring at the top of the working volume and there may be, for instance, more tartrate deposits on the tank wall where a cooling jacket has been active. It also seems that rough surfaces, such as weld lines that have not been ground flush, are significantly more likely to accumulate dirt. When choosing or designing a fermentor, care should be given to deciding on how the tank will be cleaned (e.g. automated, manual, from the top, from the bottom, etc.) so that the tank can be designed appropriately. For example, tanks cleaned from the bottom should not be overly tall. All tanks should have an appropriate drain valve diameter to allow the tank to be free draining as a drain valve that is too small will cause puddling and incomplete cleaning. Cleaning is described in greater detail in a latter chapter.

6.2 SPECIAL CONSIDERATIONS FOR WHITE WINE FERMENTORS

6.2.1 MIXING

Mixing in an active white wine fermentation is usually nearly complete due to rapid carbon dioxide evolution, as well as convection caused by temperature gradients. However, fermentations are not necessarily well mixed at the time of inoculation and for some amount of time after inoculation. The degree of mixing is a function of how well hydrated the yeast is, as well as the temperature of the fermentation. Incomplete hydration of yeast (according to manufacturer's instructions) can lead to significantly more rapid settling of the yeast proportional to the diameter of the particle squared. A paper by Vlassides and Block (2000) illustrates this issue and the magnitude of the gradients of cell concentration possible in a small pilot-scale tank because of this. In one case, it took 4 days after inoculation for the fermentation to become fully uniform in yeast cell concentration. If this tank had been used to initiate fermentation prior to transferring to multiple barrels for completion of fermentation, each barrel would have a different inoculum.

6.2.2 COOLING AND HEAT TRANSFER SURFACES

White wines are typically fermented at below-ambient temperature to preserve aromas associated with the grapes. This, as well as the fact that yeast generate heat as

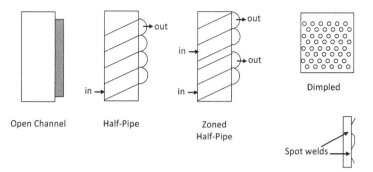

FIGURE 6.3 Diagrams of tank jacket styles. Each of the above diagrams represents half of a tank—from its center symmetry line to the outside of the tank. Overall efficiency of heat transfer increases from left to right.

they grow and convert sugar to alcohol means that white wine fermentors must have the ability to cool a fermentation and control temperature.

The heat generated by the fermentation can be removed in one of two ways, either by cooling surfaces integral to the fermentor or by periodically passing the juice or must through an external heat exchanger. The integral heat exchanger can be either a jacket on the fermentor wall or a coil or plate which is mounted within the fermentor. If a jacket is used, several styles are possible (Figure 6.3). The simplest style is an open channel jacket which is essentially a concentric cylinder or tank welded onto the outside of the fermentation vessel. These jackets are simple and have a low pressure drop, but are not, comparatively, very efficient at transferring heat. Mechanically, they also require a thicker metal to be used to maintain structural integrity. A half-pipe system is another type of jacket. This type of jacket consists of a semi-circular half-pipe welded in a coil around the outside of the tank. This arrangement is structurally sounder and gives better heat transfer. The heat transfer of this type of jacket can be further improved by using multiple inputs and outputs of coolant, usually called zones. Finally, dimpled jackets can be used. Dimpled jackets are formed by taking a piece of stainless steel and spot welding it in regular intervals to the tank. This type of jacket is the least expensive but gives excellent heat transfer, as it provides a tortuous path for coolant that promotes mixing. However, flow rates of coolant for dimpled jackets need to be less than 50 gpm or the pressure drop due to friction becomes unreasonably high (see later in this chapter for this calculation). If a coil or extra cooling plate is used inside a stainless or wood fermentor in place of a jacket, cleaning becomes an important issue, as does leaking around any penetration in the side of the tank for cooling fluid flow.

6.3 SPECIAL CONSIDERATIONS FOR BEER FERMENTORS

Beer fermentors come in two flavors—primary (atmospheric) fermentors and secondary (pressurized) fermentors. Primary beer fermentors are very similar to white wine fermentors. Secondary beer fermentors are extremely similar to sparkling wine fermentors used for the Charmat method. Beer fermentors, however, have some important distinguishing features that will be covered in this section.

6.3.1 PRIMARY BEER FERMENTORS

Primary beer fermentors (sometimes referred to as "atmospheric" or "alpha" fermentors) are somewhat similar to white wine fermentors, in that they are unmixed, jacketed fermentors. Beer fermentation temperature varies widely, but it tends to be cooler than wines. Beer fermentation can be quite cool, sometimes as low as 50°F. Three key differentiators of beer fermentors versus white wine fermentors are: (i) the use of conical bottom tanks, (ii) the increased necessity of sanitary design in fermentors, and (iii) the installation of foam-catching meshes.

Unlike in winery settings, it is common for brewing yeast to be recovered post fermentation and re-pitched into subsequent fermentations. This has the benefit of high initial inoculum rates, which greatly reduces the "lag" period in fermentation and enables high peak biomass concentrations. In order to simplify post-fermentation yeast harvesting, conical bottom tanks (Figure 6.4) are employed.

FIGURE 6.4   A conical bottom beer fermentation tank. Notice the closed top and aseptic design. Figure courtesy of Cedarstone Industry.

Once fermentation ceases, the lack of heat and carbon dioxide evolution in the fermentor ceases agitation, allowing solids (primarily yeast and yeast hulls) to settle to the bottom. The aggressive angle of the conical bottom creates a relatively high-density layer of yeast, as opposed to the thin layer you would see on a flat-bottomed tank. The yeast harvest line, located at the very bottom of the tank, enables the drawing off of the yeast slurry with minimal beer loss.

Second, note the all stainless steel, sealed design of this fermentor. Industrial beer fermentors are highly sanitary, with hard-piped lines running to and from the fermentors (as opposed to hoses), sealed tank tops, and a lack of crevices or dead zones. Additionally, beer fermentors are typically hot sanitized to deactivate potential spoilage organisms (see Section 11.1 for details).

A final interesting deviation of beer fermentors from white wine fermentors is the presence of a foam catching grill at the top of the fermentor. During primary fermentation, hop acids and proteins result in the formation of a large volume of stable foam, called *krausen* in the brewing industry. This foam is typically quite unpleasant. To prevent the re-mixing of the foam into the beer at the end of fermentation, a foam-catching grid is often included in the top of the fermentor, to allow surface area for foam deposition, rather than allowing the foam to fall back into the beer.

6.4 SPECIAL CONSIDERATIONS FOR RED WINE FERMENTORS

6.4.1 Filling

For red wine fermentations, the fermentor will be filled with must containing both juice and skins. Depending on the size of the winery and tank, the filling operation may be with a hose through the top manway or from the bottom through a drain or racking valve. If the tank is to be filled through a valve, the diameter must be chosen to be large enough to accommodate the skins and other solids in the must. It should also be a design that is not as likely to clog, such as a ball valve or diaphragm valve, as butterfly valves with the central pivot disk will create an opportunity for the solids to accumulate and may be difficult to close post-transfer. If a drain valve is to be used to remove must including solids, for instance, during sluicing to a press, it again needs to be large enough to accommodate this flow without any restrictions that could cause clogging and interrupt the operation. In the case of sluicing, the drain valve should be located at the tank low point. While this sounds straightforward, it is often not, especially if the tank has a sloped bottom. This is because the tank bottom generally slopes to a manway which also is best situated at the low point of the tank. Knowing how you are likely to fill and drain tanks can help resolve this type of design conflict.

6.4.2 Cap Management

There are several alternative ways in which the juice and skins (and seeds) are mixed during red fermentations. Cap management is typically employed to promote phenolic extraction during red wine fermentations and to cool the cap periodically down to the temperature of the liquid below the cap. The empirical (and scientific) evidence is that none of these has a bearing on color extraction directly, despite some widespread misconceptions that they do. Some indirect effects due to temperature impact on extraction are starting to become more clear, as is the impact of fermentor geometry and cap management (and are discussed in a later section of this chapter). There are also no data showing any one approach capable of capturing more flavor or aromas than the other. The origins of these approaches are usually regional, and they may affect the final sensory characteristics of the wine and its typicity for that region. The alternatives include these fairly common options:

(1) Pumping Over of Juice
(2) Punching Down of Skin Caps
(3) Using a Rotating Fermentor
(4) Ganimede Tanks
(5) Rack and Return
(6) Using a Submerged Skin Cap

6.4.2.1 Pumpovers

Perhaps the most common means of cap management is called a pumpover. In a pumpover, fermenting juice is taken from below the cap and pumped over the top of the cap, with the aid of a spray device that can target the liquid (Figure 6.5). While the frequency and duration of pumpovers vary, it is common to pump between half and two times the fermentation volume over the cap once or twice per day during fermentation. The apparatus for this type of pumpover can be fixed in place and dedicated to a particular fermentor, or alternatively, hoses and pumps can be moved from tank to tank sequentially. As can be imagined, there is a tradeoff here between one-time capital costs and ongoing labor costs. If the decision is made to have dedicated, fixed equipment, liquid will be hard piped from a racking arm to a pump situated next to the tank. The pump will draw the liquid from above the lees layer containing

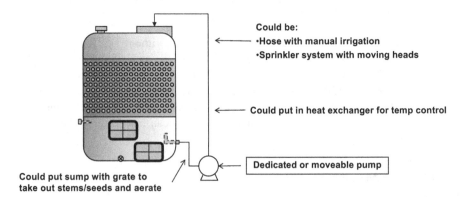

FIGURE 6.5 Diagram of a red wine fermentor set for a pumpover. In general, this method involves taking fermenting juice from the bottom of the tank and spraying it over the top of the cap. Several variations are possible including whether the pump is dedicated to the fermentor or mobile, and whether the spray is directed manually or with a fixed distribution device.

mostly yeast and grape solids and then send the fermenting juice up a riser (vertical pipe) and into the top of a tank to a spray device. It is common for these spray devices to rotate just from the force of the pumped over liquid impinging on the device. However, some of these devices employ external power sources to provide the rotation necessary to evenly cover the cap. The key with any of these automated devices is to specify and install them so that the entire cap is evenly covered to avoid channeling and hot spots in the cap. With a dedicated pumpover pump, the process can be automated on a timer or even linked to other process parameters like cap temperatures. Of course, it is still common to decide to purchase a small number of pumps that can be used for pumpovers sequentially. In this case, a flexible hose is connected to the racking valve on one end and the pump on the other. From the pump, flexible hose can be used to send the liquid to the top of the tank. Alternatively, some companies will install stainless risers from just below the tank to the top of the tank to speed up operations. Either way, the hose or riser can be connected to an automated spray device as with the dedicated system or an operator can stand on a catwalk at the top of the tank holding the hose and manually spray the cap for as long as necessary.

At smaller scales, it is common to have the option of including a screened sump between the racking arm and the pump to allow splashing and aeration of the fermenting must during pumpovers. At larger scales, it becomes difficult, if not impossible, to control temperature with a jacket alone, as the surface area to volume ratio is too small. Larger wineries will often take advantage of the necessary pumping action during a pumpover to place a heat exchanger/chiller inline to chill the fermenting juice during pumpovers.

6.4.2.2 Punchdowns

The punching down of the skin cap either manually or by a hydraulic or pneumatic plunger aims at breaking up the skin cap rather than mixing the juice. A complete punchdown does, however, thoroughly wet the cap and distribute the heat effectively through the entire fermentation. Truly complete pumpovers become increasingly difficult as the diameter of the tank increases and the center of the tanks becomes nearly impossible to reach. Punchdowns usually require open top fermentors, and these cannot be used for wine storage unless the fermentors are variable capacity tanks with moveable heads (Figure 6.6). These heads can be lowered into the tank using a hoist down to the liquid level and maintained in place using an inflatable O-ring. While use of these moveable heads can be cumbersome in practice, their use increases the utility of the open-top tank in the winery past the fermentation phase and into the storage phase, precluding the need to purchase an extra set of storage vessels.

Punchdowns can be scheduled based on the winemaker's wishes and it is practically limited to smaller fermentor volumes. In addition to the difficulty of reaching the center of larger fermentors for punchdowns, at some point the cap becomes so thick that excessive force and expensive, specialized equipment would be needed to punch down effectively. It is important to note that care should be taken in manual punchdowns to provide sufficient safe standing area for workers (as opposed to just standing on the rim of the tank) from which they can reach all areas of the cap. If sufficient space is not given, accidents involving falling into the fermentation or onto the ground below could occur causing serious injury or death. For hydraulically

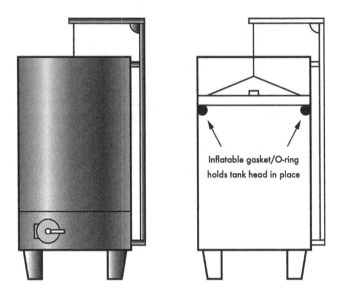

Inflatable gasket/O-ring
holds tank head in place

FIGURE 6.6 Diagram of a variable capacity tank (a) outside and (b) in cross section (M. Ogawa).

assisted punchdowns, one punchdown device typically is mounted on a rail above a bank of tanks and can be used for each of the tanks sequentially. These devices are normally operated from the catwalk located near the top of the tanks (see Figure 6.7 for examples). They can also be built into the fermentor at small to medium scale (Figure 6.8) and operated manually or in an automated fashion.

FIGURE 6.7 Punchdown device. Two examples of punchdown devices shown here. (a) This hydraulic punchdown device rides on a rail above a bank of open top tanks and is operated from the catwalk. (b) Punchdown device from a Burgundy winery.

FIGURE 6.8 Automated punchdown installation. (a) These tanks are closed top fermentors with the ability to use automated punchdowns. The cylinders coming out of the top are supports for the internal punchdown devices as seen in (b).

6.4.2.3 Rotary Fermentors

Rotary fermentors (Figure 6.9) are horizontal cylinders with one conical end. They are mounted on external roller bearings and are driven by an electrical motor at the rear. The use of internal helical baffles and the conical end enable pomace to be removed automatically at the end of fermentation. Cap management is accomplished by rotating the tank by several revolutions on a regular basis. Through experience, many wineries have found that adding additional chains mounted across the diameter

FIGURE 6.9 Cutaway drawing of a rotary wine fermentor. Notice the helical baffle that allows an automated removal of skins through the end-mounted manway. The fermentor is mounted on rollers and rotated with a chain rive. A vent is opened when it passes through the 12:00 position. Courtesy Paul Mueller Company.

of the tank aid in breaking up the cap to get sufficient contact during rotation, though this arrangement does make cleaning more challenging. The cylinder has a gas venting port, which is activated at the 12 o'clock position during each rotation. The major limitation is the dependence on electricity and the need to rotate them in order to vent the carbon dioxide. Having multiple vents mounted around the tank so that one vent is always positioned above the liquid level would increase the safety of these tanks.

6.4.2.4   Ganimede Tanks

A variation of pumping over is the "accumulated flooding" fermentors which have a large open top holding volume above the fermentor. The juice is pumped up into this vessel by temporarily preventing carbon dioxide release and the short-term build-up of pressure. When the juice volume approaches a designated height, its pressure causes a large valve in the base of the vessel to open up and to discharge the contents onto the skin cap below. This also vents the carbon dioxide and the cycle begins all over again. The frequency of pumping is then related to the rate of fermentation (and of gas evolution) with most vigorous mixing at mid-fermentation and less at the beginning and end. The main limitations to this approach are the lack of control of the timing of pumpovers and the lack of sanitation and the pick-up of airborne organisms during the pumpover events.

6.4.2.5   Rack and Return

Skin contact can be accomplished by simply draining liquid from the tank to allow the cap to fall, followed by the return of the liquid to the tank to break up the skin cap. This is called "rack and return." While generally this is done manually by a cellar worker with a pump and additional tank, there are installations that forego the pump by draining the tank by gravity and then using a mechanical hoist to lift the drained liquid tank over the fermentor with a subsequent gravity return of the juice back to the cap. Having sloped walls to the tank (smaller diameter at the top than the bottom) is said to promote mixing of the cap as it drops during draining.

6.4.2.6   Submerged Cap

The use of a floating raft or a fixed grid to keep the skin cap from rising above a certain level as the must is introduced, usually from the bottom of the tank, provides a submerged cap. The claimed advantage is not having a need for any intervention, but the limitation is a tightly compacted skin cap which has no mixing or extraction capability. Phenolic extraction in a submerged cap fermentor proceeds in a manner nearly identical to a fermentation in which the cap and liquid are closely controlled at the same temperature, usually by periodic pumpovers.

6.4.3   Skin Removal

Removal of skins from a red wine fermentor can be accomplished several ways. Perhaps the most common means of skin removal is to drain the tank of free-run wine, open the two manways, and shovel the skins manually into bins that can be transported to a press. Eventually, when there is enough room and the carbon dioxide has been evacuated from the space, a person is sent into the tank to shovel the remaining skins into bins outside the tank. In addition to being a lengthy process, this method for skin removal can be **extremely** dangerous. Several winery workers have died in California over the last 30 years by entering a tank that did not have sufficient oxygen. If this method is chosen, special care must be taken to assure evacuation of the carbon dioxide, and workers should not be alone, should have the appropriate harnesses and oxygen meters, and proper training. Another method for skin removal is "sluicing." Sluicing involves moving the free-run wine to another tank and then, in conjunction with a must pump, pushing the skins to the press while pumping the wine back into the tank with the skins. This method requires extra tank capacity, more pumping, and potential for more air contact, but does not require tank entry.

Several more automated methods have been explored for skin removal over the last few decades. A vacuum/pneumatic conveyance system was evaluated commercially, as was an automatic shoveling device using a conveyer system. Both of these have the advantage of not requiring tank entry, but both require a capital outlay and still take time. Perhaps the best solution to date has been the use of retort conical bottom tanks (like drainers) that are mounted over a press (Figure 6.10). After the free run is drained, a guillotine valve is opened at the bottom of the fermentor and the skins fall into the press below, usually with the aid of a sweep arm mounted in the bottom of the tank. Tanks on the order of 10,000 gal or more can have their skins removed in a matter of minutes, again with no tank entry, using simple gravity.

6.5   PRESSURIZED FERMENTORS

Beers and sparkling wines require dissolved carbon dioxide. While this can be achieved via inline CO_2 injection, it is often more economical (not to mention traditional) to perform secondary fermentations in a *pressurized fermentor*. A pressurized fermentor is one which is sealed to the outside environment, and capable of maintaining a pressure greater than 1 atm (14.7 psia).

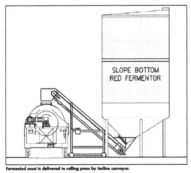
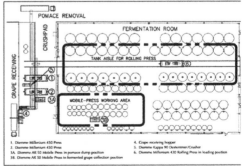

FIGURE 6.10   Automated skin removal setup for red fermentors. This system is sometimes called the Scheid Slide after the winery that developed the system. It allows skins to fall onto an elevator by gravity to be transported to a press that moves around the winery. Courtesy of Wines & Vines, Jan/Feb 2007.

Pressurized fermentations are almost always secondary fermentations. This means that once-fermented beer or wine is transferred to a pressurized fermentor, and fresh juice or wort is blended into the tank, providing a fresh sugar source. This sugar is then fermented, liberating carbon dioxide, which is trapped in the pressurized fermentor. As more carbon dioxide evolves, pressure increases, forcing more carbon dioxide to dissolve into the wine or beer.

The relation between fermentor pressure and carbon dioxide concentration can be expressed in several ways, but the simplest is via Henry's Law:

$$C_i = H_i \times P_i$$

where C_i is the concentration of a dissolved gas "i" in a liquid (expressed typically in g/L), H_i (in units of g/L/bar) is the Henry's constant for gas "i," and P_i is the partial pressure of the gas above the liquid. The Henry's constant is a strong function of temperature. Henry's constants for different gasses are readily accessible via the NIST database. At 25°C, CO_2 has a Henry's constant of 1.54 g/L/bar.

Example 6.1:

You are planning a secondary fermentation for Granite State Lager, where you wish to reach a dissolved CO2 level of 2.1 g/L. Calculate the absolute pressure that the pressure gauge in the secondary fermentors headspace will reach when this is achieved. Assume that the fermentation takes place at 25°C, that the fermentor is initially at atmospheric pressure, and that the beer is flat.

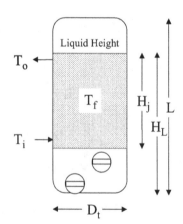

SOLUTION

We start with Henry's Law, Ci = H × Pi, and solve for pressure, Pi = Ci / Hi
We have a target Ci = 2.1 g/L
Solving for Pi, Pi = (2.1 g/L) / (1.54 g/L/bar) = 1.36 bar = 19.73 psi
At an initial fermentor pressure of 14.7 psia, the fermentor will now read 34.43 psia, or 19.73 psig

By measuring the rise in pressure due to the secondary fermentation and attributing this rise in pressure solely to carbon dioxide, Henry's law can be used to determine the concentration of carbon dioxide in solution, as seen in Example 6.1. Note that P_i is the partial pressure of species "i," in this case CO_2. If no CO_2 is initially present in the fermentor headspace, and the fermentor headspace is at starting pressure P_0 (typically 1 atm, or 14.7 psia), then

$$P(t) = P_0 + P_i$$

By definition, pressurized fermentors are *Pressure Vessels*, and have unique safety and design requirements. The American Society of Mechanical Engineers (ASME) publishes guidelines for the design of a pressure vessel (Section 8, Div. 1), which is formally defined as any vessel with an operational pressure of greater than 15 psig and an internal diameter of greater than 6 in. While the ASME code is not a code enforced on a federal level, certain states, municipalities, and insurers require that all pressure vessels match ASME code. While this can be more expensive upfront, it is critical to ensure both equipment safety and to minimize plant liability.

Pressure vessels also require some form of pressure relief to prevent catastrophic failure due to overpressure. There are two common forms of pressure relief devices: relief valves and rupture (or burst) disks. Relief valves are a mechanical, spring-loaded valve, which opens when the backpressure compresses the spring keeping the valve closed. Relief valves are reusable, and typically automatically reset. They are employed where overpressure may be somewhat common, and transient venting is an acceptable solution. A rupture disk is a physical membrane designed to fail at a certain back pressure, allowing the vessel to vent to atmosphere. These do not automatically reset and must be installed at every failure. While mechanically simpler, they should not be employed when routine overpressure is expected.

Vessels designed and rated for pressure are usually marked conspicuously on the tank itself, including the design pressure and the maximum allowable working pressure (MAWP). For vessels that are officially rated, it is common to have them tested at a pressure of 1.5 × MAWP and then inspected to assure no obvious failures in structure. Vessels that have been inspected this way and followed all of the ASME guidelines can be stamped with an official stamp. This official stamp may cost extra but proves compliance with code that is accepted in many governmental organizations and insurance carriers.

6.6 THE EFFECT OF FERMENTOR GEOMETRY AND DESIGN ON WINE CHEMISTRY

Very little practical data exist on the effects of fermentor design choices on the final chemical and sensory characteristics of wine. However, there are a small number of academic studies that examine this issue. In a 2-year study at UC Davis (J. Duane and M. Larner, M.S. Theses, with D. Block) some of these design factors were examined with Cabernet Sauvignon wines. This study examined fermentors with the same aspect ratio (height to diameter ratio, 1.3:1) at different scales including 1 gal, 10 gal, 40 gal, and 200 gal all at UC Davis, and 10,000 gal at Gallo Sonoma, different aspect ratios (0.5:1, 1.3:1, and 3:1) at the same volume, and punchdowns versus pumpovers for cap management. Results from these studies indicate that the most important effect is that from temperature—either directly through temperature control or indirectly through other factors like geometry. We found that integrated heat input (temperature held for a given time) was very well correlated with extraction of large polymeric pigment and tannin (both positively) and anthocyanins (negatively) as can be seen in Figure 6.11. Therefore, when choosing or designing a fermentor, it is likely most important to think about how your choice will affect your ability to control temperature. These experiments are exceedingly difficult to do in a controlled manner, especially at commercial scale. However, we have also been able to create a mathematical model that lets us examine the effects of geometry at any scale on fermentation kinetics and phenolic extraction (Miller et al., 2019a, b, c, d; Miller and Block 2020; Miller et al., 2020a, b). With this model, we have demonstrated that at large enough scale, the surface area to volume ratio becomes so small that a tank jacket will no longer be sufficient to control fermentor temperature and external cooling will be necessary.

Using this latter modeling approach, we were also able to investigate temperature control and mixing in concrete egg-shaped fermentors. Here, we were able to demonstrate that the concrete acted more as an insulator than a heat sink and therefore made temperature control more difficult than in a jacketed stainless steel tank of the same size. We also predicted that mixing is better in a regular cylindrical tank than an egg shape. We did not study the effects of concrete contact on the wine, which could have its own unique impact.

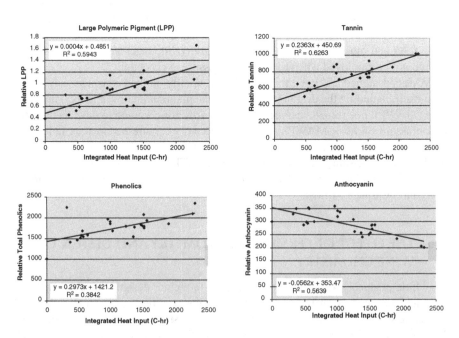

FIGURE 6.11 The effects of integrated heat input on phenolic extraction in Cabernet Sauvignon wines. Large polymeric pigment, total tannin, and phenolics are directly correlated with heat input, while anthocyanin extraction is negatively correlated with heat input.

6.7 THE APPEARANCE OF TEMPERATURE GRADIENTS IN RED WINE FERMENTORS

Unfortunately, temperature in red wine fermentors is not constant, but varies across the height and diameter of the tank, even when a cooling jacket is employed. Measurements of these temperature gradients, however, have not been widely published. In collaborative work with the University of Adelaide (Schmid et al., 2009), we measured these gradients in fermentations from commercial scale (5,000 L) down to research scale (50 L) in an attempt to quantify the gradients and determine whether fermentor design and scale contributes to temperature heterogeneity. Some important observations resulted from this research. First, as can be seen in Figure 6.12 for a 5,000 L Pinot noir fermentation, the temperature gradients in this tank are very large. A gradient of approximately 14°C (25°F) existed between the center of the cap and the liquid below. This gradient is even more important given that the temperature probe in this case (and in most cases for commercial wine fermentors) was at the bottom of the tank in the liquid. It is therefore possible to have very large gradients in a wine fermentor without knowing it. A gradient of 6–8°C also existed between the center of the cap and the side of the cap. In addition, this figure also demonstrates the fate of the temperature gradients after a manual punchdown (between the first two graphs). It can be seen that some heat is immediately dissipated in the fermenting juice below the cap. By nearly 2 hr after the punchdown, the heat left in the cap has spread out to cover a wider area of the cap. By 3 hr after the punchdown event, the gradient has begun to reform. Because the data in Figure 6.12a were taken 3 hr after a manual pumpover and Figure 6.12d were taken 3 hr after a manual punchdown, these two figures can be compared to examine the efficacy of these two cap management methods in dissipating the heat. It seems from these data that the manual punchdown may have been more effective at dissipating the heat, though it is also clear from Figure 6.12b that this particular punchdown event was not as complete as it could have been in mixing the cap and the liquid below.

Figure 6.13 illustrates how the temperature gradient changes over the course of the fermentation. Figure 6.13e gives the sugar utilization curve for the fermentation with

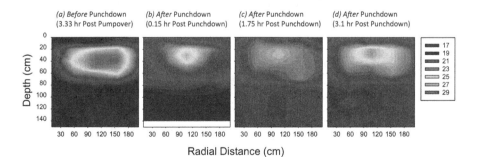

FIGURE 6.12 Progression of temperature gradient just prior to and after a punchdown event, as measured in a 5,000 L Pinot noir fermentation at a commercial winery in the Adelaide Hills. (a) First gradient measured 3 hr after the last pumpover, and prior to manual punchdown, (b) 0.15 hr post punchdown, (c) 1.75 hr, (d) 3.1 hr.

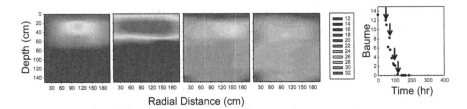

FIGURE 6.13 Progression of temperature gradient just prior to punchdown event over the course of a 5,000 L Pinot noir fermentation at a commercial winery in the Adelaide Hills. Each of the gradients (a–d) represents a time in the sugar utilization curve shown in (e). The largest gradient is apparent at the peak of fermentation.

arrows placed corresponding to the points of the temperature contour plots above. It can be seen in this figure that the gradient builds to a maximum where the sugar utilization rate is at the highest, which also corresponds to the point in the fermentation where heat generation is the greatest. As the sugar utilization begins to slow down (Figure 6.13c), the gradient is reduced to a few degrees. By the end of the fermentation (Figure 6.13d), the temperature gradient ceases to exist. A second (now Zinfandel) fermentation in the same commercial tank exhibited a significantly smaller gradient (about 5°C), while a small 50 L fermentation research fermentor that we also followed reached a gradient of 12–14°C. Therefore, we can conclude that the gradients formed are more a function of the must and processing conditions than the tank

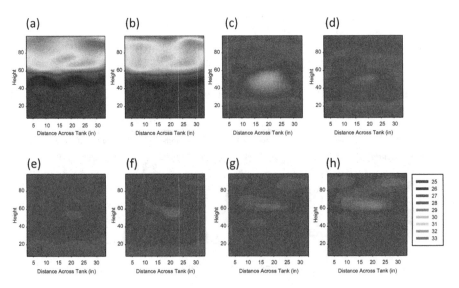

FIGURE 6.14 Progression of temperature gradients in a 1,600 L Grenache fermentation in a UC Davis pilot winery fermentor equipped with 63 Templine temperature sensors. Gradients were measured every 5 min from just before a pumpover (a) to during a pumpover (b–d) and then after (e–h).

itself. We did find, however, that it was easier to have a complete cap management event, with total dissipation of the heat throughout the fermentor, at the small scale.

With significantly more sensors and automated data acquisition, we were able capture temperature data during a pumpover and after at a much higher resolution during a Grenache fermentation. Figure 6.14 shows these data. The first plot shows the gradient just prior to pumpover with an automated pumpover/irrigation device. The second through fourth plots show the progression of the gradient during the pumpover. It can be seen here that the heat is pushed down into the liquid below the cap within the first 10 min or so of the pumpover. After the pumpover, the heat begins to reform in the cap within about 20 min and reaches the same extent by approximately 4–5 hr after the pumpover event. At large scales or with less efficient cap management techniques, heat may not be as completely dissipated which could affect both fermentation kinetics and extraction.

6.8 THE EFFECTS OF CAP MANAGEMENT ON WINE CHEMISTRY

The role of cap management in determining the final phenolic profile is likely dependent on fermentor scale and geometry. We found experimentally (Lerno et al., 2018) that pumpover volume and frequency have very little effect on phenolic extraction at the 120 L scale. In fact, at this small pilot scale, even foregoing pumpovers altogether had little significant effect. Anecdotal evidence from industry revealed a mix of experience at commercial scales in the absence of cap management (usually by accident—for instance, a broken pumpover pump). Some had had a similar experience to ours in that they were able to achieve satisfactory extraction without cap management. Others reported that no extraction occurred in the absence of regular cap management. Using the modeling approach mentioned previously (Miller and Block 2020; Miller et al., 2020a), we were able to demonstrate that cap management is more important to phenolic extraction at large scales and in tall, skinny tanks (i.e. when the height to diameter ratio is large). Cap management and temperature may also affect the ratio of skin tannins to seed tannins that get extracted into the final red wine. Therefore, having fermentors designed to have as much control of temperature and cap management is critical to having the most control over phenolic extraction.

6.9 HEAT TRANSFER AND TEMPERATURE CONTROL

6.9.1 HEAT TRANSFER IN WINERIES, BREWERIES, AND DISTILLERIES

Heat exchangers are devices that facilitate the exchange of energy between a cold stream and a hot stream. The movement of that energy, or heat, is called heat transfer. This term is used generically in the engineering literature, whether you are trying to heat or to cool your process stream. Various types of heat exchangers are used throughout beverage processing. Typical applications include:

1. Must, juice, or wort cooling prior to skin contact or fermentation
2. High temperature, short time (HTST) treatment for denaturation of laccases or other oxidative enzymes in juices, treatment of some kosher wines, or for flash pasteurization of beer

3. Temperature control during and after fermentation
4. Chilling wines for bitartrate stabilization, or beers for haze filtration
5. Heat recovery from cold fluids.

There are many common types of heat exchangers. Depending on the cooling or heating task needed in winemaking or brewing, some of these heat exchangers may become obvious choices. In other cases, any convenient heat exchanger may work fine. As an example, chilling juice to just below ambient temperature may work fine with any type of heat exchanger, but must chilling of a red must with skins and seeds would preclude the use of a heat exchanger with narrow gaps or passages usually employed to increase heat transfer efficiency.

6.9.2 Common Types of Heat Exchangers

6.9.2.1 Tank Jackets

A common method of heat transfer in a winery setting is a jacketed tank. Jackets can be ordered in various configurations. As discussed earlier in this chapter, the most common types of jackets are open jackets that may have baffles to direct flow, half-pipe arrangements where a half-pipe is welded onto the tank surface in a helical fashion, and dimpled where a pattern of spot welds interrupts the coolant flow to increase heat transfer. Any of these geometries can be "zoned" (multiple inlets/multiple outlets) for more efficient heat transfer.

6.9.2.2 Tube-in-Tube

At tube-in-tube heat exchanger is one of the simplest heat exchanger designs. It consists of two concentric pipes. The process fluid to be heated or cooled usually flows through the inner pipe, while the heat exchange fluid flows through the space in between the two pipes, usually in the opposite direction of the process stream as described below. In the wine industry, these heat exchangers are used for a wide variety of purposes, though their most common use is as a juice or must chiller at the crush pad. They are particularly well suited for fluids with suspended solids like must, as the inner pipe of the heat exchanger will not offer any more resistance than the pipe leading up to or away from the exchanger (assuming the diameter does not change). There are therefore no obstructions or tortuous paths that would trap solids. It is also common to see tube-in-tube heat exchangers used in large wineries for external temperature control in large fermentors. As noted above, this type of heat exchanger would just be placed in line during a pumpover operation, and coolant from a refrigeration system would be flowed through the other side. This type of heat exchanger is very simple, yet not overly efficient, thus requiring a fair amount of heat transfer area. This corresponds to large, required lengths, which are accomplished through a series of straight pipes on a rack interconnected by u-bends. An example of a tube-in-tube must chiller is shown in Figure 6.15.

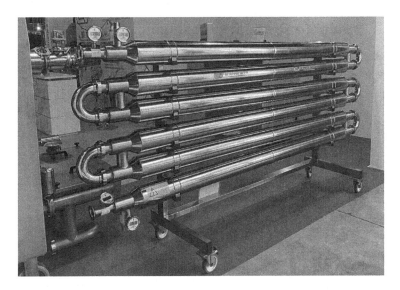

FIGURE 6.15 Example of a tube-in-tube heat exchanger. The U-bends are sometimes connected with tri-clamp fittings to allow the length to be changed as needed.

6.9.2.3 Shell-and-Tube Heat Exchangers

Shell-and-tube heat exchangers use a similar principle to tube-in-tube exchangers, but consist of many small diameter tubes in one larger cylindrical shell. Flow entering the heat exchanger on the tube side is divided equally amongst all tubes flowing through the shell. Tube diameters, lengths, and packing geometries within the shell are usually standardized for a given vendor. Therefore, increasing the heat transfer surface will correspond to a larger diameter shell. If both fluids flow in one direction (i.e. without any change of direction in the middle of the fluid path) the heat exchanger is said to be a single pass system. For some situations, a single pass system may result in a piece of equipment that is too long or awkward to fit in a process area. Therefore, some heat exchangers will have a change in direction in the middle. This tends to make the heat exchanger shorter, though less efficient as some of the flow becomes co-current (more explanation of this is given later in this chapter). This type of system would be called a dual or double pass system. Care must be taken to assess the nature of the solids in the intended processing streams, as the diameter of the tubes will likely be quite a bit smaller than those in a tube-in-tube arrangement. One of these units is shown in Figure 6.16.

Tube-in-tube and shell-and-tube heat exchangers have overall heat transfer coefficients of 50–200 Btu/hr ft^2 °F or even higher (with shell-and-tube on the higher end of this range). This is an indication of their efficiency of heat transfer as described in the next section, making them significantly more efficient than tank jackets in unmixed systems.

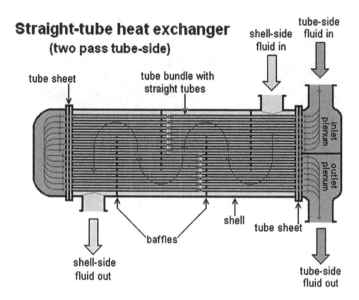

FIGURE 6.16 Sketch of a shell and tube heat exchanger. This unit is double-pass on the tube side: liquid enters the tubes at one head, exits the other head, and then flows back along the second half of the tubes. It is single-pass on the shell side. CC BY-SA 3.0, https://commons. wikimedia.org/w/index.php?curid=819523

6.9.2.4 Spiral Heat Exchanger

Spiral heat exchangers offer an alternative geometry to tube-based heat exchangers. One way to describe this geometry is to picture two pieces of paper with their long sides glued together. There is still a gap in between them through which fluid could flow. If these papers are then rolled up into a spiral starting with one of the longer ends, one has the idea for the geometry of these heat exchangers. Of course, the paper would be steel, and the ends capped. One stream would flow in between the sheets, and the other around the outside of the sheets as pictured in Figure 6.17. The two streams would also flow in opposite directions around the spiral, with one starting in the middle of the spiral and the other starting from the outside.

Commercial spiral exchangers have fluid paths (e.g. the space between the imaginary papers above) that are large enough so that they can be used with must. Because the flow in these is counter-current with good contact between the hot and cold streams, these exchangers are a good alternative to geometries like shell-and-tube where large concentrations of solids are present. Capacities could be expanded by expanding the diameter of the heat exchanger or the length.

6.9.2.5 Plate Heat Exchanger

The geometry of a plate heat exchanger is illustrated in Figure 6.18. This type of heat exchanger is made up of a series of flat rectangular plates arranged on a frame. Each of the plates has raised surfaces to promote turbulence and mixing to increase the heat transfer. Much like the geometry of a plate and frame filter discussed in a latter chapter, the fluidics are designed so that the cool fluid stream is introduced in between

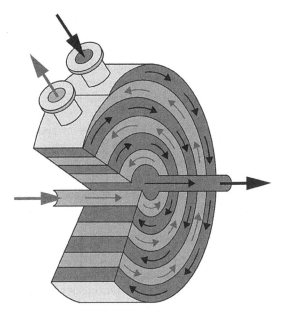

FIGURE 6.17 Diagram of a spiral heat exchanger. The cutaway view shows the two countercurrent flows (M. Ogawa).

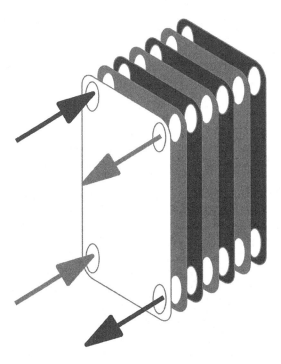

FIGURE 6.18 Diagram of a plate and frame heat exchanger. This can be visualized as an "unrolled" spiral heat exchanger; or a spiral heat exchanger can be thought of as a plate and frame burrito. Notice the countercurrent flow paths. By Plate_frame_1.png: Ubderivative work: Malyszkz (talk): Plate_frame_1.png, CC BY-SA 3.0, https://commons.wikimedia.org/w/index.php?curid=15033819

every other plate, and the hot fluid stream is introduced in between the other plates. The flow patterns are arranged so that the hot and cold streams are counter-current (in opposite directions) separated by the heat transfer plate. Increasing the area in this type of heat exchanger involves simply adding more plates. Because the fluid flow (both hot and cold) is divided equally between plates, adding more area should not appreciably change the flow characteristics or heat transfer characteristics of this equipment.

A plate heat exchanger can be used for a variety of applications from simple exchange of heat between two liquid streams to HTST treatment of juice using condensing steam as the heat transfer medium. The use of this equipment for fluids containing solids depends on the gap between plates relative to the size of the solids. It is unlikely, for instance, that must with skins and seeds (and possibly stems) will work well with this geometry, though smaller solid particles as found in lees might be fine.

6.9.2.6 Scraped-Surface Heat Exchanger

While some heat exchangers are well-suited for a wide range of applications, there are applications that require specific thought and specialized designs. Example applications include chilling wines down to below freezing conditions (<0°C or 32°F) for tartrate crystallization in order to cold stabilize wines, or to chill beers to the point of ice formation to cold concentrate the beer. If a shell-and-tube or plate heat exchanger are used, there is a risk that ice and tartrate crystals will build up in the heat exchanger until the flow channels are clogged and the system becomes over pressurized. This would be a safety risk at worst and an inefficient process at best. Wineries in California will often cool down wines for cold stabilization by simply running chilled glycol through a jacket. As described in further detail below, this practice is also far from efficient, as excess energy is fed into the production of ice on the inside and outside of the tank which only serves to make further transfer of heat less efficient. Therefore, specialized heat transfer apparatus would be needed to accomplish this function efficiently.

One type of apparatus that could be used for this application is a scraped-surface heat exchanger (Figure 6.19). This type of heat exchanger consists of an elongated cylinder that is jacketed to provide refrigerated coolant. Inside the cylinder is a rotating shaft, that may also have heat transfer capacity. As the shaft rotates, blades constantly scrape the ice off the walls of the heat exchanger so that it can be conveyed out of the cylinder and the fluid path is not occluded. This system works somewhat like home ice cream makers where ice cream solidifies at the wall of the container and is then scraped off to allow more liquid to freeze. Because ice is generated in a scraped surface heat exchanger and constantly conveyed out of the exchanger, only a portion of the wine being treated needs to be conveyed through the heat exchanger. The ice that ends up in the stabilization tank will then serve to chill the remainder of the wine. While this type of device makes sense for the unit operation of cold stabilization, it has not been widely adopted in the California wine industry, likely because of the need for additional capital outlay (above the cost of the jacketed fermentors). However, energy savings could be substantial and therefore justify the extra cost and complexity for certain wineries.

Inner pipe

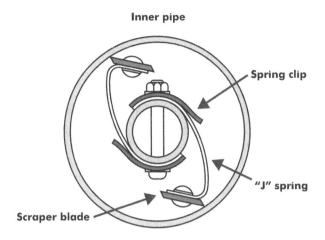

Spring clip

"J" spring

Scraper blade

FIGURE 6.19 Cross section of a scraped surface heat exchanger. The rotation of the blades scrapes ice and other solids off of the heat transfer surface (M. Ogawa after Perry's Chemical Engineering Handbook).

6.9.3 MEANS OF HEATING AND COOLING

In order to heat or cool your process stream in a winery setting, heat transfer fluids at the right conditions are needed. These fluids remain in a single phase (e.g. a liquid) and change temperature or, alternatively, they can change phase (e.g. liquid to a gas) and remain at the same temperature. As described in an earlier chapter, this corresponds to using sensible heat or latent heat. Either will work for heating or cooling. Cooling water or glycol are often used as coolants in jackets and other heat exchangers with the intention that they will change temperature, but not phase. Both of these materials (water and glycol) could also be used for heating streams, as long as the desired process temperatures remain beneath the boiling points of these heat transfer liquids. An example of a phase change for cooling would be direct expansion of the refrigerant ammonia in the jacket (from liquid to gas), while an example of phase change for heating would include condensing steam (steam condensing to liquid water). The choice will be dependent on available equipment and desired processing.

6.9.4 HEAT TRANSFER CALCULATIONS USING COMMON SCENARIOS

In order to discuss the calculations associated with heat transfer, three examples will be discussed in the next section. These examples will be (1) cooling of a fermentation using a jacket, (2) cooling of a must or wort prior to fermentation using a single-pass tube-in-tube or shell-and-tube heat exchanger, and (3) chilling of a wine or beer for cold stabilization.

6.9.4.1 Temperature Control during Fermentation

In order to control the temperature in a fermentation, all heat generated by the metabolic activity of the yeast must be removed. Most of this heat will be removed via transfer to a coolant flowing through an external jacket on the tank. The heat flux from the fermenting juice or must to the coolant can be described as:

$$\frac{\bar{Q}_{TOT}}{A} = U\Delta T_{LM} \tag{6.3}$$

where $\bar{Q}_{TOT}$ is the total heat generation rate for the fermentor, A is the heat transfer area, U is the overall heat transfer coefficient, and ΔT_{LM} is the log mean difference between the coolant temperature and the fermentation temperature. The left-hand side of this equation is called the heat flux and the driving force for the heat flux is the temperature difference between the inside of the fermentor and the jacket. To use this equation, we must further define all of the terms.

The maximum rate of heat generation during an ethanol fermentation coincides with the maximum sugar utilization rate. For white fermentations, this sugar utilization rate corresponds to 2–3 Brix/day (or Balling/day), while for red wines, 4–6 Brix/day is more typical. From a correlation in the Boulton et al. (1996), heat generation *per unit volume* at this point in the fermentation can be described by the relationship:

$$Q_{TOT} = 77.5 \times 10^{-3} \frac{W/L}{Brix/day}[R_s] \tag{6.4}$$

where R_s is the sugar utilization rate in Brix/day. The units of Q_{TOT} are then W/L. To get heat generation rate for the whole fermentor, we then use the relationship:

$$\bar{Q}_{TOT} = V_w Q_{TOT} \tag{6.5}$$

where V_w is the working volume of the fermentor.

The heat transfer area, A, can be calculated from the geometry of the tank:

$$A = \pi D_t H_j \tag{6.6}$$

where D_t is the diameter of the tank and H_j is the height (i.e. length) of the jacket. This height does come with a caveat. It should actually be called the effective height, as a jacket that reaches above the liquid level or is inactive for some other reason does not count as part of the heat transfer surface. Therefore, in deciding the heat transfer area for any given fermentor configuration, care needs to be taken to only include jacket surface area actively cooling the fermentation.

Since the temperature difference between the fermentation and the jacket is the driving force for heat transfer, this difference needs to be calculated. However, if the coolant enters the jacket at the bottom at its coldest and leaves at the top of the jacket at its warmest, the temperature difference is not constant along the side of the tank. While we could just use an arithmetic average of the two temperature differences, engineers have shown that a better approximation the characteristic temperature difference is the log mean temperature difference. The log mean of any two values, X_1 and X_2, is defined as:

$$X_{LM} = \frac{X_2 - X_1}{ln\dfrac{X_2}{X_1}} = \frac{X_1 - X_2}{ln\dfrac{X_1}{X_2}}$$

Therefore, the log mean temperature difference, ΔT_{LM}, can be calculated using the jacket inlet (T_i) and outlet (T_o) temperatures, along with the fermentation temperature, T_f (see figure in Example 6.1 for notation). The form of this mean temperature difference is:

$$\Delta T_{LM} = \frac{\left(T_f - T_o\right) - \left(T_f - T_i\right)}{\ln \dfrac{\left(T_f - T_o\right)}{\left(T_f - T_i\right)}}$$

or rearranging to:

$$\Delta T_{LM} = \frac{\left(T_i - T_o\right)}{\ln\left[\left(T_f - T_o\right)/\left(T_f - T_i\right)\right]} \tag{6.7}$$

Finally, the overall heat transfer coefficient, U, is a function of jacket geometry, coolant flows, fermentation mixing, materials of construction, and expected jacket fouling. It is a property of the fermentor design and the fermentation system. The vendor of the fermentor should be able to give you an idea of this coefficient for the tank they are selling.

The necessary (or minimum) coolant flow through the jacket can be calculated using the overall heat to be removed and the properties of the coolant as shown in Equation (6.8):

$$\bar{Q}_{TOT} = \dot{m} \times C_{p,c} \times \left(T_o - T_i\right) \tag{6.8}$$

where $\dot{m}$ is the coolant mass flow rate and $C_{p,c}$ is the heat capacity of the coolant. Care must be taken with dimpled jackets to assure that coolant volumetric flow is less than 50 gpm. Higher flow rates will cause an unacceptably high pressure drop for this type of jacket. If higher flow is necessary, a different jacket geometry should be utilized.

Example 6.2: Sizing a Fermentor and Its Jacket for the Commonwealth Winery (Mara Couch)

Commonwealth is a mid-sized winery located in the Alexander Valley AVA of Sonoma County. Commonwealth is focused on crafting exciting premium wines that speak to the storied viticultural tradition of the US. Fruit for the Old Vine Zinfandel, Prieto Field Blend, and Roosevelt Chardonnay are sourced from Estate vineyards. The Pinot Noir and Catawba used in the American Sparkler are purchased from vineyards within Sonoma County.

The American Sparkler is a dry sparkling wine (brut) produced according to the méthode champenoise. All wine (except for the American Sparkler which is inoculated) undergoes indigenous fermentation in stainless steel tanks and is then transferred to specific oak programs. The Roosevelt Chardonnay is whole cluster pressed and eventually aged in 20% new and 80% neutral American oak for 12 months. Varieties in the Prieto Field Blend are fermented in stainless steel separately, then blended in varying percentages depending on quality and sensory properties. The best Petit Sirah lots are reserved for the Old Vine Zinfandel. All Old Vine Zinfandel and Prieto Field Blend lots are aged in neutral American Oak for 18 months and allowed to undergo (inoculated) ML fermentation. All wines are sterile filtered prior to bottling for microbial control.

PROBLEM

The largest fermentors in the Commonwealth Winery are 5,500 gal (= 735 ft³) and have a L/D$_t$ = 2. The fermentors will have a working volume of 4,400 gal (= 588 ft³) and be used for red wine fermentations at 86°F. At this temperature, the fermentations are expected to use 6 Brix/day at their maximum sugar utilization rate. Chilled water at 40°F will be used in the jacket and will emerge from the jacket at 75°F. The manufacturer of the tanks states that the overall heat transfer coefficient, U, is 15 Btu/hr ft² °F. If the tank has 10 ft (i.e. height) of jacket in contact with the must, is there enough heat transfer surface area to maintain the temperature in the tank at the setpoint?

SOLUTION

First, we can calculate the dimensions of the tank. To do this, we use the equation for the volume of a cylinder:

$$V = \frac{\pi D_t^2}{4} L = \frac{\pi D_t^3}{2}$$

when L/D$_t$ = 2. Therefore,

$$D_t = \left(\frac{2V}{\pi}\right)^{\frac{1}{3}} = 7.8 \, \text{ft}$$

and L = 15.6 ft (height).

Now, we use the equation for heat flux:

$$\frac{\bar{Q}_{TOT}}{A} = U \Delta T_{LM}$$

rearranged to solve for area, A:

$$A = \frac{\bar{Q}_{TOT}}{U \Delta T_{LM}}$$

Next, we need to find the $\bar{Q}_{TOT}$:

$$Q_{TOT} = 77.5 \times 10^{-3} \frac{\dfrac{W}{L}}{\dfrac{Brix}{day}} \left[6 \frac{Brix}{day} \right]$$

$$= 0.465 \frac{W}{L}$$

Therefore,

$$\bar{Q}_{TOT} = Q_{TOT} \times V_W = 0.465 \frac{W}{L} \times 4400\,\text{gal} \times 3.78 \frac{L}{\text{gal}}$$

$$\bar{Q}_{TOT} = 7734\,W$$

$$= 10,564\,W \times \left[56.87 \frac{\text{Btu/min}}{\text{kW}} \right] \times \frac{\text{kW}}{1000\,W} \times 60 \frac{\text{min}}{\text{hr}}$$

$$= 26,390 \frac{\text{Btu}}{\text{hr}}$$

We then calculate the log mean temperature difference:

$$\Delta T_{LM} = \frac{T_i - T_o}{\ln\left[\dfrac{(T_f - To)}{(T_f - T_i)}\right]} = \frac{40°F - 75°F}{\ln\left[\dfrac{(86°F - 75°F)}{(86°F - 40°F)}\right]}$$

$$= 24.5°F$$

Therefore,

$$A_{req} = \frac{\bar{Q}_{TOT}}{U \Delta T_{LM}} = \frac{26,390\,\text{Btu/hr}}{\left(15 \dfrac{\text{Btu}}{\text{hr ft}^2 °F}\right)\left(24.5°F\right)} = 71.8\,\text{ft}^2$$

This is the required area. The available area is:

$$A_{avail} = \pi D_t H_j = \pi(7.8\,\text{ft})(10\,\text{ft}) = 245\,\text{ft}^2$$

We have plenty of area to control our fermentation temperature with this design. However, as we demonstrated earlier in this chapter, practically no heat is removed from the cap by the jacket. Because of this, we should only count the jacket height that is in contact with the liquid part of the fermentation. In this case, if we estimated that half of the fermentation was cap, our effective jacket height would be more like 3.75 ft (assuming the jacket starts 2.5 ft from the bottom of the tank). This would give us an available area of 91.9 ft², which is still more than is required.

6.9.4.2   Use of a Single Pass Heat Exchanger for Must/Juice/Wort Chilling

In designing a tube-in-tube or shell-and-tube heat exchanger for chilling juice or must, the calculations will be similar to the jacketed fermentor case described above. In this case, enough heat will be taken from the juice or must to reduce the temperature to the desired level and transferred to a coolant usually flowing in the opposite direction (i.e. countercurrent flow). The analysis of this heat transfer problem again begins with the equation for the heat flux:

$$\frac{\bar{Q}_{TOT}}{A} = U\Delta T_{LM} \tag{6.9}$$

In this case, however, the total heat to be removed will be defined in terms of the juice (or must) as in Equation (6.8):

$$\bar{Q}_{TOT} = m_j C_{p,j}\left(T_{j,i} - T_{j,o}\right) \tag{6.10}$$

where m_j is the mass flow rate of juice, $C_{p,j}$ is the heat capacity of the juice, $T_{j,i}$ is the temperature of the entering juice, and $T_{j,o}$ is the temperature of the exiting juice. The heat transfer area for each inner tube will be:

$$A = \pi D_i L \tag{6.11}$$

where D_i is the average diameter of the tube and L is the length of the tube. For a shell-and-tube heat exchanger, the area for all tubes will be summed together. The log mean temperature difference for countercurrent flow will be given by:

$$\Delta T_{LM} = \frac{\left(T_{j,i} - T_{c,o}\right) - \left(T_{j,o} - T_{c,i}\right)}{\ln\left[\left(T_{j,i} - T_{c,o}\right)/\left(T_{j,o} - T_{c,i}\right)\right]} \tag{6.12}$$

For multiple pass shell-and-tube heat exchangers, this temperature difference will be multiplied by an efficiency term, F_T, that will correct for the non-countercurrent

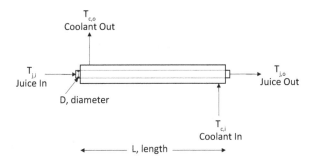

FIGURE 6.20   Diagram of a countercurrent tube-in-tube heat exchanger with notation.

flow in part of the exchanger. The overall heat transfer coefficient for a tube-in-tube or shell-and-tube heat exchanger like this one will generally be considerably higher than for the jacketed fermentor, especially when no mixing exists in the fermentor. A diagram of a heat exchanger with notation is shown in Figure 6.20.

Figure 6.21 explains why countercurrent flow in a heat exchanger is usually more efficient than co-current flow. As can be seen in this figure, with countercurrent flow, the temperature difference or heat flux driving force remains constant along the length of the heat exchanger, thus evenly using the capacity of the heat exchanger. In the co-current case, there is a large temperature difference on one end of the heat

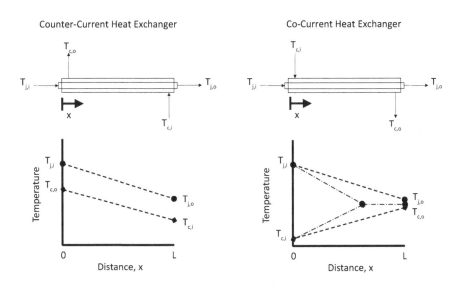

FIGURE 6.21   A comparison of counter-current and co-current flows in a heat exchanger. The counter-current configuration maintains a more constant temperature difference across the length of the heat exchanger, making this configuration more efficient. In the co-current configuration, there is a large temperature difference driving force for heat flux at the entrance and then very little to no temperature difference at the end, L, of the heat exchanger. This is less efficient and means that the some portion of the heat exchanger near the juice outlet may not be actively exchanging heat at all.

exchanger and a smaller or non-existent temperature on the opposite end as heat is exchanged. In this case, a large heat flux will occur on the left end of the exchanger and none will happen on the right side. If the temperatures meet in the middle of the heat exchanger, no further heat exchange will occur. For this reason, countercurrent heat exchange is more desirable when it is possible.

Example 6.3: Sizing a Juice Chiller for the Commonwealth Winery

PROBLEM

To examine the use of a juice chiller for white juices in the Commonwealth Winery, we look at a design for a counter-current tube-in-tube heat exchanger to chill juice coming from the press pan into a tank below. The juice is leaving the press pan and entering the heat exchanger at 85°F. We would like to chill the juice down to 60°F prior to settling and fermentation. To do this, we will use glycol at 23°F. The approach temperature (the temperature of the coolant leaving the heat exchanger) is 75°F. The volumetric flow rate of the juice is 31.1 gpm (approximating 1,400 gal of juice from 16 tons of fruit in the first 45 min of pressing), and the diameter of the inner tube/process line is 2 in. The density of the juice is 70 lb_m/ft^3. The heat capacity of the juice is 0.91 Btu/lb_m °F. A typical overall heat transfer coefficient for a tube-in-tube heat exchanger is 200 Btu/hr ft^2 °F. What heat transfer area is necessary for this temperature change (at this flow rate)? What length of pipe is needed to achieve this area?

SOLUTION

To find the length of the juice chiller, we first start with the equation for heat flux:

$$\frac{\bar{Q}_{TOT}}{A} = U\Delta T_{LM}$$

To calculate $\bar{Q}_{TOT}$, we use the equation:

$$\bar{Q}_{TOT} = \dot{m_j}C_{P_j}\left(T_{ji} - T_{jo}\right)$$

$$= 31.1\frac{gal}{min} \times \frac{ft^3}{7.48\,gal} \times 70\frac{lb_m}{ft^3} \times 60\frac{min}{hr} \times 0.91\frac{Btu}{lb_m\,F} \times 25°F$$

$$\bar{Q}_{TOT} = 401,310\frac{Btu}{hr}$$

For the log mean temperature, we use the equation:

$$\Delta T_{LM} = \frac{\left(85°F - 75°F\right) - \left(60°F - 23°F\right)}{\ln\left[\dfrac{\left(85°F - 75°F\right)}{\left(60°F - 23°F\right)}\right]} = 20.6°F$$

Now we can return to the heat flux equation rearranged to calculate area:

$$A_{req} = \frac{\bar{Q}_{TOT}}{U\Delta T_{LM}} = \frac{401,310\,\text{Btu/hr}}{\left(200\dfrac{\text{Btu}}{\text{hr}\,\text{ft}^2\,°F}\right)\left(20.6°F\right)} = 97.4\,\text{ft}^2$$

To find the length:

$$A_{req} = \pi DL$$

$$L = \frac{A_{req}}{\pi D} = \frac{97.4\,\text{ft}^2}{\pi\left(\dfrac{2}{12}\,\text{ft}\right)} = 186\,\text{ft}$$

This seems like a large length, but this operation amounts to a huge amount of cooling in a very short period of time. In fact, as we will see later, this type of operation can easily take up half of the expected refrigeration in a winery. Heat exchangers like this, as pictured earlier in the chapter, are made up of a series of long lengths of heat exchanger connected by u-bend connectors to accommodate the need for these types of long lengths. It should be noted that picking the grapes early in the morning can preclude the need for this type of unit operation, in addition to saving money and energy on refrigeration.

6.9.4.3 Use of a Jacket and Unstirred Tank for Cold Stabilization of Wine

The third example to be discussed is cold stabilization of a wine in a tank using glycol in the cooling jacket. Because tanks are generally not stirred during this process, the cooling occurs through the process of conduction (heat transfer through a stationary solid or liquid) instead of through convection (heat transfer aided by currents and mixing). Heat transfer through conduction is a much slower process. When the coolant is run through the jacket, the wine near the wall of the tank will begin to cool down to the temperature of the coolant, while the temperature in the center of the tank will remain higher thereby forming a parabolic temperature profile in the tank. Many times when this is done in a winery setting, one can notice ice on the outside of the tank. Unfortunately, ice also typically forms on the inside of the tank creating a further insulating barrier to heat transfer. Overall, this is a very inefficient process and wastes energy. More efficient methods for heat transfer would include using mixing in the tank (e.g. using a Guth mixer) while chilling, pumping the wine over with a heat exchanger in the pumpover line, or staggering the cold stabilization of the tanks in a winery and using one cold tank to pre-chill the next tank (using a heat exchanger). This latter option is likely the best, as it avoids wasting expensive refrigeration. While it is straightforward, it is very rarely performed in practice in the US industry. For wineries or breweries looking to reduce their energy utilization, this idea, then, provides a straightforward means to reduce energy waste. Hot streams can be used to heat or pre-heat other streams that need to be heated, and likewise, cold process streams can be used to chill or pre-chill process streams that need to be chilled—as opposed to just allowing these streams to lose energy to the environment and move toward ambient temperature. This concept is called heat recovery or heat integration and is practiced extensively in facilities like refineries to save energy and reduce costs.

REFERENCES

Boulton, R.B., Singleton, V.L., Bisson, L.F., and Kunkee, R.E. *Principles and Practices of Winemaking.* Chapman and Hall, New York, 1996.

Lerno, L.A., Panprivech, S., Ponangi, R., Hearne, L., Blair, T., Oberholster, A., and Block, D.E. Effect of Pump-over Conditions on the Extraction of Phenolic Compounds during Cabernet Sauvignon Fermentation. *American Journal of Enology and Viticulture,* **69,** 295–301, 2018.

Miller, K. and Block, D.E. A Review of Wine Fermentation Process Modeling. *Journal of Food Engineering,* **273,** 109783, 2020.

Miller, K. V., Noguera, R., Beaver, J., Medina-Plaza, C., Oberholster, A., and Block, D.E. A Mechanistic Model for the Extraction of Phenolics from Grapes During Red Wine Fermentation. *Molecules,* **24,** 1275, 2019a.

Miller, K., Noguera, R., Beaver, J., Oberholster, A., and Block, D.E. A Combined Phenolic Extraction and Fermentation Reactor Engineering Model for Multiphase Red Wine Fermentations. *Biotechnology and Bioengineering,* **117,** 109–116, 2020a.

Miller, K. V., Oberholster, A., and Block, D.E. Creation and Validation of a Reactor Engineering Model for Multiphase Red Wine Fermentations. *Biotechnology and Bioengineering,* **116,** 781–792, 2019b.

Miller, K. V., Oberholster, A., and Block, D.E. Predicting the Impact of Red Winemaking Practices Using a Reactor Engineering Model." *American Journal of Enology and Viticulture,* **70,** 162–168, 2019c.

Miller, K. V., Oberholster, A., and Block, D.E. Predicting fermentation dynamics of concrete egg fermenters. *Australian Journal of Grape and Wine Research,* **25,** 338–344, 2019d.

Miller, K., Oberholster, A., and Block, D.E. Impact of Fermentor Geometry and Cap Management on Phenolic Profile Using a Reactor Engineering Model. *American Journal of Enology and Viticulture,* **71,** 44–51, 2020b.

Schmid, F., Schadt, J., Jiranek, V. and Block, D.E. Formation of Temperature Gradients in Large and Small Scale Red Wine Fermentations During Cap Management. *Australian Journal of Grape and Wine Research,* **15,** 249–255, 2009.

Vlassides, S. and Block, D.E. Evaluation of Cell Concentration Profiles and Mixing in Unagitated Wine Fermentors. *American Journal of Enology and Viticulture,* **51,** 73–80, 2000.

PROBLEMS

1. Your brewery has decided to purchase stainless steel tanks for beer fermentations. The total volume for each tank will be 3,000 gal with a 2,400 gal working volume during fermentation. The dimensions for the tank are shown in the

figure below. The tank will have a dimpled jacket. The bottom approximately 25% of the tank cannot be jacketed because of the dual manway design.

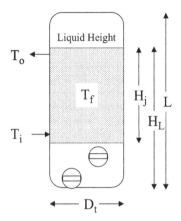

(a) The beers will be fermented at 55°F. During the maximum sugar utilization period of fermentation, the fermentation is dropping 3°Brix/day. The coolant will be chilled water entering the jacket at 40°F and leaving the jacket at 45°F. If the manufacturer tells you that the overall heat transfer coefficient, U, will be 10 Btu/hr ft² °F for the tanks, what is the heat transfer area that will be needed? What fraction of the potential jacket zone does this correspond to?

(b) What is the corresponding flow rate of chilled water through the jacket? Is this reasonable for a dimpled jacket?

(c) Many years later, you are looking to sell this tank off to a salvage equipment company, who is looking for a red wine fermentor. Could this tank be used for red wine fermentations? Assume that you will control the red fermentations at 80°F, the cooling water will enter the jacket at 40°F, and leave the jacket at 70°F. The maximum sugar utilization rate for these fermentations will be 6°Brix/day.

2. After graduating from Davis, your first job is as an assistant winemaker for Domaine de la Ingeniere in charge of white wine production. The owner of Domaine de la Ingeniere has asked you to produce a Sauvignon blanc wine in a stainless steel tank (8,000 gal working volume). When you are done with your fermentation (at 65°F), you decide to chill the wine down for cold stabilization. Instead of chilling it down in the fermentation tank, you decide to run the wine through a shell-and-tube heat exchanger at 67 gpm (= 8.96 ft³/min) using glycol on the shell side prior to doing the remainder of the chilling and stabilization in a second tank. The flow will be counter-current.

(a) If the maximum capacity of the glycol system allows you to supply 170 gpm (= 22.7 ft³/min) glycol at 23°F and the glycol emerges from the heat exchanger at 45°F, what is the minimum temperature of the wine exiting the heat exchanger?

$$\rho_{wine} = 61.8 \ lb_m/ft^3$$

$$\rho_{glycol} = 68.6 \ lb_m/ft^3$$

$$Cp_{wine} = 1.07 \ Btu/lb_m \ ^\circ F$$

$$Cp_{glycol} = 0.24 \ Btu/lb_m \ ^\circ F$$

(b) How much heat transfer surface area is needed for this shell-and-tube heat exchanger? The heat exchanger manufacturer tells you that the overall heat transfer coefficient, U, should be 400 Btu/hr ft^2 °F.

(c) During the setup of the heat exchanger, one of the cellar workers mixes up the glycol inlet and outlet for the heat exchanger. Therefore, the heat exchanger unexpectedly becomes a co-current system. Assuming the same overall heat transfer coefficient, how much heat transfer surface area is needed in this case?

3. **[Integrative Problem]** Immediately prior to crush this year, your Foothills winery receives a new stainless steel jacketed tank to use for white wine fermentations. Unfortunately, you find out too late that your glycol cooling system is already at its limits and cannot be used for the new tank. Luckily, the municipal water source (i.e. city water) is relatively cool in this region at 55°F, and you think you can just about control the fermentation at the desired temperature of 65°F using this water (assuming that the water leaves the jacket at 62°F). The tank is 10,000 gal (= 1,337 ft^3) total volume with a working volume of 8,000 gal (= 1,070 ft^3).

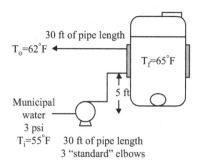

(a) For 2.5 Brix/day maximum fermentation rate, the fermentation will generate heat at a maximum rate of 2.5 Btu/hr gal. If the tank has a diameter of 9.5 ft, what is the height of jacket necessary to control the temperature at the desired level? The tank manufacturer tells you that the overall heat transfer coefficient for this fermentor should be U = 15 Btu/hr ft^2 °F.

(b) What is the necessary flow rate for the municipal water to maintain temperature? Assume that the heat capacity of the water is Cp = 1 Btu/lb$_m$ °F and the density of the water is 62.4 lb$_m$/ft^3.

(c) For the flow rate calculated, find the best size for the pipe entering and leaving the jacket and the circulation pump. Be sure to verify that you have turbulent flow and a "reasonable" pressure drop. The pipe can be assumed to be "smooth tube." The viscosity of the water is 1 cp (= 0.000672 lb_m/ft s).

(d) Now that the tank and pipe are in place, you need to choose the size of the centrifugal circulation pump. The tank manufacturer tells you that the jacket will have a pressure drop of 10 psi. The centrifugal pump used for circulation is at a height equal to the bottom of the fermentor, and the jacket begins 5 ft above this. The pipe connecting the municipal water to the tank is a total of 30 ft in length with three "standard" elbows. The return line exits the top of the jacket and goes back to a holding tank 30 ft away (at the same height, no elbows). The municipal water source is supplied at 3 psi pressure. Select an appropriate pump and pump curve online.

4. **[Integrative Problem]** You get a job with a large winery and are put in charge of their custom crush business. In your first harvest, you get several lots of machine-harvested Chardonnay with a high degree of mold. This mold secretes a polysaccharide that causes a gel-like substance to form during fermentation. To avoid this problem in future years, you decide to install a pasteurization system to kill the mold. As shown below, this system consists of a tube-in-tube heat exchanger for heating up the juice to 170°F from an initial 70°F, a tube-in-tube heat exchanger that maintains the juice at 170°F for the pasteurization (killing) step, and a third heat exchanger for chilling the juice back down to fermentation temperature.

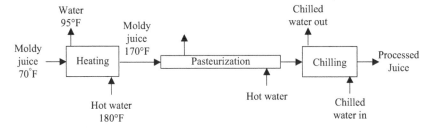

(a) To minimize the effects of this pasteurization on scheduling, you decide that you need the pasteurization unit to handle 10,000 gal (= 1,337 ft^3) in 2.5 hr. The density of the juice is approximately 70 lb_m/ft^3 and the viscosity is 2.6 cp. What size should the pipe be that runs through the pasteurizer in order to assure turbulent flow and a reasonable pressure drop. The inner surface of the pasteurizer can be assumed to be "smooth tube." (1 cp = 0.000672 lb_m/ft s)

(b) In the initial "heating" heat exchanger used for heating the juice, what is the necessary volumetric flow rate (i.e. gpm) for hot water? How long should the exchanger be (assuming a simple tube-in-tube design)? The heat capacity of the juice is 0.94 Btu/lb_m °F and the heat capacity of the water is 1 Btu/lb_m °F. The overall heat transfer coefficient for the exchanger is 500 Btu/hr ft^2 °F. The densities of the juice and hot water are 70 lb_m/ft^3 and 62.4 lb_m/ft^3, respectively.

5. You are helping to design a tube-in-tube juice chiller that will be used for your proprietary white wine, *Venology*. You know that your juice usually leaves the press at 80°F, and you would like the juice to be chilled to 60°F for the initial settling. Assume that the maximum flow rate through the chiller will be 33 gpm, and that the inner "tube" will have an average diameter of 2 in. Again, the density of the juice is 70 lb_m/ft^3. The heat capacity of the juice is 0.91 $Btu/lb_m°F$. The overall heat transfer coefficient, U, for this chiller is 150 $Btu/hr °F ft^2$.

   (a) The system is to be designed for glycol use with an entrance temperature of 23°F and a coolant exit temperature of 70°F. For a countercurrent system, what heat transfer area would be required for this operation? What length of pipe does this correspond to?
   (b) After ordering the chiller, your winery decides to switch over to a less energy-intensive chilled water system wherever possible to do their part to reduce energy use in California. Would it be possible to switch over the juice chiller as designed in (a) to use with chilled water? The chilled water would enter the outer tube at 40°F and exit at 70°F.
   (c) How could you design the chiller so that it could easily be used with either coolant?

6. Since the next step that the Casa Ingeniero winemaker is planning after emptying all of the barrels into the blending tank is cold stabilization, you suggest to him that he might want to place a plate and frame heat exchanger in the transfer line to chill the wine from the cellar temperature of 55°F to 38°F prior to reaching the blending tank. The coolant available is chilled glycol at 28°F that leaves the heat exchanger at 46°F in a counter-current configuration. The overall heat transfer coefficient for this type of heat exchanger is typically 350 $Btu/hr ft^2 °F$. The heat capacity for the wine is 1.04 $Btu/lb_m °F$.

   (a) How many heat transfer plates are necessary if each one has 3 ft^2 of heat transfer area?
   (b) Give an explanation for why you might suggest an external heat exchanger for pre-chilling prior to cold stabilization instead of providing all of the chilling capacity through the jacket in the blending tank.

7. In search of the perfect Zinfandel, you decide to build a winery in the Sierra Foothills. To save money, you buy used 6,000 gal fermentors with a 5,000 gal (= 18,900 L) working volume. Instead of buying a glycol chiller to maintain fermentation temperature, you decide to use water from the nearby mountain stream as a coolant. Average monthly water temperatures are shown in the table below. The tank has the dimensions shown in the figure below.

Month	Mean Temperature (°F)
January	60
February	57
March	52
April	49

Month	Mean Temperature (°F)
May	51
June	57
July	62
August	68
September	72
October	74
November	67
December	64

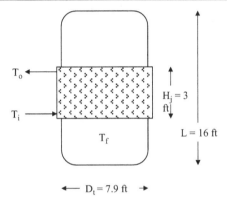

(a) To make sure the equipment will work, you get some early Zinfandel from a vineyard in Davis in August. If the maximum fermentation rate for the Davis Zinfandel fermentation is 5 Brix/day, is the heat transfer surface area in this tank sufficient to control the fermentation at 85°F (Note: T_o should not exceed 78°F)? The overall heat transfer coefficient, U, is 40 Btu/hr °F ft². (Unit conversion: 1 W = 3.41 Btu/hr)

(b) Your estate Zinfandel ripens in the middle of October, what is the maximum fermentation rate (i.e. Brix/day) that will still allow you to control the temperature at 85°F?

8. [Integrative Problem] On a recent visit to Chablis, you taste some Grand Cru Chablis and decide that your goal is to produce a great "Chablis-style" California Chardonnay (of course, in deference to your French colleagues, you will not call it "Chablis"). To do this, you decide to emulate one of your favorite wineries in Chablis in every way. This includes using 3,000 gal working volume (= 401 ft³) fiberglass fermentors with internal heat exchangers for temperature control. The internal heat exchanger consists of a series of 12 ft long plastic tubes (each with an average diameter of 0.5 in) that hang from a manifold inside the fermentor as shown below in the diagram and photo. The cooling water enters the tubes at 40°F and leaves the tubes at 50°F. The fermentation is to be controlled at 60°F. At its maximum, the sugar utilization rate at this temperature is 3.8 Brix/day, which corresponds to a heat output/gal of 3.8 Btu/ hr gal. See the photo of the inside of a fiberglass tank with heat exchanger, looking up toward the top of the tank.

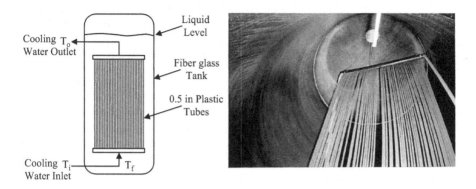

(a) If the overall heat transfer coefficient for this system is 15 Btu/hr ft^2 °F, what is the total heat transfer surface area needed and how many tubes will you need?

(b) If the heat capacity of the water coolant is 1 Btu/lb$_m$ °F and the density of water at this temperature is 62 lb$_m$/ft^3, what is the necessary total volumetric water flow through the tubes?

(c) If equal amounts of water flow through each of the tubes in the heat exchanger, what is the velocity of the water and pressure drop through each tube? Is the pressure drop reasonable? Assume an inner tube diameter of 0.25″ and a Fanning friction factor of $f_f = 0.02$.

(d) Name one concern that you might have about using this design in your normal operations.

9. After traveling around the world making wine for several years, you decide to settle in California and open up a winery that specializes in Syrah. From your experience at crush positions in Australia, you decide to purchase several 6,000 gal (= 802 ft^3) rotary fermentors with 4,500 gal working volumes. Because you don't have much money to start, you go with a low-cost stainless steel tank fabricator that has not made one of these tanks before, but is interested in starting. Therefore, you need to give him a lot of information on the design. The basic geometry is shown in the figure below.

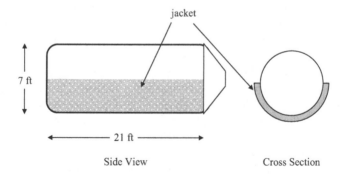

Side View   Cross Section

(a) You plan to ferment your Syrah at a maximum temperature of 86°F. To maintain this temperature, you are hoping to use a municipal water source

that is typically at 72°F during the late September harvest. The temperature at the outlet of the jacket should not exceed 78°F. At your planned
fermentation temperature, your experience tells you that you should have
a maximum sugar utilization rate of about 5.5 Brix/ day. Assuming an
overall heat transfer coefficient of 15 Btu/hr °F ft², will you have enough
heat transfer surface area if only the bottom half of the fermentor is jacketed? (12.9 Btu/hr gal = 1 W/L).

(b) What is the necessary volumetric flow rate of cooling water through the
jacket? Would it be reasonable to use a dimpled jacket with this type of
flow rate? The heat capacity and density of the cooling water are 1 Btu/lb$_m$
°F and 62.4 lb$_m$/ft³, respectively.

10. **[Integrative Problem]** You are opening a new winery in Napa on Mt. Veeder
that will be making an Australian-style Shiraz called ShirOz. You decide to
copy everything about Australian winemaking including the design of the
cooling jackets on their tanks. Therefore, you have your tanks built with a
dimpled jacket 3 ft wide that runs the entire height of the tank (as shown in the
figure below), as you have seen in your travels through Australia. This tank has
a total volume of 2,500 gal (= 9,463 L) with a working volume of 2,000 gal (=
7,570 L). You are planning to ferment the Shiraz at 82°F. You are using an
industrial water source to feed the jacket that typically does not go above 64°F.
The jacket outlet temperature is to remain below 70°F.

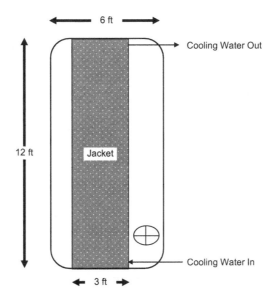

(a) If you expect the maximum sugar utilization rate in this fermentation to be
6 Brix/day and your tank manufacturer thinks that the overall heat transfer
coefficient, U, will be 25 Btu/hr ft² °F, will this be enough heat transfer
area to control the temperature?

$$1W = 3.41 Btu / hr$$

(b) During the first crush in your new winery, you use these tanks for the first lots of Shiraz. Some of your Shiraz comes in late, however, so you are able to use each tank a second time. However, you are worried about the amount of water you are wasting for cooling. Because of your mountain location, you realize that if you just open up your winery doors the ambient temperature is 50°F at the highest during the day. What is the minimum that U, the overall heat transfer coefficient, could be and still have sufficient control of the fermentation temperature if you just use ambient air to cool the tanks instead of water in the jacket?

(c) While visiting one of your favorite Shiraz producers in the Barossa Valley, you took a picture of some of their presses (see below). You are now weighing whether or not you want to use the same type of equipment/ process for your ShirOz. Briefly, give two potential advantages and two potential disadvantages of this type of press over other types that you could use.

11. After several years of successful operation, the Green Grapes Winery begins to have problems with their supplier of Gewurztraminer. The grapes are coming in with a large amount of mold. To avoid issues with the associated oxidative enzymes, you decide to do a heat treatment of the juice coming out of the press pan prior to filling the tank. Since we decided to pick grapes at night and stop using our tube-in-tube heat exchanger for must chilling after the first harvest, this heat exchanger has just been sitting there doing nothing. Therefore, we plan to see if we can use this heat exchanger for the juice heat treatment. We will use condensing steam to heat the juice from 70°F to 90°F (prior to a short hold step and a cooling step).

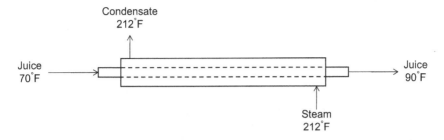

(a) The flow rate of the juice out of the press pan is 50 gpm (= 6.7 ft³/min) and the density of the juice is 68 lb$_m$/ft³. The manufacturer of the heat exchanger feels that the overall heat transfer coefficient, U, for this situation should be around 200 Btu/hr ft² °F. Calculate the necessary heat transfer area for this heat exchanger.

$$Cp_{juice} = 0.91\,Btu/lb_m\,°F$$

(b) If the existing must chiller has a 2.5 in (inner tube) diameter and a maximum length of 250 ft (broken into ten 25 ft lengths), will this work?

12. **[Integrative Problem]** After several years of successful operation, the Green winery begins to open a new winery in the Willamette Valley in Oregon and make just Pinot noir. You have travelled to some of the great Domaines in Burgundy and decide to follow their lead by using oak upright tanks for half of your production. These oak uprights will have a total volume of 2,000 gal (= 267 ft³ =7,560 L) with a working volume of 1,500 gal (= 201 ft³ = 5,670 L).

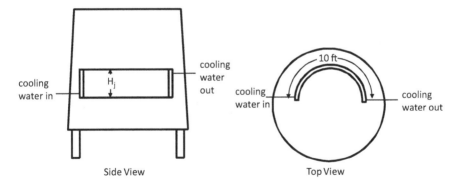

(a) You have a limited amount of money, so you talk to a low cost cooperage, Faux Freres, about buying the oak uprights. The salesman tells you that you can buy a tank made from French oak from the Vosges, Troncais, or Limousin forests or American oak from Missouri. He tells you that for Pinot noir, you should definitely use the wood from Vosges, and that it is worth the extra money over any of the other possible choices (including the other French forests). Using what you know about barrel chemistry, explain why this might not be the best choice given the extra cost. In general, what chemical differences do you expect in wine made in contact with these different wood sources, if any?

(b) After the first vintage in these tanks, you are disappointed to find that the fermentations are getting hotter than you would like. Since you cannot jacket the oak upright, you decide to add an internal stainless steel heat transfer surface (a "fin") as shown in the figure below. You plan to make the surface a rectangular shape bent into a semicircle. The length of the arc is 10 ft. Calculate how tall the surface needs to be (H_j) given that you expect your Pinot to ferment at a rate of 7.5 Brix/day controlling at a fermentation temperature of 88°F. Your cooling water enters the fin at 60°F and leaves at 78°F. You have some indication that the overall heat transfer coefficient, U, for this type of heat transfer device should be around 10 Btu/hr ft^2 °F. Ignore heat transfer from the four thin edges of the fin.

$$1\,W = 3.41\,Btu\,/\,hr$$

(c) Name two negative aspects of adding the fin, aside from the costs of purchasing and installing them.

13. After working on the winemaking staff for Arboreal Winery for 5 years, you decide that it is time to follow your dream and start your own label. You find an old property down the road from Arboreal (in Oregon) that is available and purchase it. The existing winery has lots of old equipment and several open-top, rectangular concrete tanks.

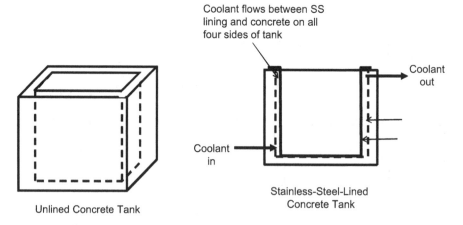

Unlined Concrete Tank

Coolant flows between SS lining and concrete on all four sides of tank

Coolant out

Coolant in

Stainless-Steel-Lined Concrete Tank

(a) While you like the idea of the open top tanks for the Pinot noir that you are planning to make, you are worried about the "cleanability" of the tanks. Give two possible problems with these concrete tanks based on our class discussion of sanitary design.

(b) The original concrete tanks are approximately 9 ft wide by 9 ft long by 10 ft deep. To fix the potential sanitation issues, you decide to install a stainless steel liner in the tanks. This liner sits directly on the bottom of the concrete tank, but there is a small 2″ gap around all of the sides between the stainless steel and concrete (but not on the bottom). You decide that this gap can be used as a jacket to provide temperature control for your fermentations. If the working volume for this tank is 80% of the total volume and heat is only *removed* through the bottom half of the must (i.e. the liquid part underneath the cap), will you have enough surface area to control the temperature at 90°F? Assume that the maximum fermentation rate for your Pinot noir fermentations is 6.4 Brix/day and your coolant is chilled water at 45°F and your approach temperature is not more than 80°F. The overall heat transfer coefficient for this configuration is likely to be approximately 10 Btu/hr ft² °F.

$$1\,W = 3.41\,Btu/hr$$

(c) What is the minimum volumetric flow rate of coolant (assume chilled water) that will allow the temperature to be controlled in this fermentor (given a coolant temperature of 45°F)?

$$C_{p,water} = 1.0\,Btu/lb_m\,°F$$

$$\rho_{water} = 62.4\,lb_m\,/\,ft^3$$

(d) What is the minimum ΔT_{LM} that would still allow the temperature to be controlled in this fermentor?

(e) To what maximum coolant temperature would that correspond if the approach temperature is never more than 80°F?

14. **[Integrative Problem]** After leaving UC Davis, you take a job as assistant winemaker at Pedunculate Oaks Winery. At this high-end winery, the consulting winemaker has chosen to use 5,000 gal cylindrical upright concrete tanks with a 4,000 gal (= 15,120 L) working volume for the Cabernet Sauvignon production. Expecting that he will still need cooling to control temperature, he decides to have a stainless steel coil installed inside the tank and hands off the project to you to figure out how long the coil needs to be.

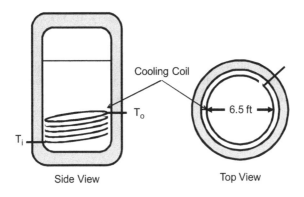

Side View   Top View

(a) The concrete tank with the stainless steel coil is pictured in the diagram above. The diameter of the stainless steel used for the coil is 1 in. Your consulting winemaker likes the fermentations to be fast and believes that the maximum sugar utilization rate in your Cabernet fermentations will be about 6.5 Brix/day at 90°F. For cooling, the chilled water entering the coil will be 46°F and exit with a temperature of 80°F. How long does the coil need to be (i.e. what is the total length and how many complete turns around the inside of the tank does the coil make)? The manufacturer of the coil thinks that the U will be 40 Btu/hr ft² °F for this type of heat exchanger.

$$1\,W = 3.41\,Btu\,/\,hr$$

(b) What volumetric flow rate of chilled water will you need to maintain these conditions?

$$\rho = 62.4\,lb_m\,/\,ft^3$$

$$c_p = 1\,Btu\,/\,lb_m\,°F$$

(c) One day before crush, your consulting winemaker is looking around the winery grounds and finds a storage room in an old winery building that has some equipment from the previous owners that has not been used for at least 40 years. This includes (1) a ratchet-type basket press with a wood-slat basket and a painted metal base (with the paint chipping off in places); (2) a reducer that has had an early type of tri-clamp flange welded on (the welding looks OK on the outside, but the inside feels a little rough to touch where the weld has been ground); (3) a ball valve that is still functional though there is noticeable rouging on the inside of the ball; and (4) several glass carboys with glass airlocks. He leaves it to you to decide what to salvage and what to toss. For each of these items, decide whether you will keep it (for use in your winery) or not and why?

15. **[Integrative Problem]** After graduating from UC Davis, you get a job as Assistant Brewmaster at Lakeshore Brewery, a successful medium-sized brewery set on the north shore of Lake Tahoe. At this winery, they are making a lager in 3,500 gal fermentors with 2,800 gal (= 10,584 L) working volumes. Your first month on the job, the refrigeration/chilled water system in the brewery goes down just as your first fermentations are taking off, reaching their maximum sugar utilization rate of 2.3 Brix/day. You would like to save your fermentations from over-heating and would like to continue to control the fermentation temperature at 65°F.

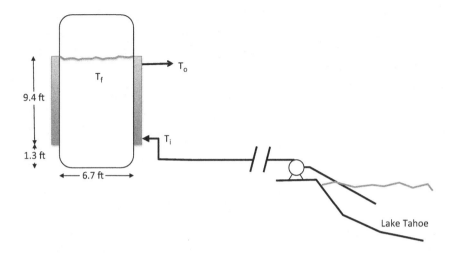

(a) Wanting to make a good impression on the Brewmaster, you come up with the idea of pumping water from the lake (at 55°F) directly through the jacket to provide cooling. However, you need a pump to do this. You look around the storage area of the brewery and find three pumps that are not being used, a used diaphragm pump, a centrifugal pump that looks brand new, and a progressive cavity pump with a stator that is made of an unknown synthetic rubber material. Based on your knowledge of pumps, which pump will you use for this application? Explain your answer.

(b) You look up the pump information on the internet and find that the maximum flow rate that the pump can produce is 10 gpm (= 1.34 ft³/min). You are concerned that using the water for cooling the fermentor will heat up the water to a temperature that is too warm to put back into the lake. For this flow rate, at what temperature do you expect the cooling water to come out of the fermentor jacket? Show your calculations.

$$c_p = 1 \, \text{Btu} / \text{lb}_m \, °F$$

$$\rho = 62.4 \, \text{lb}_m / \text{ft}^3$$

$$1 \, \text{W} = 3.41 \, \text{Btu} / \text{hr}$$

(c) If the jacket is 9.4 ft tall, the tank has a diameter of 6.7 ft, and the liquid level is just at the top of the jacket, does this tank have sufficient surface area to control the temperature. The tank manufacturer has told you that the overall heat transfer coefficient, U, is 50 Btu/hr ft² °F. Show your calculations.

16. After a few harvests at Bodega Impetu, you decide it's time to start up your own winery in the Paso Robles area and call it M Block Vineyards after your best vineyard. At this winery, you would like to produce a fruit-forward Sauvignon blanc that you ferment cold. You decide to destem the grapes and

then press them, followed by chilling the juice down in a plate and frame heat exchanger prior to cold-settling and fermentation. You find a used GEA heat exchanger on the market that has fifty 2 foot by 1 foot plates and is set up for counter-current flow.

(a) If you would like to chill your juice from 80°F to 55°F using chilled water at 40°F (and exiting the heat exchanger at 70°F), what is the highest volumetric flow rate of juice that you will be able to handle coming out of the press pan? You find literature that states that the overall heat transfer coefficient will be 100 Btu/hr ft² °F.

$$c_{p,juice} = 0.91 \, Btu \, / \, lb_m \, °F$$

$$\rho_{juice} = 70 \, lb_m \, / \, ft^3$$

(b) If you expect to get 120 gal of juice from each ton of Sauvignon blanc pressed and the pressing cycle takes 1.5 hr, what is the largest lot of grapes that can be pressed at one time to accommodate this juice chiller and not overflow the press pan? You can assume that 2/3 of the juice will be pressed out in the first 45 min of the cycle.

17. After graduation, you decide to stay around in the new winery at UC Davis to help start up a bank of fourteen 500 gal (= 1,890 L) fermentors, one of which is shown schematically below. The working volume is 400 gal (= 1,512 L).

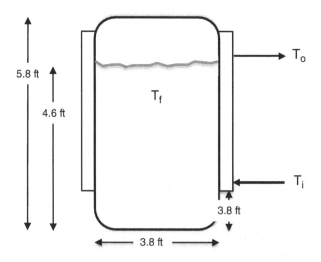

(a) The winemaker tells you that the fermentor has been constructed using 304L stainless steel. Name three features of this stainless steel that would increase its "cleanability."

(b) If the fermentor is going to be used for a red wine fermentation and you expect that the cap will be approximately half of the working volume, what is the heat transfer area available to help control the temperature?

(c) Given this working volume and heat transfer surface area, what is the maximum fermentation rate you can have in this fermentor and still control the temperature at 77°F? You find out from the fermentor manufacturer that the jacket should have an overall heat transfer coefficient of 15 Btu/hr ft² °F. The house chilled water enters the jacket at 40°F and leaves the jacket at 68°F.

$$3.41\,\text{Btu}\,/\,\text{hr} = 1\,\text{W}$$

(d) After figuring this out, the winemaker asks you to calculate the necessary volumetric cooling water flow rate (e.g. in gpm) to control fermentation temperature for one tank at this maximum fermentation rate.

$$c_{p,\text{water}} = 1.0\,\text{Btu}\,/\,\text{lb}_m\,°\text{F}$$

$$\rho_{\text{water}} = 62.4\,\text{lb}_m\,/\,\text{ft}^3$$

$$7.48\,\text{gal} = 1\,\text{ft}^3$$

18. In your first internship, you decide to head to South America and land a job at a medium-sized winery called Bodega Fria. Their specialty is aromatic, fruity white wines, which they ferment cool in stainless steel. The fermentors that they use for this are 4,000 gal (= 535 ft³) with a working volume of 3,200 gal (= 428 ft³ = 12,096 L). As shown in the diagram below, these tanks have two jackets, one near the top and one near the bottom. They can be operated individually or together.

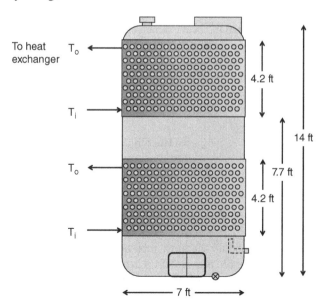

(a) The winemaker at Bodega Fria tells you that their refrigeration system is having issues. To start, she tells you that you can only use the bottom jacket and that the maximum flow rate of chilled water through this jacket will be 5 gpm (0.67 ft³/min). If the chilled water enters the jacket at 40°F and leaves at 48°F, what is the maximum rate of fermentation for which the temperature control will still work?

$$7.48\,gal = 1\,ft^3$$

$$\rho_{cw} = 62.4\,lb_m\,/\,ft^3$$

$$c_{p,cw} = 1\,Btu\,/\,lb_m\,°F$$

$$3.41\,Btu\,/\,hr = 1\,W$$

(b) If you intend to ferment a Muscat juice at 58°F, will the area of the bottom jacket be sufficient? Show your work. Assume $U = 20$ Btu/hr ft² °F for this tank/jacket and the maximum sugar utilization rate that you just calculated.

(c) Generally the top jacket is used for cold wine storage. However, when you are using the bottom jacket, you see debris coming out and decide to try just the top jacket instead. Will this work? Assume the inlet and outlet chilled water temperatures are the same as above, as is the overall heat transfer coefficient. Show your calculations.

(d) Up until now, the chilled water coming out the top of the jacket ends up warming to room temperature before being chilled again in the refrigeration system. You explain to Bodega Fria's winemaker that you think that you can help them save on refrigeration by using this stream that is still cold to chill white juice on its way from the press pan to a fermentor using a tube-in-tube heat exchanger. If you get a temperature pickup of 25°F for the chilled water going through the heat exchanger, what is the minimum temperature of the juice leaving the heat exchanger on the way to the fermentor. You can assume that you are pressing a 12 ton lot of Muscat at 80°F and that you will release about 800 gal (= 107 ft³) in the first 70 min of pressing.

$$\rho_{cw} = 62.4\,lb_m\,/\,ft^3$$

$$c_{p,cw} = 1\,Btu\,/\,lb_m\,°F$$

$$\rho_{juice} = 70\,lb_m\,/\,ft^3$$

$$c_{p,juice} = 0.91\,Btu\,/\,lb_m\,°F$$

(e) What diameter should the inner tube be for the juice to assure turbulent flow and a reasonable pressure drop in a tube-in-tube heat exchanger? You do not have to verify turbulence and pressure drop if you are confident that your choice is correct.

(f) If the heat exchanger that you find is 100 ft in length, is this sufficient to accomplish the heat transfer? Show your calculations. Assume $U = 140$ Btu/hr ft² °F.

19. **[Integrative Problem]** The winemaker at Bodega Fria is very impressed with your winemaking (and engineering) abilities and offers you a permanent position to create a new line of fruit-forward red wines for them. You have Cabernet Sauvignon on the property and decide to use that. The winemaker tells you that you have free reign to choose temperature, pumpover regimes, and oak for this project. You will use the same 4,000 gal tanks as in the previous problem.

   (a) Given the data that you have seen in class, what qualitative temperature profile and pumpover regime would you use to maximize color, but minimize seed extraction (that might lead to increased bitterness and/or astringency)? Make a graph to illustrate how these will vary over time and explain your answer.

   (b) You want to have the prestige of "oak aged" on your label, but do not think you can afford to use barrels at the $8 per bottle price point. You decide to turn to oak alternatives, but choosing between the many options available to you seems like a daunting task. Then, you find an oak alternative company that lists very specific flavors for their various products, as well as a range of shapes and sizes. You get some samples and they seem promising, but you are not sure if the large lots that you need will give the same flavor consistently. From what you know about oak and oak chemistry, name two key questions that you will ask of the salesman to assure that their manufacturing process will produce product that will be consistent from batch to batch and that you will get the right amount of extraction in the one week period you have set aside for this step. Explain your answers.

   (c) If worker safety laws were to change and make it more difficult to send someone into this tank to shovel the skins out, what would your options be for removing the skins from this tank?

20. **[Integrative Problem]** Ten years after graduation from UC Davis, you are chosen for your dream job—you are asked to startup a small high-end winery. This small winery is to be called Wine Cubed (it will say "wine³" on the label). The owner tells you that you can choose all the features of the winery, except one—all fermentors have to be cubes. Thankfully, you find a stainless steel open top fermentor commercially with a total volume of 1,460 gal and a maximum working volume of 1,100 gal (= 4,158 L). It is shown below in schematic and a photograph.

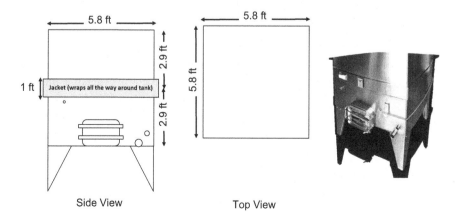

Side View   Top View

(a) The tank manufacturer tells you that the tank is made of 304L stainless steel and that all welds have been ground out both inside and out to an Ra of 25 μin. Are there any other potential issues with this tank from a sanitary design point of view? Explain.

(b) This tank will be used for a Pinot noir fermentation with manual punchdowns. If you receive 5.5 tons of fruit for this fermentation and expect the must volume to be approximately 1,100 gal (= 4,158 L), will you be able to maintain the temperature at a maximum of 86°F? As you are trying to figure this out, you find that the previous consulting winemaker has left you with some data for Pinot noir fermentations from the same vineyard. These data are shown below. You can assume that chilled water is supplied to the jacket at 40°F and exits the jacket at 70°F. You have also noticed that you are not getting any heat transfer from the cap, which is the top 35% of the working volume. The tank manufacturer tells you that this tank has an overall heat transfer coefficient, U, of 12 Btu/hr ft² °F.

$$3.41 \, Btu \, / \, hr = 1W$$

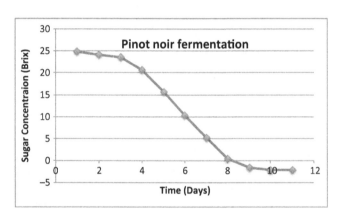

Day	Brix
1	24.8
2	24.2
3	23.6
4	20.6
5	15.6
6	10.4
7	5.3
8	0.4
9	-1.5
10	-2.1
11	-2.1

(c) You also produce a Sauvignon blanc at wine[3]. After tasting it at the end of the primary fermentation, you decide it might work well as a "Fume blanc" style wine with aging in barrels. Specifically, you are looking for added coconut aroma and a bit of smokiness. Just as you are thinking this, a barrel salesperson contacts you with new information that his company has developed. They have completed a chemical analysis of the extraction profile for their various types of barrels using a white wine. The concentrations of the key sensory impact molecules are different for each barrel type they produce. On which compounds will you focus in evaluating this sales literature given your needs? Explain your answer.

7 Post-Fermentation Processing Equipment

After fermentation is complete, an entire series of unit operations are possible from blending to stabilization to clarification. Whether for winemaking or brewing, some of these operations will take place right in the fermentation or hold tanks that we have already discussed. Other unit operations will require specialized equipment. In this chapter, we discuss key post-fermentation unit operations such as filtration, settling, and centrifugation as means of clarifying wine or beer, as well as some specialized techniques for adjusting the concentration of volatile components, such as acetic acid or ethanol, in order to achieve stylistic or regulatory goals.

7.1 TYPES AND MECHANISMS OF FILTRATION

Filtration is the first post-fermentation unit operation we will discuss here. The two main reasons for filtration are to clarify and to stabilize (microbially) the wine or beer produced. There are multiple types of filtration, basically grouped by the size of particles they will remove from the product. However, this grouping is also related to the relative flux through a given filter as can be seen in Figure 7.1. Starting in the upper right of this figure is conventional particle filters. These filters are designed to remove the largest particles from wine and beer and also allow the largest flow rate through a given area of filter material (i.e. the highest flux). Conventional particle filters are designed to remove particles usually considerably larger than 10 μm, such as grape pulp, bentonite, or trub. "Rough" filters will remove only the large particles, while a "tight" filter will remove smaller particles. Types of conventional particle filters include plate and frame, pressure leaf or diatomaceous earth filters, lenticular stacks, and rotary drum filters, among others (all to be described in more detail later in this chapter). The next type of filter would be a microfilter. These filters will remove smaller particles, down to 0.22 μm or 0.45 μm depending on their material, though at a lower flux. This includes yeast and most bacteria that might be present in wineries and breweries. These filters are said to be "absolute" or "sterile." Most microfilters will either be found in the form of cartridges or in cross-flow filtration units. As we get to the next group of filters, we are now looking to remove soluble components above a specific molecular weight. This type of filtration is known as ultrafiltration or UF. Finally, reverse osmosis (RO) is a very specialized form of filtration that will only allow uncharged compounds below a molecular weight of 100 daltons to pass. RO is frequently used for water purification, but as we will see, it also has specialized use in the wine industry for removal of volatile acidity and ethanol. Its flux is the lowest of all types of filtration.

DOI: 10.1201/9781003097495-7

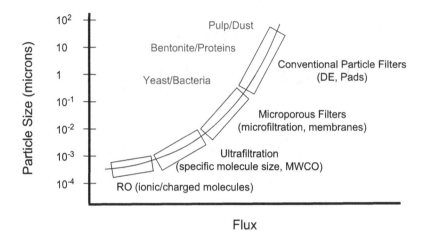

FIGURE 7.1 Four classes of filtration. This is a classification based on the size of particle removed and not the inherent flux capacity of the methods.

All of these types of filtration use one of two filtration mechanisms (shown in Figure 7.2). Filters are either depth filters or membrane filters, and they work in fundamentally different ways. All conventional particle filters are depth filters. A depth filter is made up of a series of fibers (e.g. cellulose acetate) or solid particles (e.g. diatomaceous earth) that are closely packed. When looking at the microscopic structure as depicted in Figure 7.2a, the distance between fibers or particles in a depth filter may be relatively large, but the density of the packing gives a *high probability* that a wine or beer particle traveling through the filter will stick or become trapped and thus be removed from the fluid. The density and nature of the fibers or solid particles will determine what size of fluid particle is likely to be trapped. However, these filters are not absolute, and some larger particles could and do pass through—this is why they are

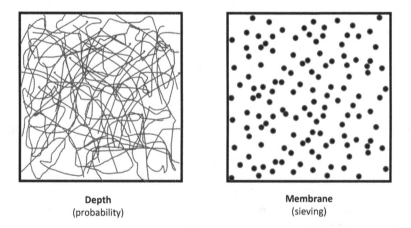

FIGURE 7.2 Two mechanisms of filtration. (a) Depth filters, based on probability of a particle hitting a filter fiber, and (b) membrane filters based on actual pore size and sieving.

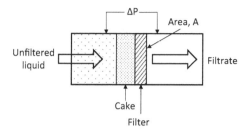

FIGURE 7.3 Schematic of a dead-end filtration.

said to have a "nominal" rating that is reported by the manufacturer of the filter material. A membrane filter is a polymer sheet with very uniformly sized holes or channels and the mechanism of action is *sieving*. This is probably more aligned with the mechanism people think of for filtration. As an example, if you put grapes in a colander and run water over them, the water will go through the holes, but the grapes will not because they are too large. Here, the pore size exactly determines the size of particle retained, so these filters are said to be "absolute" and are rated for their absolute size.

Finally, filters are also classified by the direction of fluid flow relative to the filter material. When the fluid flow is perpendicular to the filter surface, this is called dead-end filtration or, more recently, normal-flow filtration (Figure 7.3). This is the kind of filtration you probably used in an introductory chemistry lab. When the fluid flow is tangential to the filter surface, the filtration system is called tangential flow filtration or cross-flow filtration.

As we step through the various types of filters common to the wine and brewing industries, we will see all of these terms come up again. But first, it is important to understand some of the theory of filtration so we can understand how these filters operate and how to choose the right type and size for the application.

7.2 FILTRATION THEORY

A quantitative description of filtration starts with Darcy's Law which gives the following expression for volumetric flux through a filter:

$$\frac{1}{A}\frac{dV}{dt} = \frac{\Delta P}{\mu_o R} \tag{7.1}$$

where A is the filter area, V is the volume filtered, ΔP is the pressure drop across the filter and filter cake, μ_o is the viscosity of the filtrate, and R is the resistance of the filter and filter cake. The resistance, R, can be further described as:

$$R = R_m + R_c = R_m + \alpha c \frac{V}{A} \tag{7.2}$$

where R_m is the resistance of the filter, itself, and R_c is the resistance of the cake which changes over time as the cake builds up. R_c can be further described using α, the specific cake resistance, and c the concentration of solids in the liquid to be filtered. The term V/A has units of length and can be thought of as a sort of cake thickness that grows over the course of filtration. This expression for R can be substituted into the equation for flux:

$$\frac{1}{A}\frac{dV}{dt} = \frac{\Delta P}{\mu_o \left[R_m + \alpha c \dfrac{V}{A} \right]}$$

(7.3)

This equation can be integrated to give the following:

$$\frac{t}{V/A} = \frac{\mu_o \alpha c}{2\Delta P}\left(\frac{V}{A}\right) + \frac{\mu_o R_m}{\Delta P}$$

(7.4)

This means that if t/(V/A) is plotted versus V/A for a small-scale filtration trial, the result will be straight line with a slope equal to $\dfrac{\mu_o \alpha c}{2\Delta P}$ and an intercept equal to $\dfrac{\mu_o R_m}{\Delta P}$ (Figure 7.4). From these two values, the R_m and α can be calculated. Then, the following equation can be used to calculate filtration time for a given area and pressure differential (or any other combination of these variables).

$$t = \frac{\mu_o V}{\Delta P A}\left[\frac{\alpha c}{2}\left(\frac{V}{A}\right) + R_m \right]$$

(7.5)

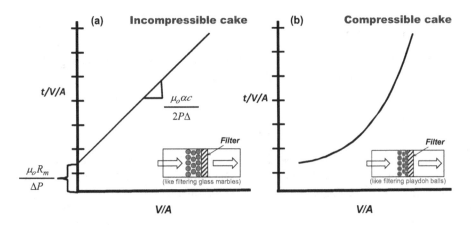

FIGURE 7.4 Graphs of filtration data. (a) A noncompressible/ideal filter cake. (b) A compressible filter cake. For the ideal case, the slope and intercept can be used to calculate the specific cake resistance and filter resistance, respectively.

This description assumes an ideal, incompressible cake (think water flow past glass marbles, for instance). However, this level of incompressibility is not often observed in practice. The solids filtered out in wineries and breweries will likely be at least somewhat compressible (think water flow past playdoh spheres that deform with increasing pressure, closing the fluid paths). Therefore, the plot idealized in Figure 7.4a will actually look more like the one in Figure 7.4b. Several methods can be used to incorporate compressibility into these equations. However, one straightforward way to look at this is:

$$\alpha = \alpha' \left(\Delta P \right)^s \tag{7.6}$$

where s = 0 for an incompressible filter cake and s approaches 1 for highly compressible cakes.

Example 7.1: Sizing the Pad Filter for the Peak Winery

PEAK WINERY (ANNIE GALLOIS)

Peak is wine for people who "love wine but don't know anything about it." It is a Colorado winery that makes quality, dry wine at a reasonable price in straightforward, easy to understand packaging. They offer three wines at the base level— Maroon Bell Red Blend, Blanca White Wine, and Quandary Rosé—each named for one of Colorado's 14,000 ft peaks. They are packaged in 250 mL cans for retail and distributed in kegs to bars and restaurants. The labels and tap pulls feature a line drawing of the peak for which it is named. The can labels also include a short description of the wine with pairing suggestions and an ingredients list. Peak also makes two premium wines under the Snowcapped label: Mountain Cru, a Bordeaux blend, and a Chardonnay—with the goal of increasing the reputation of Colorado wine generally. The Snowcapped wines are packaged in traditional 750 mL glass bottles and together constitute 10,000 cases of production. All of the base level wines are aged in stainless steel while the Snowcapped wines are aged in small French oak barrels after malolactic fermentation.

An environmentally conscious winery based in Palisade, Colorado, Peak Winery is solar powered. Cans and kegs are chosen for the base level wines, because they are greener than bottles: cans are virtually infinitely recyclable while kegs can be refilled. Cans also offer the advantage of being cheaper to store and ship, more portable than bottles, and appealing to younger consumers used to buying craft beer in cans. All Peak wine is made from grapes purchased from Colorado's Grand Valley AVA.

PROBLEM

The bottled Chardonnay produced by Peak will be blended from barrels back into a 3,000 gal tank. The winemaker would like to know how much filter area is necessary to filter 2,900 gal of Chardonnay in a reasonable amount of time. This will be a pad (or plate and frame) filtration using MTX-35 Cellulose pads with a 4 µm nominal pore size.

SOLUTION

The volume of wine to be filtered is:

$$V = 2,900 \, \text{gal} \times \frac{3.78 \text{L}}{\text{gal}} = 10,962 \text{L} = 10.96 \text{m}^3$$

The viscosity of the wine, μ_0, is 0.0016 kg/m s. A preliminary bench filtration shows that the total solids load in the wine, c, is 25 g solids/L or 25 kg/m^3. We will control the ΔP across the filter at 25 psi (= 172,321 Pa =172,321 kg/m s^2).

First, we will need to do an experiment where we filter a small amount of the Chardonnay through a filter disk of the material used in MTX-35 pads. We choose to use an 8 cm diameter disk. Data are recorded for volume as a function of time. The data are shown here.

Filtration Trial Data

Time (s)	Volume (mL)
8	78
24	214
50	343
132	542
260	847

These data can be used to calculate t/(V/A) and (V/A) which can be plotted against each other as shown here:

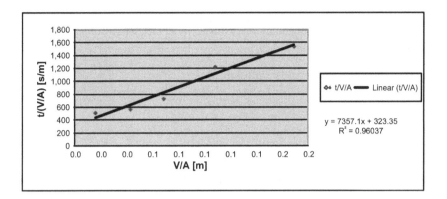

The intercept and slope can then be used to calculate resistance of the filter medium, R_m, and the specific cake resistance, α, respectively as follows:

$$intercept = \frac{\mu_0 R_m}{\Delta P}$$

Therefore,

$$R_m = \frac{\Delta P \left(intercept \right)}{\mu_0} = \frac{\left(172,321 \dfrac{Kg}{ms^2} \right) \left(323 \dfrac{s}{m} \right)}{\left(0.0016 \dfrac{Kg}{ms} \right)}$$

$$R_m = 3.48 \times 10^{10} \frac{1}{m}$$

Similarly,

$$slope = \frac{\mu_0 c \alpha}{2 \Delta P}$$

or

$$\alpha = \frac{2 \Delta P \left(slope \right)}{\mu_0 c} = \frac{2 \left(172,321 \dfrac{Kg}{ms^2} \right) \left(\dfrac{7,357s}{m^2} \right)}{\left(0.0016 \dfrac{Kg}{ms} \right) \left(\dfrac{25 Kg}{m^3} \right)} = 6.3 \times 10^{10} m/Kg$$

Once we know these values, we can use the equation in Section 7.2 to calculate the time needed for the filtration of this wine. At first, let's assume an area, A, of 5 m².

$$t = \frac{\mu_o V}{\Delta P A}\left[\frac{\alpha c}{2}\left(\frac{V}{A}\right) + R_m\right]$$

Substituting,

$$t = \left[\frac{\left(0.0016\frac{Kg}{ms}\right)\left(10.96\,m^3\right)}{\left(172,321\frac{Kg}{ms^2}\right)\left(5m^2\right)}\right]\left[\frac{\left(6.3\times\frac{10^{10}m}{Kg}\right)\left(\frac{25Kg}{m^3}\right)\left(10.96\,m^3\right)}{2\left(5m^2\right)} + 3.48\times10^{10}\frac{1}{m}\right]$$

This calculation gives a time of t = 35,841 s, which is equal to t = 9.96 hr. This would be too long to do in one 8-hr workday, especially if we want to include time for set up and clean up. Therefore, since the pads are rated for 45 psi, what if we increased the pressure from 25 psi to 40 psi? Let's calculate that time by substituting 40 psi (= 275,714 kg/m s²) instead of 25 psi. If we do this, the time calculated is 6.2 hr. This will still be too long if it will take an hour or more to set up and clean up. We could then try 25 psi and 8 m² area. This gives t = 3.94 hr and should work well. Since each pad is 25″ × 25″ or 4.34 ft² (= 0.4 m²), we would need 20 pads for this filtration. It should be noted that the equation for time above can be solved for area or volume (for a given time) using a quadratic equation.

7.3   COMMON FILTER CONFIGURATIONS

7.3.1   PLATE AND FRAME FILTERS

A plate and frame filtration system (sometimes called a "pad" filtration system) consists of a series of distribution and collection plates and disposable filter pads or sheets that are placed on a frame (Figure 7.5). At the ends of the frame, there is a capstan mechanism that allows the pads and plates to be secured in place and sealed. The plates provide the fluid path so that every other plate becomes a distribution plate, distributing feed to filters on both sides of the plate, and the remaining plates act as collection plates to take in the filtrate from the two adjacent filter pads. Any volume element of fluid only passes through only one filter pad and then exits as filtrate. By adding more pads and plates in parallel, throughput can be increased. On some units, it is possible to add a "crossover" plate that allows the user to load a second series of tighter pads that the fluid will pass through after exiting the rougher pads. In this way, one might start with, for instance, 10 μm nominal pore size pads, and then crossover to a 3 μm nominal pore size pad. By filtering out the larger particles first and then going to the tighter pads, it is often possible to increase the volumetric flow rate and time to filter blinding.

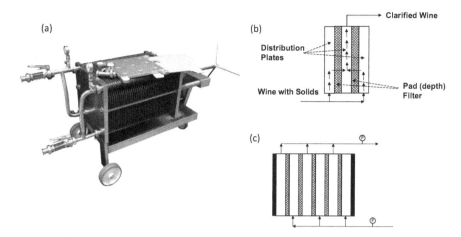

FIGURE 7.5   Pad filtration. (a) Photo of a Velo Acciai plate filter, courtesy of Scott Labs. (b) Fluid paths through an individual filter element in a frame. (c) Flow through multiple elements in parallel elements.

During filtration, the operator monitors the pressure drop across the filter pads using pressure sensors on the feed and filtrate sides of the filter skid. While according to Darcy's Law, a higher pressure drop will result in a larger flux, filter pads are rated for a maximum pressure that should not be exceeded, as pad failure will allow a greater number of particles to pass. Depending on the pad used, fluxes for pad filtration units could be anywhere from 20 gal/hr ft^2 to 100 gal/hr ft^2 or more at pressure drops of 15–30 psi (1–2 bar or atm). Pads come in different sizes, depending on the size of the plates and frame. Filtration units vary in size from that of a shoe box (for home winemakers or brewers) to larger than a shipping container.

7.3.2   Kieselguhr/K/Pressure Leaf Filters

Another type of filter commonly used in the wine and beer industries is a pressure leaf filter (see Figure 7.6a). These filters are often called Kieselguhr Filters or K-Filters in the brewing industry. While geometries differ with manufacturer, all of these filters use a series of screens to support solid material of relatively uniform particle size deposited on the screens to act as a depth filter. Therefore, the two key elements of these units are the screens and the particulate material. The screens in these systems are organized into "leaves" that are packed into the filter housing. An individual "leaf" is the supporting element which holds two screens (one on each side of the leaf) and provides the path (in between the screens) for the filtrate into the outlet manifold. The leaves are generally rectangular in shape although circular units are also utilized, especially in brewing. The fluid to be filtered is pumped into the cylindrical filter housing holding the leaves and surrounds the leaves, passing through the filter cake, particulates, and screen to be collected as filtrate. Housings can be vertical or horizontal cylinders and the leaves are found in different orientations, as well, depending on the manufacturer and norms in the industry. In the wine industry,

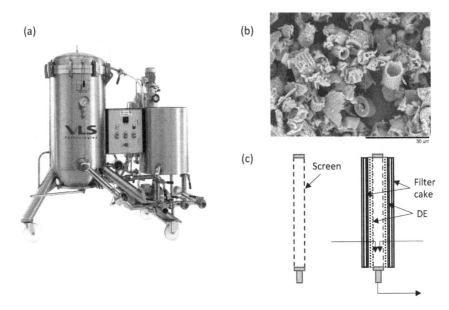

FIGURE 7.6   Pressure leaf/K-filtration. (a) A VLS Technologies horizontal pressure leaf filter, image courtesy of Scott Labs. (b) An SEM image of diatomaceous earth (DE) is the skeletal remains of diatoms (aglae). It is used to create a fairly incompressible depth filter on a series of screened "leaves." By Dawid Siodłak: Template:Own/Opole University, CC BY-SA 4.0, https://commons.wikimedia.org/w/index.php?curid=39632693. (c) Schematic of flow patterns in a single leaf.

it is common to use diatomaceous earth as the basis for the particulate depth filter. Diatomaceous earth (DE) is an assortment of dried, calcined, skeletal remains of diatoms (algae) that can vary in size and shape (Figure 7.6b), allowing variation in the nominal rating of the filter. Other filter aids, more commonly seen in brewing, include perlite and bentonite.

To operate a pressure leaf filter, the two screens on each leaf need to be coated with DE. Once a thin coating is in place, the fluid to be filtered will be sent into the filtration unit and solids will build up on the DE in the form of a filter cake and the filtrate will pass through the cake and screen to the receiving tank. Because typical solids in wine and beer are compressible to some degree, an extra step is usually taken. This is called a body feed. A slurry of DE is actually metered into the fluid to be filtered as it enters the unit. This may seem somewhat counterintuitive at first. Why would we actually want to add solids, which would increase c (our solids concentration), and thus increase time to filter? We would do this because the body feed, when mixed with the solids in the wine or beer, will actually make the cake less compressible and more ideal. Therefore, the specific cake resistance, α, will decrease significantly. If α decreases a lot more than c increases, then the filtration will be more efficient with the body feed. When the filter cakes of adjacent leaves grow together, filtration must stop, the leaves are cleaned and recoated, if necessary, and filtration resumes.

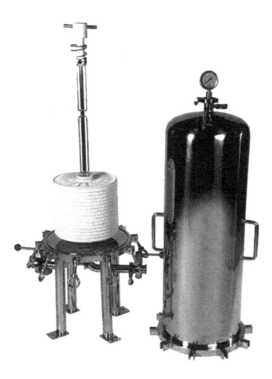

FIGURE 7.7 A lenticular stack filter, with the filter housing open and the stack visible. A series of stack of filter disks are mounted in a stainless steel housing. Solids-bearing liquid is filtered from the outside of the stack. Filtered material flows through each filter disk into an inner collection cylinder. Image courtesy of Scott Labs.

7.3.3 LENTICULAR STACKS

Lenticular stacks (Figure 7.7) are made up of preformed disks of DE or cellulose supported on plastic, all around a central plastic channel. The stacked disks provide a large amount of filtration surface area in a manner similar to having many pads in a pad filter or many leaves in a pressure leaf filter. The stacks are placed in a stainless steel housing similar to the cartridge described below. Wine or juice go through the outside of the disks to the inside where they are collected in a central channel that is mounted on the outlet of the filter housing. The convenience and performance of these stacks in the wine industry has driven adoption of this technology in a number of related industries (such as the biotech industry).

7.3.4 ROTARY DRUM FILTERS

Rotary drum filters have been used since at least the 1940's in the antibiotic production industry. The principle of the rotary drum filter is illustrated in Figure 7.8a with a photo of a unit in Figure 7.8b. The filter is a cylindrical drum that has a screen on its

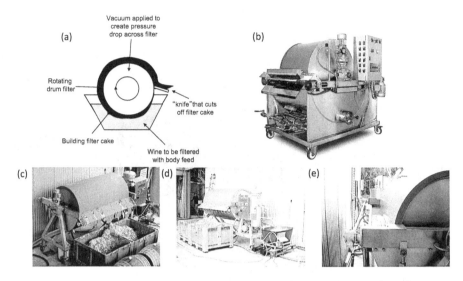

FIGURE 7.8 Rotary drum filters. (a) This schematic illustrates the operation of a rotary drum filter. (b) A Cellar-Tek rotary drum filter. Image courtesy of Cellar-Tek Supplies Ltd. (c) Operation of the knife, (d) body-feed reservoir, and (e) feed trough.

surface. As the drum rotates through a trough holding the liquid to be filtered, a vacuum is pulled on the inside of the drum, driving liquid through the screen and depositing solids on the screen that has been precoated with a layer of DE. Just prior to the drum re-entering the trough, a "knife" or adjustable metal bar slices off the filter cake to the same thickness that it was on the last rotation (Figures 7.8c–e). In this way, the filter cake never builds up in thickness, thus precluding the need for a large pressure drop as the filtration proceeds. This type of filter is very well suited for filtering lees. One downside to this type of filter is that as the drum leaves the trough, air may be drawn into the wine, which could lead to oxidation.

7.3.5 Cartridge and Normal Flow Membrane Filters

Cartridge filters are widely used in the winery and brewery settings, usually just prior to packaging. Cartridge filtration systems (Figure 7.9) are composed of an outer stainless steel housing and an inner, disposable filter element. The housings come in various sizes that hold one or multiple filter elements that are 10″, 20″, or 30″ tall. The housings have an inlet and outlet, along with a vent valve and potentially a pressure gauge on the top of the unit.

The cartridge elements, themselves, can be fabricated with filter material that is either a depth filter (e.g. cellulose acetate) or an absolute membrane filter. For the latter, this is typically used just prior to the filler on a bottling line and is chosen to be a "sterile" filter, which is defined as either 0.45 μm (most common in the wine industry) or 0.22 μm absolute pore size. The most common materials for these membrane filters are polyvinylidene fluoride (PVDF) or polyethersulfone (PES), though other materials can be found. The filter membrane itself is highly pleated to

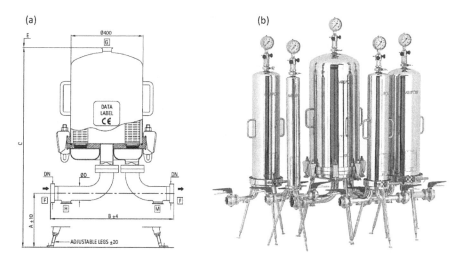

FIGURE 7.9 Cartridge/membrane filters. (a) Schematic of a Millipore multiple cartridge filter housing. Wine or beer flows in from the central line and filters from the outside of the filter stacks in. Once inside the filter stack, the clarified liquid leaves the element. (b) Photo of sample filter housings. Images courtesy of Millipore Sigma.

fit a large surface area in a compact volume. This surface area is distributed evenly around a central, hollow plastic core and enclosed in a plastic cage for support. This filter cartridge or element is constructed to be inserted in the stainless steel housing using o-rings as a seal and does not reach the top of the housing. Fluid flow goes from the outside of the cartridge to the inside, where it is collected in the hollow portion of the cartridge and then continues out of the housing and on to the filler bowl (described in more detail later). Because sterile cartridges clog easily and are expensive, it is important to protect them by filtering out all large particles prior to this final filter. This can be accomplished using a "filter train," two filter housings in series with the first one being a depth filter and the second the sterile absolute filter. Alternatively, some manufacturers make cartridges that have the nominal depth filter built into the outer layer of the cartridge, with a sterile absolute filter underneath.

When starting the flow through cartridge filters, is important to make sure that the housing is completely full so that the cartridge is surrounded by fluid to its complete height. To accomplish this, the vent valve at the top of the housing is opened and a valve downstream of the filter is closed until fluid comes out the vent valve. Then, the vent valve is closed and the downstream path opened to begin filtration. Pressure is monitored across the filters to assure that maximum operating pressures are not exceeded.

It is impossible to inspect a filter cartridge visually for holes that may allow microbes through. Therefore, it is imperative that there be some sort of experimental method for assuring the integrity of the cartridge filter both before **and** after use. Filtration companies have developed three types of integrity tests whose results they have correlated with the ability to filter out microorganisms. These are the forward flow test, pressure hold test, and the bubble point test. In all three types of tests, the

FIGURE 7.10 A Millipore automated filter integrity tester. Automated testers allow choice of automated forward flow, pressure hold, or bubble point tests. Image courtesy of Millipore Sigma.

filter is first wetted to fill the pores with liquid. For the forward flow test, a pressure is maintained upstream of the cartridge using an inert gas. The flow rate of the gas is monitored downstream of the filter. A large flow rate corresponds to a lack of integrity. For the pressure hold test, a pressure is applied upstream of the filter using inert gas and then the source of the gas is shut off. Pressure decay upstream of the cartridge is then monitored with a rapid pressure decay corresponding to a loss of integrity. In the bubble point test, an inert gas is applied upstream of a wetted filter. The pressure is gradually increased until enough pressure is present to displace the water from the membrane pores allowing bubbles of gas to escape. The lower this pressure, the more likely the filter has lost integrity. In the biotech industry where integrity testing is done frequently, the process has been automated. Many filter companies sell an automated integrity-testing machine (see Figure 7.10).

7.4 TANGENTIAL FLOW FILTERS

All of the types of filtration described to this point were normal-flow (or dead-end) filtration. In these cases, the bulk flow of fluid was normal or perpendicular to the filter itself, leading to the buildup of a filter cake on the filter and an increase in resistance over time. Eventually, in these cases, the cake will provide enough resistance, that flow will stop or pressure drops across the filter will rise past the rupture point for the filter. Tangential flow filters (TFF, sometimes called cross-flow filters in the wine industry) operate in a fundamentally different manner. With TFF, the flow is tangential to or across the filter as seen in Figure 7.11. This provides a continuous sweeping action across the filter that precludes the buildup of a filter cake with its associated resistance. However, because a pressure drop is maintained across the filter, particles smaller than the pore size will still pass through to the permeate side. All particles that can fit through the pores will not flow through on one pass by the filter though. In order for TFF to work well, the particles that are retained are recycled back to the feed, where they pass by the filter repeatedly until passing through to the filter unit outlet. One of the issues with a TFF unit is this need for recycle. In fact, the ratio of the feed flow rate to permeate is often on the order of 10:1. This is not a

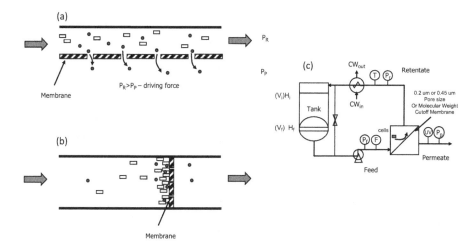

FIGURE 7.11 Tangential flow filtration. The principle of TFF is illustrated in (a). In TFF, flow is parallel to the membrane filter, as opposed to normal or perpendicular flow, as seen in (b). A schematic of a TFF process is shown in (c).

problem, though, if the throughput capacity is sized dependent on the permeate flow rate. In other words, if you need a 10 gpm filtration rate, your recirculation pump will need to provide on the order of 100 gpm and you will need the appropriate filter area as discussed further below.

TFFs are typically either ceramic or polymeric in design. Polymeric filters are typically cheaper and "swappable," they are lightweight, and they can be changed out in the filter housing for different applications, or when a filter becomes worn out and needs to be replaced. Polymeric filters are not particularly robust, and are prone to damage at high transmembrane pressures (TMPs), in hostile solvent or cleaning solutions, or at very high temperatures.

Ceramic filters are typically made by doping a metal oxide (i.e. titanium oxide) onto a stainless steel filter chassis. This allows for good control over filter pore size and means that ceramic membranes are extremely robust. Ceramic membranes can typically be steam sterilized without issues, operate at high TMPs without issue, and can handle a wide array of environments. However, ceramic membranes are expensive, bulky, and are typically only seen in shell-and-tube style membrane configurations.

7.4.1 Terminology and Flow Patterns

The general terminology for TFFs is to refer to fluid entering the filter as the "feed," the filtrate leaving the filtration unit as the "permeate," and for the fraction which is retained or recycled as the "retentate." During filtration, a feed pump sends the feed into the filter. The permeate leaves the unit filtered, while any fluid rejected by the filter returns to a hold tank as the retentate and then is recycled repeatedly through the filter until the desired volume passes into the permeate stream. Flow across the membrane filter is controlled by the pressure across the filter, more commonly known as TMP (described further below).

7.4.2 APPLICATIONS

When TFF was first introduced into the wine industry, the use was limited to mainly a final polishing step just prior to bottling. This is because the hollow fibers used initially were easily clogged by larger particles. However, advances in the technology have overcome this issue and now TFF units can be used to replace many of the types of filtration discussed above. In addition, TFF is also the basis for ultrafiltration and reverse osmosis—just with different membrane material. Therefore, the remainder of this section is also applicable for these applications that are discussed later in this chapter.

7.4.3 TYPES OF TFF MEMBRANE GEOMETRIES

There are at least three common geometries for TFF membranes. These are illustrated in Figure 7.12. First, flat membrane sheets can be used, usually stacked parallel to each other. This is sometimes called a flat plate arrangement. This geometry is least susceptible to clogging, but it is the least efficient means of packing filter area into a given volume. The second geometry is a spiral wound geometry. Very much like a spiral heat exchanger in the previous chapter, this geometry uses two flat rectangular sheets, fused along the long sides and then wound in a spiral and placed in a long cylindrical housing. This geometry is intermediate in both clogging potential and efficiency of packing. It is commonly used for ultrafiltration and reverse osmosis applications. Finally, hollow fibers are like hollow pieces of spaghetti made out of membrane material. The feed enters the lumen of each fiber and the permeate gets collected around the outside of the fibers in a cylindrical housing. This gives the most efficient packing of area per volume, but also the highest clogging potential. If a particle approximately the size of the lumen of the fiber lodges in the end of the fiber, that fiber will be lost to flow. Only regular backflow through the fibers solves this issue and makes hollow fibers a viable alternative. Many of the TFF units used in the wine industry are hollow fiber geometry with automated backflow to avoid rapid clogging.

7.4.4 SIZING AND SCALE UP OF TANGENTIAL FLOW FILTRATION

Tangential flow filtration sizing and scale up is distinct from traditional dead-end or normal-flow filtration. One of the benefits of TFF is the simplicity of scale up from

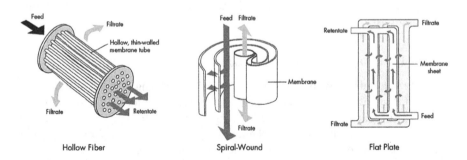

FIGURE 7.12 Three common TFF geometries. (a) Hollow fiber, (b) spiral wound, and (c) flate plate (aka "cartridge"). Images courtesy of Millipore Sigma.

laboratory or pilot scale data direct to industrial scale application via the concept of *similitude*—for a given membrane type and a given liquid to be filtered, the relationship between the *intrinsic* quantities of flux and TMP should be the same regardless of scale. This analysis is equally applicable to solids removed (microfiltration), protein removal (ultrafiltration), or nanofiltration/reverse osmosis.

Recall that we define flux as the rate of liquid filtered per area per time (L/min/m²). In a Tangential Flow Filter, we define TMP as the pressure difference between the unfiltered material (feed or retentate) and the filtered material (permeate).

TMP is the "driving force" of cross-flow filtration (Figure 7.13). At very low TMPs, little to no liquid will pass through the filter. As TMP is increased, liquid will be forced through the filter, and liquid flux will increase linearly with TMP. As TMP is further increased, flux will level out as the filter becomes "choked" at its maximum possible flux. Further TMP increases will not increase flow and may either decrease flow (by causing particles to become trapped in the filter's flow channels), or even damage the membrane.

Note that cross-flow filtration is a pseudo-continuous process. Over the course of operation, the pores in the membrane which allow for liquid to flow through can become plugged, either with solids in the case of microfiltration, or with colloids or protein aggregates in ultrafiltration. High TMPs exacerbate this phenomena. As a result, flux in a clean membrane starts out high and slowly decays with time. To combat this, backflushes (where liquid is pressurized from the retentate through the permeate) are performed periodically to push debris out of the pores. Backflushes may occur as frequently as every 20 min, or as rarely as every few hours. Membrane cleaning is also essential, with a combination of surfactants and caustic chemicals used to clean out the membrane pores and any scum layer that forms on the outside of the membrane.

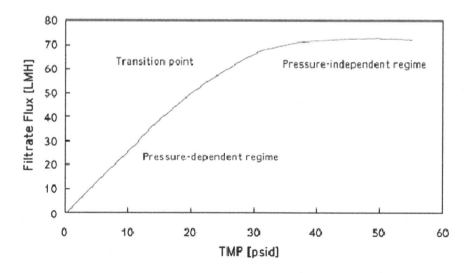

FIGURE 7.13 Example of the relationship between TMP and Flux in TFF. At low TMPs, flux increases linearly with TMP. As TMP increases, flux levels off and may even drop. Image courtesy of Millipore Sigma.

Example 7.2: Sizing of an Industrial TFF from Laboratory Data

You are running laboratory trials to see if you want to invest in a microfiltration TFF for beer clarification. The vendor loans you a bench top TFF with a membrane area of 0.1 m2 and recommends a feed pressure of 70 psig. You perform a bench-top filtration experiment, collecting the filtered permeate and returning the solids-rich retentate back to the feed. You find that this produces an initial flow rate of 0.2 L/min permeate at the beginning of filtration, which decays linearly to a flow rate of 0.05 L/min at the end of filtration. Additionally, you are collecting the permeate in an open-air container, with a pressure of 0 psig.

(a) *What is the average flux over the filtration operation?*
(b) *What is the TMP of the TFF?*
(c) *You are now imagining the piping to an industrial TFF. If the permeate was maintained at 10 psig, and the retentate has a 5 psig pressure drop from the feed, what would be the average TMP?*
(d) *If the relationship between flux and TMP is linear, what will be the new flux at this new TMP?*
(e) *Your brewery needs a TFF system that will filter a 20,000 L tank in 12 hr. From the above laboratory data, what size filter will you need?*

SOLUTIONS

(a) The average flow rate over this operation is (0.2 + 0.05)/2 = 0.125 L/min, for an average flux of (V/t)/A = 1.25 L/min/m2
(b) The TMP is the difference in feed side pressure and permeate pressure. In this case, 70 − 0 = 70 psig
(c) We must first determine the average feed side pressure, which is the average of the feed and retentate pressures. Our feed enters the TFF at 70 psig, and leaves as retentate at 65 psig, for an average feed-side pressure of 67.5 psig. Our permeate is maintained at 10 psig. TMP is defined as feed side pressure − permeate side pressure. This leads to an average TMP of 67.5 − 10 = 57.5 psig.
(d) If we are in the linear regime for flux versus TMP, then we can take our flux at a TMP of 70 psig and scale it linearly to 57.5 psig: (57.5/70) × 1.25 L/min/m2 = 1.03 L/min/m2
(e) The required flow rate comes to 20,000 L / 12 hr ~ 28 L/min. The average flux of the unit is 1.25 L/min/m2. If Flux = (V/t)/A, then A = (V/t)/Flux = 27.2 m2. Note that this is the minimum required area, membranes are often oversized by 10–25% to account for process upsets, downtime for cleaning, or variables feeds.

Cleaning efficacy can be assessed by measuring the "clean water flux"—the flux of pure, solids free water through the membrane at a given pressure. The manufacturer will typically report this value. A drop in clean water flux indicates a fouled or incompletely cleaned membrane.

7.5   GRAVITY SETTLING (SEDIMENTATION) AND CENTRIFUGATION

As discussed in the previous sections, clarification of wines and beers is often accomplished by filtration. However, this is not the only means of clarification. There are times when clarification can be performed more simply or with higher throughput, for example, and other methods may make more sense. Two alternative clarification methods will be discussed here, gravity settling and centrifugation, two methods that rely on density differences to effect separation between unwanted particulates and the liquid product.

7.5.1   SETTLING THEORY

On a fundamental level, gravity settling in a tank or settling pond is the same operation as centrifugation: the separation of denser-than-water solids via buoyancy separation. In gravity settling, the gravity provides the driving force for separation. You can see this if you have ever seen a beach: sand rapidly falls through water to the ocean floor. In a centrifuge, the internal rotation of the centrifuge imparts a centrifugal force which hurls high density particles to the centrifuge periphery.

Both gravity settling and centrifugation can be analyzed via Stoke's Law, which dictates the velocity of small particles in a liquid media:

$$v = \frac{a \cdot d^2 \cdot \left(\rho_p - \rho_L \right)}{18\mu}$$

where a is the acceleration imposed on the particle (gravity or centrifugal force), d is the particle diameter, ρ_p is the particle density, ρ_L is the liquid density, and μ is the fluid viscosity. This is an intuitive result: particle settling is faster for larger, denser particles in low viscosity liquid (think a bowling ball settling in water versus a pebble settling in honey).

In gravity settling, $a = g = 9.8$ m/s^2. For centrifugation, $a = \omega^2 r$, where ω is the rotational rate of the rotational frequency of the centrifuge, and r is the centrifuge radius.

Example 7.3:   Chill-Settle-Racking of White Wine

Freshly fermented white wine is to undergo gravity settling to remove particulates prior to fermentation. Solid particles are evenly distributed throughout the liquid. 10,000 L of wine is placed in a cylindrical tank 2 m in diameter. The particulates have a settling velocity of 150 cm/hr.

(a) How long will it take for the wine to be solids free?
(b) What would be the settling time if a tank 4 m in diameter (with the same volume) was used?

(a) *In order to determine the settling time, we must determine the maximum settling path length—that is, the height of the liquid—such that a particle at the very liquid top has enough time to hit the bottom. 10,000 L = 10 m3. The volume of a cylinder is V = H × pi × (d/4)2, so the height of the liquid in this tank is 12.73 m. At a settling velocity of 150 cm/hr = 0.15 m/hr, this settling operation will take 12.73 m/ 0.15 m/hr = 85 hr, or 3.5 days.*

(b) *For a tank 4 m in diameter, the liquid height would be 3.2 m. In this case, the settling time is only 21.3 hr, or less than a day. Note that by doubling our diameter, we cut our settling time by a factor of 4 (22)!*

In order for gravity settling to be effective, all particles must settle to the bottom of the holding vessel in the required time, or else particles will still remain in suspension.

As we can see from this example, more area for settling makes for a faster sedimentation operation. We can imagine a continuous settling process, where a feed of flow rate Q is fed into one side of a settling pond, and clarified liquid is continuously removed from the other end of the pond. For this to be effective in removing solids, we must balance the time the liquid spends in the settling pond (the residence time, τ) with the time it takes for a particle to settle, t_s. For a rectangular settling pond, $V = H \times A$, so:

$$\tau = \frac{V}{Q} = \frac{H \cdot A}{Q}$$

$$t_s = \frac{H}{v}$$

$$\tau = t_s, \frac{H \cdot A}{Q} = \frac{H}{v}$$

$$A_{required} = \frac{Q}{v} \tag{7.8}$$

This tells us the required settling area for a continuous gravity sedimentation process is a function of the flow rate of liquid to be clarified and the settling velocity of the particles to be settled!

7.5.2 CENTRIFUGATION THEORY AND SIZING

For a given particle removal duty, there will be some feed rate limit beyond which a centrifuge cannot remove all the solids from the liquid. Above this feed rate, solids will "break through" into the product stream, as the liquid will not have a long enough residence time for the solids particles to settle out of the liquid.

The approach of sizing a settling pool for continuous clarification is the basis for practical centrifuge sizing. All centrifuge configurations can be assigned a "Sigma Value," Σ, which relates the effectiveness of a centrifuge to an equivalently sized settling pond. Sigma values range from as "low" as 10,000 ft^2 up to 1,000,000 ft^2 for different centrifuge configurations. Large centrifuges with high internal surface areas and fast rotational speeds have high sigma values, while smaller, slower centrifuges have smaller sigma values.

The beauty of a sigma value is it allows for the linear scaling of centrifuge sizing. A centrifuge with twice the sigma value of a smaller centrifuge can perform the equivalent level of clarification on a feed twice the flow rate of the smaller centrifuge.

When scaling centrifuges, we keep the ratio of flow rate/sigma constant, such that for system 1 and 2,

Example 7.4: Sizing of an Industrial Centrifuge from Pilot Data

You are an engineer at a mid-size brewery, looking to install a centrifuge for yeast removal after fermentation. The vendor you are working with brings a small, skid mounted centrifuge to your brewery for pilot trials. You vary the feed rate of beer to the centrifuge and measure the solids content of the corresponding product at each feed rate. You find that beyond 5 L/min, the product contains an unacceptable amount of solids. The vendor states that the pilot centrifuge has a sigma value of $\Sigma = 12,000$ ft^2. You need a unit which can process 70 L/min with adequate solids removal. What is the minimum sigma required for this process?

Since we know the sigma value of the pilot centrifuge, the acceptable flow rate in the pilot centrifuge, and the desired flow rate in the industrial centrifuge, we can scale the pilot sigma value by the ratio of the flow rates to find the minimum sigma value required for the industrial unit:

$$\Sigma_{industrial} = \Sigma_{pilot} \cdot \frac{Q_{industrial}}{Q_{pilot}} = 12,000\,ft^2 \cdot \frac{70\,L/min}{5\,L/min} = 168,000\,ft^2$$

The industrial unit must have a sigma value of at least 168,000 ft2.

$$Q_1 / \Sigma_1 = Q_2 / \Sigma_2 \tag{7.9}$$

7.5.3 CENTRIFUGATION EQUIPMENT

The simplest form of continuous centrifuge is a *tubular bowl centrifuge.*

In a tubular bowl centrifuge, liquid is fed into a rotating cylinder. The cylinder's rotation imparts a centrifugal force on the liquid and solids. Solids migrate to the periphery of the tube, leaving a clarified liquid at the center of the tube. These types of centrifuges are rarely encountered in wineries but are the simplest and most intuitive type of continuous centrifuge to visualize. A typical sigma value for a tubular bowl centrifuge would be 20,000 ft^2.

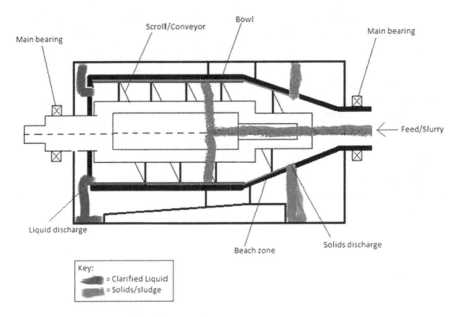

FIGURE 7.14 Decanter Centrifuge. Solids-rich liquid is fed into the middle of the rotating bowl. The screw along the central axis provides additional area for solids settling, and serves to actively remove settled solids. By Log.smith: Own work, CC BY-SA 3.0, https://commons. wikimedia.org/w/index.php?curid=29011003

The next simplest form of continuous centrifuge is a *decanter centrifuge* (Figure 7.14). A decanter centrifuge is essentially a tubular bowl centrifuge with a rotating screw along the center shaft. This screw serves to both provide more surface area for separation (enhancing separation), and to actively remove the solids-rich phase in a matter similar to an augur. Decanter centrifuges are usually used for feeds with very high solids (>5%), such as recovering wine from tank bottoms or used filter cake.

Decanters are often used in breweries for wort extraction from trub slurry and/or lauter tun solids. Typical sigma values for a decanter centrifuge are in the 10,000–100,000 ft^2 range.

Finally, the most common type of centrifuge encountered in wineries is the "disk-stack" centrifuge (Figure 7.15). This is a rotating bowl with a stack of disks, which serve to greatly increase the available surface area for solids settling. Solids accumulate at the periphery of the bowl and are periodically purged from the unit via a discharge cycle, where a gap is opened in the bowl periphery to allow for solids to bleed out.

Disk-stack centrifuges are the centrifuge configuration most commonly encountered in a winery or brewery environment. They excel at high-throughput clarification with low solids feeds (<5 % solids v/v). Product streams of <0.5% solids are often attainable via disk-stack centrifuges. Large facilities will often couple a single decanter centrifuge with multiple disk-stacks, which allows for further liquid recovery from the solids discharge of the disk-stack units by running the discharge through the decanter. The bowl geometry and the addition of disk-stacks greatly enhances the

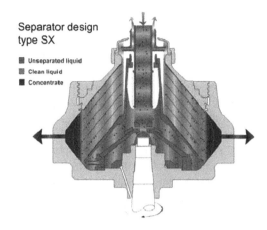

FIGURE 7.15 Schematic of a disk stack centrifuge. Solids-rich liquid is fed to the center of the bowl, and then passes through the spinning disks, which provide surface area for settling. Clarified liquid stays in the center of the bowl, while solids migrate to the periphery and are ejected by hydraulically dropping the bowl. Disks are present inside the bowl to add additional surface area for settling. Wibawa (2018).

separating power of disk-stack centrifuges, with large industrial units having sigma values upward of 1,000,000 ft^2.

The operation of a disk-stack centrifuge depends on a clear channel between the disk-stacks. At high solids feeds (>5%), large solids particles, or flocculating particles, these channels can clog, hindering separation and potentially damaging the centrifuge.

Disk-stack centrifuges encountered in wineries and breweries are typically "solids discharge" centrifuges. This means that rather than constantly removing solids, solids are periodically removed via a discharge cycle. Between cycles, solids accumulate at the bowl periphery. The bowl housing is then briefly pushed down via hydraulics or pneumatics, which allows solids to discharge from the bowl interior to a catch-pan where the solids are then removed. These discharge cycles are typically controlled by a timing cycle, a turbidity feedback cycle, or a combination of both. A timing cycle is very simple: after a set time of operation (say, 20 min), the bowl undergoes a solids discharge step. Turbidity control utilizes a turbidity meter in the clarified liquid line. Clarified liquid turbidity will increase as solids accumulate in the bowl. When the turbidity reaches some threshold value, the bowl is discharged to remove the accumulated solids.

7.6 ADJUSTING VOLATILE PRODUCT COMPONENTS POST FERMENTATION: REVERSE OSMOSIS AND SPINNING CONE

Clarification is not the only unit operation utilized post fermentation. Winemakers will sometimes decide to adjust the volatile profile of their product to meet stylistic goals or regulatory limits. While other volatiles may be adjusted, here we will focus on reduction of ethanol and acetic acid using reverse osmosis or spinning cone technology.

7.6.1   Reverse Osmosis

Osmosis is a physical phenomenon, common in nature, in which water will pass through a membrane from a region of low salt concentration (osmolarity) to a region of high salt concentration. In this way, water flow works toward evening the solute concentrations on the two sides of the membrane. When the salt or solute concentrations are the same on both sides of the membrane, flow ceases. The potential for flow, or osmotic pressure, drives this process (see Figure 7.16a).

If a pressure is applied to an osmotic cell across the membrane in the opposite direction of the osmotic pressure (Figure 7.16b), it is possible to get fluid to move across the membrane against the osmotic potential. This process is known as "reverse osmosis" and can be used as a separation technique for very small solutes. In reverse osmosis, a polymer membrane is chosen with an extremely small molecular weight cut off (MWCO), essential that of 100 daltons. The surface of the membrane is also treated to have a charge layer that will reject any charged moiety, either negatively or positively charged. The wine to be treated that will have a high solute concentration is passed over the membrane using a TFF system (as described earlier in this chapter) with a pressure differential that creates flow toward the dilute side of the permeable membrane against the osmotic potential. The permeate in this system will thus contain molecules smaller than 100 daltons that are not charged. This is a very small number of molecules in a wine matrix as can be seen in Figure 7.17. Essentially, the permeate will contain water (18 daltons), ethanol (46 daltons), acetic acid (60 daltons, if present and undissociated), and potentially ethyl acetate (86 daltons, though some will be rejected by the membrane). Unlike other kinds of TFF, the retentate is not usually continually recycled in RO. Because not all molecules like ethanol and acetic acid will permeate the membrane on the first pass, however, multiple passes using membranes in series can be used to increase recovery.

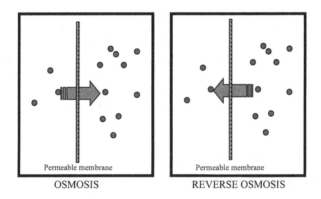

FIGURE 7.16   Osmosis and reverse osmosis. In normal osmosis (a), water will move across a semi-permeable membrane to the high solute concentration side in order to "even out" the solute concentration through dilution. For reverse osmosis (b), a pressure greater than the osmotic potential is applied in the direction opposite the osmotic gradient to drive water out of the salt solution.

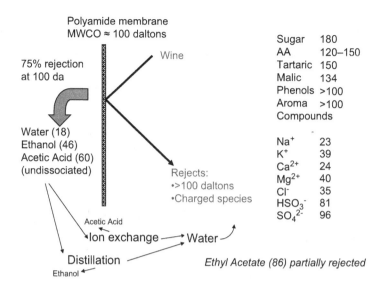

FIGURE 7.17 Properties of a reverse osmosis membrane. Components with a molecular mass greater than 100 will be rejected by the membrane, as well as charged species. This means that water, ethanol, and undissociated acetic acid will pass through the membrane. Sugars, hop acids, esters, phenolics all fail to permeate. Acetic acid can be removed via ion exchange, while ethanol can be removed via distillation. The remaining water can be added back to the wine.

With the permeate separated from the remainder of the wine, it can then be treated with little worry of affecting wine quality further, as sensory impact molecules other than ethanol and acetic acid will have been rejected. If the goal of the processing is to reduce ethanol in the wine, the next step will be to distill the permeate and discard the ethanol. The remaining water can then be added back to the treated wine. If the goal of processing is to reduce acetic acid, the permeate is typically passed over an ion exchange column that will separate the acetic acid from water. The acetic acid would then be discarded, and the water returned to the product. For alcohol reduction, wine must be transported to a bonded facility to be treated in order to carry out the distillation in a bonded facility. Treatment for acetic acid is more flexible. In either case, all of the wine could be treated to reduce the ethanol or acetic acid concentrations, but it is more likely that a smaller portion will be treated to remove most of the ethanol or acetic acid and then blended back with the bulk of the wine to achieve the desired concentrations and minimize any unintended consequences of further treatment.

7.6.2 SPINNING CONE

An alternative means to ethanol reduction in wine or beer is to use distillation. The entire next chapter of this text is dedicated to discussing this idea in the context of distillery design. However, here we discuss a very specialized means of distillation for the purpose of reducing or removing ethanol or other volatiles from wine. The main difference is that here, the product is what remains after alcohol is removed, as opposed to the alcohol itself. This means that special care needs to be taken to avoid the potential effects of heating wine or beer that would not meet with the stylistic goals of the product. In the previous section on RO, this was achieved by separating out the water and alcohol first prior to distillation. With spinning cone technology, this is achieved by distilling at low temperatures with specialized equipment.

Spinning cone technology was developed in the Australian industry in the 1980's. The technology has two key features. First, it uses a vacuum to reduce the boiling point of ethanol so that the distillation can be accomplished at temperatures just above ambient (e.g. 40°C) instead of closer to 100°C as would be necessary at atmospheric pressure. Second, it uses a stack of spinning and stationary cones in the distillation column (Figure 7.18) to make the heat transfer and mass transfer more efficient. Wine (or beer) is first introduced into the column under vacuum and heat. The liquid runs down the series of cones, spread out into a thin film and mixed by the spinning ones. At the same time, the vapor moves up through the column in the opposite direction. The close contact between the phases allows for volatiles to be stripped out of the liquid and for water vapor and other less volatile components to condense back into the liquid. In this first pass through the column, the most volatile components of the wine are collected and saved. The wine is then treated a second time, this time to remove the ethanol. After the ethanol is stripped, the most volatile fraction from the first pass is added back to the wine. In this manner, the wine is only

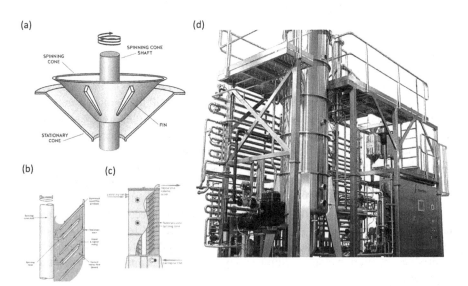

FIGURE 7.18 Spinning cone technology. The spinning cone is a vacuum distillation unit operation, constructed of a series of rotating cones along a central shaft, and stationary cones attached to the walls of the column. (a) Wine is introduced at the top of the column and moves downward in close contact with the vapor moving up the column. (b) The vapor strips the volatile components and carries them out the top, where they can be condensed, captured, and later added back. (c) Cut away diagram of the entire column in operation. (d) Spinning cone unit, with the column in the foreground. All images courtesy of Flavourtech 2021.

at elevated temperature for a matter of seconds, which is meant to minimize any negative "cooking" reactions or loss of desirable volatiles.

This type of approach has also been used to remove other undesired volatile compounds, such as sulfur-containing ones associated with a negative sensory impact. As with RO, all wine can be treated to remove a small amount of ethanol, or a fraction can be treated to remove much of the alcohol and blended back into the untreated wine. Because this is a distillation process, it must be conducted in a specialized, bonded facility, so wine must be transported to and from the treatment facility.

REFERENCES

Boulton, R.B., Singleton, V.L., Bisson, L.F., and Kunkee, R.E. *Principles and Practices of Winemaking.* Chapman and Hall, New York, 1996.

Harders, T., Sykes, S.J. and Prince, R.G.H. Spinning Cone Distillation in Wine Treatment. *Proc. Chemeca, 95,* 32–36, 1995.

Lydersen, B.K., D'Elia, N.A., and Nelson, K.L. (Eds.). *Bioprocess Engineering: Systems, Equipment and Facilities.* John Wiley & Sons, New York, 1994.

Wibawa, D.S., Nasution, M.A., Noguchi, R., Ahamed, T., Demura, M., Watanabe, M.M. Microalgae Oil Production: A Downstream Approach to Energy Requirements for the Minamisoma Pilot Plant. *Energies,* 11(3), 521, 2018. https://doi.org/10.3390/en11030521.

PROBLEMS

1. You are testing out a new pad filter medium for eventual rough filtration of beer after centrifugation. To do this, you will use a laboratory Buchner funnel with an 8 cm disk of the pad material. The viscosity of the material is 3.0 cp. The data below were obtained from your filtration in which a 600 mg Hg (80.0 kPa) pressure drop was applied across the filter. At the end of your lab-scale filtration, you find that you have 14.0 g of solids in a total of 690 mL of beer.

TABLE 7.1
Data for Lab-Scale Filtration

Time (s)	Volume of Filtrate (mL)
26	100
96	200
197	300
342	400
537	500
692	600
989	690 (end)

(a) Determine the specific cake resistance α and the filter medium resistance R_m.

(b) Using these values, estimate how long it would take to obtain 10,000 L of filtrate of this beer using a filter with a surface area of 10 m^2 and a pressure drop of 500 mm Hg.

(c) What surface area would you need to accomplish this filtration in 4 hr assuming the same pressure drop (500 mm Hg)?

2. After the cold stabilization is complete in the second tank, you need to filter the beer to remove cold-temperature turbidity (often referred to as "chill haze"). To get ready for this operation, you complete a small-scale test with the same filter material. In this test, you find that the specific cake resistance, α, is 1.1×10^{10} m/kg, and your filter resistance, R_m, is 1.3×10^{10} 1/m.

(a) If you do the full-scale pad filtration at a 50,000 Pa (= kg/m s^2) pressure differential and have an initial solids concentration of 0.3 g/L (= kg/m^3), how long will it take to filter 8,000 gal (= 30.2 m^3) of the beer using 3 m^2 of filter area? Assume a filtrate viscosity of 2.5 cp (= 0.0025 kg/m s).

(b) You decide that this time is too long. Therefore, you decide to reduce the time by increasing the pressure drop across the filter. What pressure drop, DP, would you need to finish processing in 1 hr? Will this work if the filter apparatus is rated for 15 psi?

(c) After completing these calculations, you have another idea to reduce the processing time. You decide to look into changing the filter medium to one with a lower resistance. What fraction of the resistance is due to the filter (as opposed to the resistance due to the filter cake)? Will this do any good?

3. Prior to bottling, you decide with your winemaking and cellar staff to perform a rough filtration of your 20,000 L (= 20 m^3) of wine to facilitate the sterile membrane filtration. You decide to use a diatomaceous earth (DE) pressure leaf filter. You use a precoat of DE and then meter in a DE body feed to the wine being filtered. From your experience, the resistance of the precoated pressure leaves is $R_m = 4 \times 10^{10}$ m^{-1} and the specific cake compressibility with the body feed is a = 2.2×10^{10} m/kg. With the body feed, the solids concentration in the wine is 40 kg/m^3. The viscosity of the filtrate is 0.0016 kg/m s (= 1.6 cp).

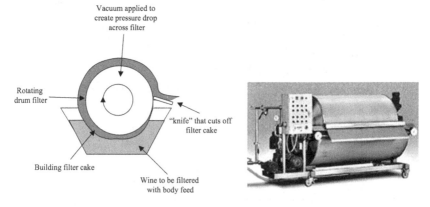

Vacuum applied to create pressure drop across filter

Rotating drum filter

"knife" that cuts off filter cake

Building filter cake

Wine to be filtered with body feed

(a) Assuming an incompressible filter cake, how long will it take to filter the wine through 8 m^2 of filter area using a pressure differential of 30 psi (= 206,786 Pa = 206,786 kg/m s^2)?

(b) A filtration salesman visits you soon after this filtration to tell you about a new type of wine filter that they are marketing called a rotary drum filter (this type of filter has been used in antibiotic production for about 60 years). As pictured below, this filter consists of a mesh rotating cylinder that is precoated with DE. A vacuum is maintained in the inner portion of the cylinder to draw wine up through the coated cylinder in the portion of the cylinder that is submerged (about 8 m² in this case at any given time). As the drum rotates out of the wine, the cake is dried and then sliced off with a knife blade. This last step maintains a constant cake thickness and resistance throughout the filtration. If the resistance, R, of the submerged surface remains constant at $R = 1 \times 10^{11}$ m⁻¹ during the entire filtration, how long will this filtration take? Assume that the vacuum maintains a steady pressure drop across the filter of 10 psi (= 68,929 kg/m s²).

4. Prior to bottling, all beer at the Falling Rocks Brewery will be filtered with a pad (plate and frame) filtration system. From your experience, you are expecting your hefeweizen to be the toughest to filter, so you decide to do some small-scale tests with it using characteristic beer and a small piece of the cellulose pad you will use. You come up with the following data table of resistance, R, versus V/A.

V/A (m)	R (m⁻¹)
0.00	2×10^{10}
0.08	8.4×10^{10}
0.20	18.0×10^{10}

(a) If the solids concentration of the hefeweizen in both the test and actual wine is 40 kg/m³, the pressure drop across the filter is 20 psi (138,000 Pa = 138,000 kg/m s²), and the filter area of the process unit is 15 m², how long would you expect it to take to filter 35,000 L (= 35 m³) of beer? The viscosity of the beer is 2.55 cp (= 0.00255 kg/m s).

(b) How would your answer change if you were able to add a new, rigid filtration aid to the beer prior to filtration that decreased the specific cake resistance to half of its original value when it is added at a rate of 8 g/L?

5. **[Integrative Problem]** At Hetch Hetchy Vineyards, you plan to use a cartridge filter train just prior to bottling. This includes a 1 mm depth filter followed by a 0.45 mm membrane filter. Both are 20 in cartridges.

(a) What is the difference between a depth filter and a membrane filter?

(b) The wine is fairly clear by this point and has a maximum solids concentration of 0.5 kg/m³. If we would like to filter and bottle an entire tank (= 4,000 gal =15,120 L =15.12 m³) in 3 hr, what should the pressure drop be across the depth filter if the filter area is 1.2 m²? The viscosity of the wine is 1.55 cp (= 0.00155 kg/m s). From preliminary filterability tests, we have found the specific cake resistance to be 1.4×10^{10} m/kg and the depth filter media resistance to be 1.6×10^{10} m⁻¹.

$$\left(1\,Pa = 1\,kg / m\,s^{2}\right)$$

(d) You decide to sterilize these filters with hot water prior to using them. You find from some testing that you typically have a maximum of 10,000 live bacteria on the membrane filter prior to sterilization. You want your probability of contamination to be 1/100 to assure no problems after the wine is in the bottle. Initially you start with 20 min of 80°C water from the beginning of the filter train, but find that you are only getting 99.9% kill in the membrane filter because the temperature of the water is falling by the time it gets to the second filter in the train. To what temperature has the filter fallen? Assume bacteria with $a = 7 \times 10^{22}$ min^{-1} and $E = 37,200$ cal/mol.

(e) How much time would it take at this same temperature to get all the way to a 1/100 probability of contamination?

(f) How would this answer change if you insulated the cartridge housing during sterilization to maintain the temperature at 80°C?

6. **[Integrative Problem]** Given that many of your products will be off-dry at Eagle's Pride, filtration will be an important unit operation to stabilize the product microbiologically prior to bottling. To do the final filtration prior to bottling, you decide to use 0.45 μm membrane cartridge filters after 1 μm depth filters. You are filtering and bottling 12,000 gal (= 45.4 m³ = 45,360 L) at a time. For the filtration, you will be using 30″ membrane cartridges, each with 2.4 m² of filtration area.

(a) If you use a bottling line that can fill 120 standard (i.e. 750 mL) bottles/min, how much time will it take to bottle this volume?

(b) To calculate the number of cartridge filters you need to accomplish the filtration part of this process in the amount of time calculated above, you do a small-scale trial with a disk of the membrane filter material and a small amount of the wine to be filtered/bottled. You find that the resistance of the membrane, R_m, is 1.5×10^{10} m^{-1} and the specific cake resistance, α, is 3.3×10^{10} m/kg. There is an initial solids concentration entering the membrane filter of 4 kg/m³ and you choose to maintain a pressure drop of 10 psi (= 68,929 kg/m s²) across the filter. How many cartridge filters do you need to use in parallel to keep up with the bottling rate. Assume a wine viscosity of 1.5 cp (= 0.0015 kg/m s).

(c) Prior to using the tangential flow filtration unit, you plan to sterilize it with hot water at 85°C. If you start the sterilization with 10^5 viable cells and run the hot water for 30 min, what is the probability of contamination? Assume that the cells that are in the filter prior to sterilization have an Arrhenius constant of 7.4×10^{21} min^{-1} and an activation energy of death of 36,200 cal/mol.

7. In order to make your bottling go more smoothly at Breakfast Winery, you decide that you will do a rough pad filtration prior to the cartridge filtration at the bottling line. The plate and frame filter that you purchase can hold up to sixteen 0.6 m × 0.6 m (= 2 ft × 2 ft) filters. You choose to use "Matrix MTX-35" filter sheets from Gusmer that have a nominal pore size of 4.0 μm. The filter sheets are rated for a maximum differential pressure of 45 psi (= 310,500 kg/m s²). Using a lab-scale filtration trial on your Chardonnay with a small pad filter disk, you find the specific cake resistance, α, and filter resistance, R_m, for

your system are 0.8×10^{10} m/kg and 0.5×10^{10} 1/m, respectively. You measure your solids concentration to be 12 kg/m³. The filtrate has a viscosity of 1.8 cp (= 0.0018 kg/m s).

(a) If your goal is to filter 9,000 gal (= 34 m³) of wine in a 5 hr period, what pressure would be necessary to accomplish this with just ten filter sheets? Will this work?

(b) If you use all 16 sheets, how long will the filtration take if you use a pressure differential of 30 psi (= 207,000 kg/m s²)?

8. **[Integrative Problem]** After chilling your Riesling and allowing the yeast to settle at the cold temperature, you then plan to filter it to stabilize the wine on the way to bottling it. You use a filter train with a 1.5 µm nominal pore size depth filter followed by a 0.45 µm pore size membrane filter in line on the way to the filler of the bottling line.

(a) Will the mechanism of filtration be the same for these two kinds of filters? Explain.

(b) The depth filter that you decide to use is a Cellupore 1300SD Cellu-stack from Gusmer Enterprises. This is a lenticular stack that will fit in a stainless steel housing (see attached information sheet). You decide to use a bench-scale filtration trial to pick a cartridge size for the 1 µm depth filter. To do this, you take 1.5 L of your settled Riesling and filter it through a circular filter disk with an area of 0.01 m² (about 5 cm diameter) using a pressure drop of 5 psi (= 34,500 kg/m s²). Your initial solids concentration is 4 kg/m³ and the viscosity of the filtrate is 0.002 kg/m s. You get the data shown in the following graph.

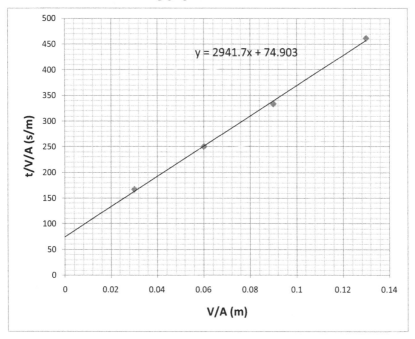

Use these data to calculate the resistance of the filter medium and the specific cake resistance, α.

(c) Your bottling line can fill 750 mL bottles at an average of 60 bottles/min. Your Riesling tank holds 2,400 gal (9,072 L = 9.072 m³). If you want to bottle this whole tank at this average rate, what is the minimum size cartridge that you should use given the areas shown in the attached information sheet (that is, should you use a stack with 0.63, 0.95, 1.37, 1.68, or 3.6 m²). Be sure to show your calculations. Assume that you will run at a pressure of 8 psi (55,200 kg/m s²).

(d) Prior to filtering your wine, you decide that you need to sterilize the transfer path from the filters through to the filler tips by running hot water through the path. When you try this out the first year in your new winery, you find that the cold Oregon winters combine with your uninsulated lines to give you a 2°C drop in temperature from the hot water going into the filter to the water coming out of the filler tips. From some initial swabbing of the filler, you find a typical live microbe count of 10^4 microbes in the filler bowl prior to sterilization. If you want to achieve a 1/1,000 chance of contamination during your 20 min sterilization, at what temperature should your water enter the filter train? Assume that the typical live microbe in your filler bowl has an Arrhenius constant, α, of 1×10^{23} min^{-1} and an activation energy of death, E, of 38,400 cal/mol.

(e) If you fully insulated the transfer line so that the temperature did not go down at all throughout the sterilization pathway, how much shorter could the sterilization be?

9. **[Integrative Problem]** Just prior to your Chardonnay entering the filler on your bottling line at Yi-Di, it is filtered using two cartridge filters, a 3 μm depth prefilter and a final PVDF 0.45 μm sterile filter. You think the sizing of the prefilter is going to be more critical, so you start with this.

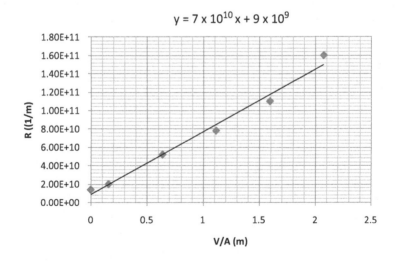

$$y = 7 \times 10^{10}x + 9 \times 10^9$$

(a) To size the prefilter, you use a 4 cm (= 0.04 m) diameter filter disk and filter some of your wine. Specifically, you take 3 L of your Chardonnay from the tank to be filtered and filter it using a pressure drop of 8 psi (= 55,200 kg/m s^2). Your initial solids concentration is 5 kg/m^3 and the viscosity of the filtrate is 0.0016 kg/m s. In your experiment, you get the data shown in the graph above. Use these data to find the filter resistance, R$_m$, and the specific cake resistance, α. The equation for the regression line is shown at the top of the graph.

(b) Based on these data, how much time will it take to filter the 10,000 gal (= 37,800 L = 37.8 m^3) of Chardonnay, if the cartridge filter has an area of 7.2 m^2. Assume that a pressure across the filter of 10 psi (= 69,000 kg/m s^2) is maintained.

(c) What would the average bottling speed be (i.e. bottles/min) with this filter? Would this be considered a low-, medium-, or high-speed line for wine packaging?

(d) If you can insert a centrifuge in the line prior to the filters, you think that you can get the solids concentration down to 1 kg/m^3 prior to entering the filters. What effect would this have on the average bottling speed possible? Explain and show your calculations.

(e) Prior to bottling, it will be necessary to sterilize the two filters and the bottling line. If you have done some swabbing and plating to find that there are typically 2 × 10^8 live cells left in this equipment after cleaning, how long would you have to sterilize with steam at 100°C to reach a 1/10,000 probability of contamination. Some lab testing indicates that likely contaminants would have an Arrhenius constant for death, α, of 1.5 × 10^{23} min^{-1} and an activation energy of death, E, of 38,200 cal/mol.

10. As a VEN 135 student at UC Davis, you remember going on a tour of a large brewery where they recently started to use Perlite with a pressure leaf filter-to-filter large volumes of beer. Perlite, a volcanic glass material, is generally less dense than diatomaceous earth (DE) and therefore acts as a less expensive alternative to DE. You decide to try this with one of your red wines from Vigneto Impulso that has recently finished ML fermentation. Data from a bench-scale filtration of your wine using an 8 cm disk coated with Perlite using a pressure of 20 psi (= 137,857 Pa = 137,857 kg/m s^2) are shown in the table and figure below.

Time (s)	Volume of Filtrate (mL)	Volume (m³)	V/A (m)	t/V/A (s/m)
5	60	6.00E-05	0.01	418.88
20	180	1.80E-04	0.04	558.51
42	300	3.00E-04	0.06	703.72
258	900	9.00E-04	0.18	1,440.95

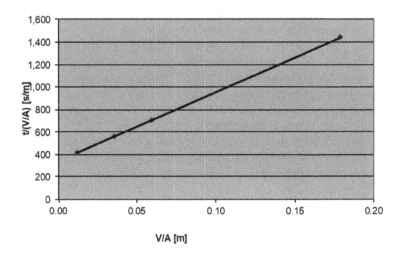

(a) Using the data above, find the resistance of the media, Rm, and the specific cake resistance, α. You can assume that the wine has 20 g/L (= 20 kg/m³) of solids and the viscosity of the filtered wine is 1.8 cp (= 0.0018 kg/m s).

(b) Given this information, you go to scale this process up. The pressure leaf filter that you will use has a series of mesh disks ("leaves") that are each 1 m in diameter as shown in the diagram below. How many of these mesh disks will you need in order to filter all 1,600 gal (6.04 m³) of wine in less than 4 hr (= 14,400 s) using 10 psi (= 68,929 kg/m s²)?

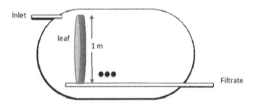

11. The winemaker at Ekaya would like to membrane filter the Syrah just prior to bottling. However, using some bench-scale trials, the winemaker decides the flow rate through the membrane filter will be too small without a prefilter of some type. For this prefiltration the winemaker decides to use a cellulose lenticular stack depth filter with a 3 μm nominal pore size.

(a) What is the difference between membrane filters and depth filters? Explain briefly.

(b) To size the lenticular stack, you get a 3 cm disk of the filter material to do a bench trial. You conduct the trial using a pressure of 25 psi (= 172,321 kg/m s^2) and get the following data. You can assume that the wine has 12 g/L (= 12 kg/m^3) of solids and the viscosity of the filtered wine is 1.6 cp (= 0.0016 kg/m s).

Time (s)	Volume of Filtrate (mL)	Volume (m³)	V/A (m)	t/V/A (s/m)
2	23	2.3E-05	0.033	61.5
10	62	6.2E-05	0.088	114.0
24	104	1.04E-04	0.147	163.1
120	225	2.25E-04	0.318	377.0

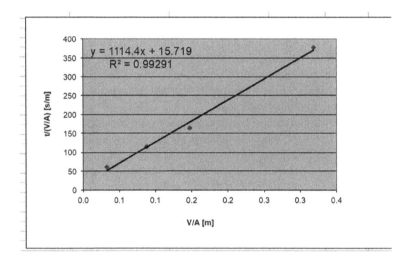

Find the specific cake resistance, α, and filter resistance, R_m.

(c) If we want to finish the filtration of 800 ft^3 (= 22.6 m^3) in 5 hr, how many cartridge stacks do we need? One filtration company sells 12″ 17.7 ft^2 (= 1.6 m^2) stacks and 16″ 38 ft^2 (= 3.5 m^2) cartridges. You can fit multiple stacks of the same diameter in the same stainless steel housing.

12. **[Integrative Problem]** At Weeping Mountain Vineyards and Winery, you decide to filter all of your wines through a 0.45 μm pore size membrane cartridge filter just prior to bottling to make them more microbiologically stable. You have done extensive racking on your Pinot noir, so you feel that rough filtration will not be necessary prior to the sterile filtration

Time (s)	Volume of Filtrate (mL)	Volume (m³)	V/A (m)	t/V/A (s/m)
16	32	3.20E-05	2.01E-02	7.95E+02
48	64	6.40E-05	4.02E-02	1.19E+03
120	100	1.00E-04	6.29E-02	1.91E+03
500	140	1.40E-04	8.80E-02	5.68E+03
1,000	154	1.54E-04	9.68E-02	1.03E+04

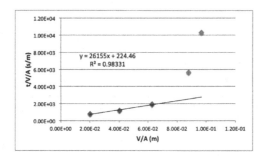

(a) To evaluate the filtration characteristics of the Pinot noir, you use a small 0.45 µm pore size PVDF filter disk with a 0.045 m diameter. An initial filtration indicates that you have 4 g solids/L (= 4 kg solids/m³). The viscosity of the filtrate is 1.4 cp (= 0.0014 kg/m s). Your filtration apparatus controls the pressure upstream of the filter at 8 psi (= 55,143 kg/m s²). You measure the volume filtered over time and get the data above. Calculate the resistance of the filter media, R_m, and the specific cake resistance, α, using the first three data points.

(b) You notice that the last two data points deviate considerably from the trend of the first two. What is your interpretation of what is happening during this part of the filtration?

(c) Given the parameters that you calculated in part (a), you decide to size the cartridge filters for the commercial scale filtration. In this filtration, you plan to run with a pressure drop of 30 psi (= 206,785 kg/m s²) and would like to filter 4,000 L (= 4 m³) of wine in 4.5 hr or less. You find one filtration company that sells four sizes of cartridge, 10″ (0.78 m² area), 20″ (1.56 m²), 30″ (2.34 m²), and 40″ (3.12 m²). Will any of these be sufficient? Show your calculations.

(d) After getting the filter train set up, you decide to sterilize from the filter through the pipe to the filler heads of the bottling line prior to filtration and bottling. Microbial screening tells you that the most difficult to kill bacteria in your filler heads has an Arrhenius constant of 6.44×10^{26} min^{-1} and an activation energy of death of 44,000 cal/mol. If your screening indicates that each of the 16 filler heads on your filler is likely to have 200 live cells, how long should you run your 80°C water through the system to have a 1/10,000 chance of contamination?

13. At first, you expect that you will be allowing all of your wines at Gunrock Winery to settle for clarification. However, they are not settling fast enough and you decide to use pad filtration to do a rough filtration. The nominal pore size for the Seitz K800 pads is 7–8 µm. To size the production scale filtration, you complete a lab scale filtration using a 3 cm × 3 cm (0.03 m × 0.03 m) square sample of the filter, along with 1 L (0.001 m³) of your unfiltered wine. You end up with 32 g (= 0.032 kg) of solids at the end of the filtration. The viscosity of the filtered wine is 1.6 cp (= 0.0016 kg/m s). The pressure upstream of the filter is maintained at 10 psi (= 68,928 kg/m s²). The data are shown below.

Time (s)	Volume of Filtrate (mL)	Volume (m³)	V/A (m)	t/V/A (s/m)
4	50	5.00E-05	0.06	72.00
30	200	2.00E-04	0.22	135.00
55	300	3.00E-04	0.33	165.00
250	700	7.00E-04	0.78	321.43
475	1,000	1.00E-03	1.11	427.50

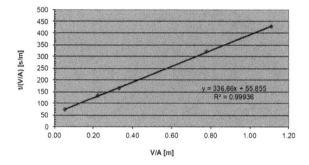

(a) Calculate the resistance of the filter medium, R_m, and the specific cake resistance, α.

(b) How many 0.6 m × 0.6 m square sheets do you need to complete the filtration 50,000 gal (= 189 m³) of wine in 7 hr (= 25,200 s)? These sheets are rated for a maximum pressure of 45 psi (= 310,178 kg/m s²).

(c) Could you use hollow fiber cross-flow filtration for this purpose? What is the advantage of cross-flow filtration over conventional filtration?

14. In order to make your bottling go more smoothly at Breakfast Winery, you decide that you will do a rough pad filtration prior to the cartridge filtration at the bottling line. The plate and frame filter that you purchase can hold up to sixteen 0.6 m × 0.6 m (= 2 ft × 2 ft) filters. You choose to use "Matrix MTX-35" filter sheets from Gusmer that have a nominal pore size of 4.0 μm. The filter sheets are rated for a maximum differential pressure of 45 psi (= 310,500 kg/m s²). Using a lab-scale filtration trial on your Chardonnay with a small pad filter disk, you find the specific cake resistance, α, and filter resistance, R_m, for your system are 0.8×10^{10} m/kg and 0.5×10^{10} 1/m, respectively. You measure your solids concentration to be 12 kg/m³. The filtrate has a viscosity of 1.8 cp (= 0.0018 kg/m s).

(a) If your goal is to filter 9,000 gal (= 34 m³) of wine in a 5 hr period, what pressure would be necessary to accomplish this with just 10 filter sheets? Will this work?

(b) If you use all 16 sheets, how long will the filtration take if you use a pressure differential of 30 psi (= 207,000 kg/m s²)?

8 Distillation

8.1 CONCEPTS IN DISTILLATION

8.1.1 Separation by Volatility

Distillation is the separation of liquids from a solution by exploiting differences in volatility. What does this mean? Consider a pot of water on a stove, boiling at 100°C. The vapor leaving the pot is pure water, so both the liquid mass fraction of water (called x) and the vapor mass fraction of water (called y) are both equal to 1 (i.e. 100% water).

Now consider wine boiling on a pot, perhaps to make a delicious sauce. We can consider this wine to be a mixture of ethanol and water. Pure ethanol boils at 78.4°C (at 1 atm of pressure). This means that our wine will boil at some temperature between 78.4°C and 100°C. This boiling point is not a function of the heat input from the stove, but purely a function of the composition of the wine: pure ethanol would boil at 78.4°C, pure water at 100°C. We call the condensed and collected vapor "distillate."

But what about the vapor leaving this boiling pot of wine? Let us say our mass fraction ethanol, x_{EtOH}, equals 0.1. Ethanol is more volatile than water, so the vapor leaving the pot should have $y_{EtOH} > 0.1$. In fact, the vapor should have a mole fraction of $y_{EtOH} = 0.4$, a four-fold enrichment! We express this enrichment as "k," *volatility*, defined as:

$$k_i = \frac{y_i}{x_i} \tag{8.1}$$

If we know the volatility and the liquid fraction, we can solve for the vapor fraction. A "T-X-Y" diagram (Figure 8.1) is a practical way to determine boiling point and vapor composition for a given liquid solution. T-X-Y diagrams can be presented in mass fraction, mole fraction, or volume fraction. The T-X-Y diagram for ethanol/water in Figure 8.1 is presented in ethanol percent.

8.1.2 Relative Volatility

Volatility by itself is a clumsy way to analyze distillations. Volatility of a component is not constant and is a strong function of composition. For example, when $x_{EtOH} = 0.1$, $k_{EtOH} = 4$, but when $x_{EtOH} = 0.6$, $k_{EtOH} = 1.1$.

When trying to separate liquids, a vastly more useful expression is *relative volatility*, α, which describes how readily two liquids split when boiled. Species with a high relative volatility separate readily (for example, ethane and n-pentane have a $\alpha \sim$

DOI: 10.1201/9781003097495-8

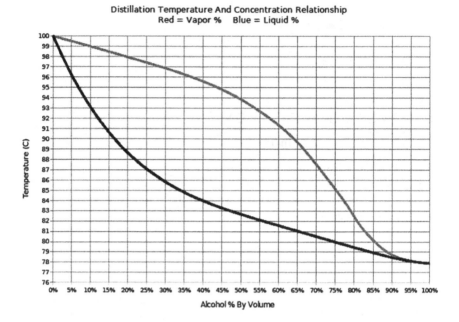

FIGURE 8.1   T-X-Y diagram of ethanol/water at 1 atm. X-axis is ethanol composition, either % volume of proof. Y-axis is boiling point, °F or °C. Figure courtesy of "Zymurgy" Bob Brunjes, Cultus Bay Distillery.

28,000), while species with a relative volatility of 1 will never separate. Relative volatility is defined as:

$$\alpha_{i,j} = \frac{k_i}{k_j} \tag{8.2}$$

where by convention, species "i" is the more volatile component and species "j" is the less volatile component, such that α is always greater than 1. It is important to

Example 8.1:

At what temperature will a 30% ethanol by volume liquid boil? What would be the ethanol composition of the resulting vapor?

We start at the 30% by volume tick mark on the bottom of the plot (Figure 8.2), and draw a line up to the liquid composition curve (1). We then read the temperature (2) at the intersection of 30% ethanol and the liquid line: ~86°C. Next, we draw a line at this temperature from the liquid curve to the vapor curve (3), and then trace down (4) to read the % ethanol composition of the resulting vapor: ~73% ethanol v/v.

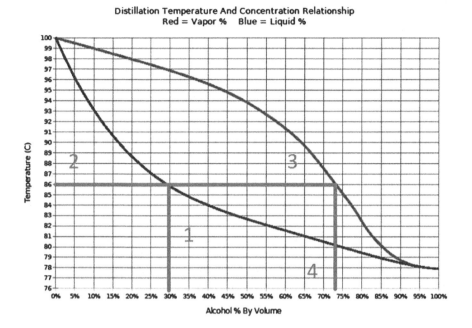

FIGURE 8.2 Example 8.1 solution.

note that while relative volatility can also change with composition, assuming constant relative volatility (unlike constant volatility) is often a good assumption.

Liquid solutions can be conceptualized as "ideal" solutions and "non-ideal" solutions. Ideal solutions have constituents where the interactions between molecules of the pure components are similar to the interactions between the constituents—that is, two primary alcohols will likely form an ideal solution, as would two straight-chain alkanes of similar size; while a primary alcohol and a straight-chain alkane would not form an ideal solution. For example, ethanol and 1-propanol are very similar molecules, with a single alcohol functional group at the end of an alkane chain. These species will interact with themselves in a very similar manner to the way they interact with each other. A liquid solution of n-Butane, n-Pentane, and n-Hexane will also behave like an ideal liquid for the same reason. In ideal liquids, the relative volatility between solutions stays essentially constant over a wide range of liquid compositions.

This is sadly not the case in water–ethanol systems. Water's extremely strong hydrogen bonds interact strongly with alcohols and lacks an alkane core to enable Van der Waals interactions. As such, relative volatilities in water–ethanol systems range from 10 (very dilute ethanol) to 1 (at 95.63% ethanol). You might ask yourself—why is the relative volatility equal to 1 when the system is not 100% ethanol? That leads us to our next section. The ethanol–water system exhibits an *azeotrope*, a composition beyond which purification is not possible by distillation. At 1 atm, the ethanol–water azeotrope is at $x_{EtOH} = 0.9563$. This azeotrope can be changed by changing the pressure or addition additional components, but azeotropic distillation and the solutions thermodynamics underlying azeotropes are outside the scope of this text.

8.1.3 CLASSES OF DISTILLATION COMPOUNDS

Spirit distillation involves dozens, if not hundreds of compounds. Major species, along with their boiling points, are identified in Table 8.1.

Boiling points are given as an approximate guideline for relative volatility. In the water–ethanol system, which is highly non-ideal, water is almost always the least volatile species. In distillation parlance, ethanol is the "Light Key" (LK) in this system—the compound that we are trying to enrich specifically via distillation. Water is the "Heavy Key" (HK)—the compound that we are trying to exclude from the product via distillation.

"Heads" are the compounds which are more volatile than ethanol. Some heads—such as methanol—are toxic. Heads in general contribute to the perceived fruitiness or lightness of a spirit. In distillation engineering parlance, heads are referred to as "Lighter than Light Key" or LLK components. Conversely, "Fusels" are compounds which are more volatile than water but less volatile than ethanol, referred to as "Intermediate Key" or IK components. Fusels are typically higher order alcohols, though distillation feeds rich in fatty acid esters will see those compounds in the fusels as well. Fusels are typically not toxic. Fusels at low levels (100s–1,000s of mg/L) contribute to a feeling of fullness and mouthfeel. Fusels are typically immiscible with water and only slightly miscible in water–ethanol solutions. Larger fusels like active amyl alcohol have a (relatively) large carbon backbone, and undergo substantial Van der Waals interactions, making water a poor solvent for them. This can lead to issues with organic phase formation in continuous distillation. Note that while several of these fusel have higher boiling points than water as pure components, in water–ethanol systems they are actually more volatile than water. Any component less volatile than water, the heavy key, would be termed "Heavier than Heavy Key" or HHK. For example, fatty acid esters, which are present in trace amount in distillation feeds, would be HHK components.

TABLE 8.1
Important Distillation Compounds and Boiling Points

Species	Boiling Point (°C)	Boiling Point (°F)	Class
Acetaldehyde	20.2	68.4	Heads
Acetone	56.1	132.9	Heads
Methanol	64.7	148.5	Heads
Ethyl acetate	77.1	170.8	Heads
Ethanol	78.4	173.1	-
tert-Butyl alcohol	81–83	179–181	Fusels
Isopropyl alcohol	82.5	180.5	Fusels
1-Propanol	97	206.6	Fusels
Water	100	212	-
n-Butanol	117.7	243.9	Fusels
Active-amyl alcohol	128.7	263.7	Fusels
Isoamyl alcohol	131.2	268.2	Fusels

8.2 DISTILLATION MODES

There are two classes of stills encountered in the spirits industry: batch stills, where a single fixed amount of wine or beer is charged to a still and then distilled, and continuous (often referred to as "column" or "Coffey") stills, where wine or beer is continuously fed to the still and distillate is continuously produced and recovered as product.

In general, batch stills are better suited to more flavorful distillates—by their nature, batch stills are not powerful separations tools, so many (potentially desirable) impurities beyond ethanol will find their way to the distillate. While lighter spirits (vodka, light rum, Canadian whiskey, etc.) can be produced via batch still, production will suffer from poor yields. Batch stills are better suited to brandies (including cognacs), heavy rums, bourbon, scotch, and tequila.

While column stills can also be used to produce high end, robust spirits, their fractionation power lends them to neutral spirit production. The overwhelming majority of the world's volume of vodka, light rums, and light whiskeys are produced via column still, though excellent character can still be achieved for more complex spirits via column distillation.

8.2.1 BATCH STILLS

Batch stills are the older and more traditional of the two. The "typical" batch still is a single-stage pot still. In a pot still, wine is fed into a vessel and then boiled, typically by either steam heat exchange or a direct-fired burner underneath the vessel. The resultant vapor is piped to a condenser, where the vapor is cooled and condensed to a liquid. This liquid is enriched in the more volatile components. Figure 8.3 shows a schematic of a pot still:

There are several important ramifications of using a pot still. First, overall separation is quite poor compared to a continuous still. Typically, two to four distillations

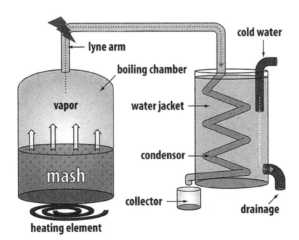

FIGURE 8.3 Schematic of a pot still. By Redrex—Own work, CC BY-SA 4.0, https://commons.wikimedia.org/w/index.php?curid=35425666

are required to hit the desired ethanol concentration, where the product from the prior distillation is collected and re-distilled for further enrichment. By the same token, yield is poor compared to a column still, with 50–60% recovery of ethanol from the original feed to the final product not uncommon (for comparison, well designed continuous column distilleries can often have 98% + integrated yields).

Second, batch processes change with time. The liquid and vapor concentrations at the beginning of a pot distillation are very different from the concentrations at the end of a pot distillation. Over the course of distillation, the pot is being depleted in volatile components, as shown in Figure 8.4. This makes batch distillation a moving target, where the distiller must decide when to begin and when to stop collecting product.

Due to the changing nature of the vapor leaving the still, common operation of a pot still involves taking several product fractions, typically referred to as "cuts." An early cut, highly enriched in heads, might be taken and then discarded to purge methanol. Subsequent cuts, richer in ethanol and with desirable heads and fusels concentrations, will be collected for either re-distillation or blending. Finally, cuts toward the end of the run excessively rich in fusel oils may be collected for blending, re-distillation, or destruction. Typically, the liquid runoff from the condenser will run to a manifold where product can be routed to tanks containing these different cuts.

There are several ways to control cuts in a pot still. The most common method is temperature control, where thermometers are inserted into the liquid and/or the vapor. Cuts are controlled by temperature, with low temperatures indicating a vapor rich in

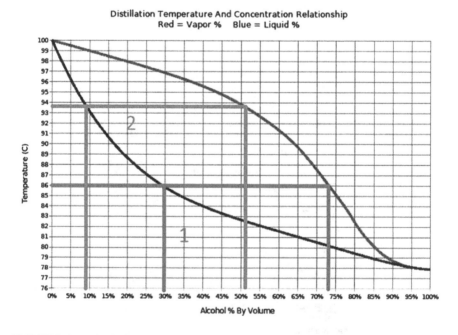

FIGURE 8.4 A feed of 30% v/v ethanol is charged in a pot still (1), which boils at 86°C and produces a vapor at 73% ethanol v/v. Over time, ethanol is depleted, until the liquid contains only 9% ethanol (2), which boils at 93.5°C and produces a vapor at 51% ethanol v/v.

heads and ethanol, and higher temperature rich in fusels or water. Another common method involves mass or volume fractions, where flow meters on the condensed distillate line or a scale on the pot still will determine the total amount distilled and route the distillate to different cuts as a function of distillation progress. This second method is typically used in tandem with temperature control when a process is well established and standardized. Another method, often used in tandem with other control schemes, is to measure the density of the condensed liquid, which can be used to determine the ethanol concentration in the distillate. A typical configuration for this control configuration is to have a hydrometer floating in an open loop of the distillate line.

Pot distillation usually requires multiple distillation steps to hit a sufficient ethanol concentration for common spirits (whiskey, brandy, tequila, etc.). Figure 8.5 shows the effect of multiple distillations on a T-X-Y diagram. Intermediate products (distilled material to be distilled again) is often referred to as "low wine" if multiple more distillations will be needed, and "high wine" if only one more distillation is required.

Finally, pot design and operation differ substantially from continuous distillation. Pot stills are typically built entirely of copper, which reacts away volatile sulfides and improves product quality. Copper is both an expensive metal and difficult to work with, which results in industrial sized pot stills being quite expensive (multi-million-dollar units are common). Second, pot stills are amenable to high solids loading in the charged liquid. While solids cooking and fouling can be an issue at high (>5%) solids loading, pot stills handle 2–4% solids, typical for unfiltered wines, without any serious issues. In contrast, even continuous distilleries designed to handle heavy fouling service will struggle with a feed of 2% solids.

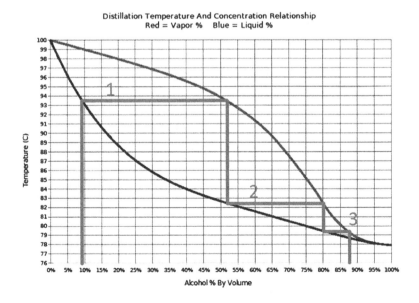

FIGURE 8.5 A T-X-Y diagram, with three distillations of a 9% ethanol feed. This allows a theoretical maximum distillate concentration of 85% v/v ethanol.

Example 8.2:

A pot still is charged with 20,000 lb of feed. The feed has the following composition:

Species	Mass Fraction
Ethanol	0.232
Heads	0.0004
Fusels	0.0004

At the end of distillation, 6,000 lb of 68% ethanol by mass distillate have been collected.

(a) *How many pounds of spent feed remain in the pot?*

(b) *What was the percent ethanol recovery of the distillation?*

(c) *What is the maximum possible mass fraction of heads in the distillate?*

(d) *Would you expect the same concentration of fusels as heads in the distillate? Why or why not?*

SOLUTION

(a) *By material balances, 14,000 lb remain in the pot*

(b) *First, we calculate the amount of pure ethanol removed by multiplying the distillate mass times the distillate ethanol mass fraction of ethanol: 4,080# × 0.68 4,080 lb pure ethanol are recovered. Divide this by the 4,640 lb of pure ethanol fed (20,000# × .232), for a recovery of (4,080#/4,640#) × 100 = 87.9%.*

(c) *The initial feed contained 8 lb (0.0004 × 20,000#) heads. If all the heads were distilled into the product, the mass fraction in the distillate would be 0.0013 (8# heads/6,000# total)*

(d) *No. Fusels are less volatile than ethanol which is less volatile than methanol. If ethanol was left behind in the pot, than some fusels were too. The recovery of fusels in the distillate will be less than the recovery of heads.*

8.2.2 CONTINUOUS DISTILLATION

Continuous distillation is a conceptual extension from batch distillation. Imagine a pot still, where ethanol enriched vapor is leaving the still. Now imagine that a pump is drawing liquid, depleted in ethanol, from the pot. We now have two streams leaving the still: vapor and liquid. To keep the operation continuous, we pump in fresh feed into the pot at the same rate that material is leaving the still from the vapor and the liquid (Figure 8.6). In this configuration, our still is now known as a "flash," the vapor leaving is the distillate, and the liquid leaving is known as "bottoms," "stillage," or "slop."

Now imagine that the vapor from this flash is condensed and redistilled in a subsequent flash, with the vapor collected as a distillate product, and the liquid returned to the original flash along with the feed. Additionally, image the bottoms fed to a third flash, where the vapors from flash #3 are condensed and returned to the original flash, and the bottoms are removed (Figure 8.7):

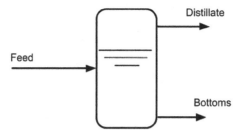

FIGURE 8.6 Schematic of a flash separation.

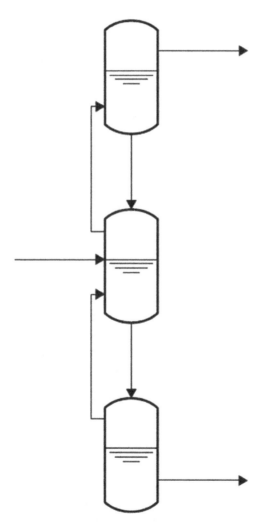

FIGURE 8.7 Three flashes arranged to produce a higher purity distillate and lower lost ethanol in bottoms.

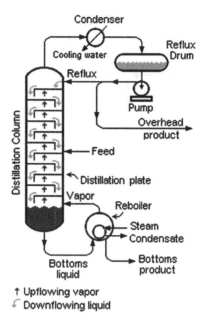

FIGURE 8.8 Schematic of a distillation column. Liquid is fed on a central plate somewhere in the column. Liquid falls down tray to tray, and is enriched in less volatile components as it falls. Vapor boils off the bottom and rises through each tray, and is enriched in the more volatile components. CC BY-SA 3.0, https://commons.wikimedia.org/w/index.php?curid=1315927

This conceptual system is the exact arrangement of continuous distillation, where multiple flashes are "staged" to produce a superior overall separation. Since it is clumsy to employ multiple pot stills operating in a continuous fashion, distillation columns are used instead.

A distillation column is essentially a vertical cylinder filled with multiple devices called "trays" or "plates" every several feet along the column's height. These trays serve the same function as the flashes in Figure 8.7—the tray is filled with liquid falling down from the stage above it, and vapor from the stage below bubbles up through the standing liquid level. The liquid at the bottom is boiled by heat exchanger to drive the separation, and each tray has liquid at its boiling point. In this way, multiple single stage distillations are carried out in a single unit operation, as seen in Figure 8.8.

Example 8.3:

A column still is fed 20,000 lb/hr of feed. The feed has the following composition:

Species	Mass Fraction
Ethanol	0.232
Heads	0.0004
Fusels	0.0004

The distillate is drawn at 6,000 lb/hr of 68% ethanol by mass distillate have been collected.

(a) What is the flow rate of bottoms leaving the still?

(b) What was the percent ethanol recovery of the distillation?

SOLUTION

(a) By material balances, 14,000 (20,000 − 6,000) lb/hr leave the still as bottoms

(b) 4,080 lb/hr pure ethanol are recovered (6,000 × .68) of the 4,640 lb/hr fed (20,000 × .232), for a recovery of 87.9%.

Note that this analysis is the same as Example 8.2, where mass (pounds) has been changed to mass flow rate (lb/hr)!

8.3 CONTINUOUS DISTILLATION EQUIPMENT AND OPERATION

Distillation columns are comprised of three key pieces of equipment: the trays, the reboiler, and the condenser.

8.3.1 DISTILLATION TRAYS

Distillation trays (or "plates") provide a place for vapor and liquid to come into intimate contact and enable separation by relative volatility differences. Liquid from above falls on the tray and vapor from below bubbles through the liquid on the tray, resulting in a new equilibrium. Liquid leaves enriched in less volatile components and falls to the tray below. Vapor leaves the tray enriched in more volatile components and rises to the next tray. A tray is comprised of three parts: a "downcomer," which supplies liquid flowing from the tray above, the "active" tray, where vapor and liquid comingle, and the "weir," a wall that maintains a liquid level on the tray and allows some liquid to flow down to the next tray (Figure 8.9).

There are two common geometries for the tray active area used in sprits distillation.

Sieve trays are little more than a perforated metal plate, where the vapor rises up the perforations and the liquid flows across them. Sieve trays are extremely cheap to manufacture and handle solids very well. As such, stainless steel sieve trays are commonly employed for trays below the feed point, as they will see any solids that are in the feed. The main drawback of sieve trays is their poor efficiency and turndown as defined here.

We define efficiency (ϵ) for a tray as:

$$\epsilon = \frac{N_{ideal}}{N_{actual}}$$

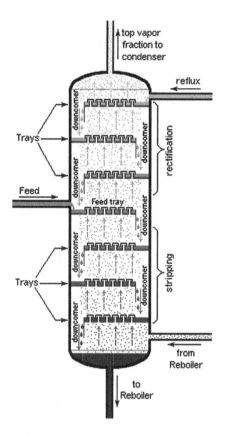

FIGURE 8.9 Schematic of trays inside a distillation column. Note the liquid falling from the tray above, over a weir and through the downcomer. The liquid enters the active tray area, where it is mixed with vapor from the tray below. Vapor rises from the tray to the tray above. Liquid flows over the active area, hits the weir, which maintains the liquid level, and then flows to the tray below. CC BY-SA 2.5.

where N is the number of trays needed to achieve a given separation, "ideal" refers to the number of perfect equilibrium stages, and "actual" is the number of real trays needed to achieve separation. The typical efficiency of a sieve tray is ~50%, meaning that two trays are needed to achieve the separation of one ideal stage.

We define turndown ratio (τ) for a piece of equipment as:

$$\tau = \frac{Maximum\ Flow\ Rate}{Minimum\ Flow\ Rate} \tag{8.4}$$

For example, a pump may have a maximum flow rate of 100 gal/min, but it can be operated as low as 10 gal/min. This pump would have a turndown ratio of 10. Sieve trays have a particularly poor turndown ratio, often as low as 2. This is because the only thing that stops liquid from leaking down the sieve holes ("weeping") is the pressure of the vapor from the stage below. At low gas velocities, sieve tray efficiency drops rapidly due to weeping.

The other common type of tray geometry is the "bubble cap" tray, which is also shown in Figure 8.9. These trays utilize a small, inverted metal cup, called a "riser," placed above each hole in the tray. This riser serves to ensure good vapor–liquid mixing, and to prevent liquid from weeping to the tray below. In this manner, bubble cap trays are the opposite of sieve trays: they have high efficiency (>90%), excellent turndown (>10), are very expensive to manufacture, and do not tolerate solids at all. Sieve trays are almost always installed above the feed point to prevent any solids from fouling the risers, and they are often constructed from copper to react away sulfides in a similar manner to copper pot stills.

Finally, it is important to note that trays in their entirety can be replaced by a class of vapor–liquid contacting equipment called "packing," as illustrated back in Figure 5.8. Unlike the discrete equilibrium stages offered by a tray, packing is used to provide a constant area for mass transfer between vapor and liquid streams. While common in the chemicals industry, packing is rather rare in distilled spirits, often relegated to hobby distillers. Analysis of packed columns is much more complex than tray columns, so readers are directed to the reference sections for further reading.

8.3.2 REBOILERS AND DIRECT STEAM INJECTION

Distillation is an energy intensive process. In a pot still, the pot serves as the boiler: heat is applied via steam or fire, the liquid boils, and the vapor leaves. In continuous distillation, we impart heat to the liquid that falls to the bottom of the distillation column via a type heat exchanger called a "reboiler." We add the "re-" since liquid is constantly circulating through the still, so an element of liquid might be boiled multiple times. Heat addition is an absolute requirement for the operation of a column still. Without heat, the liquid fed into the middle of the still will simply drop from tray to tray and out the bottom of the column.

A reboiler is typically a shell-and-tube style heat exchanger, with steam on the shell side and bottoms on the tube side. Increasing the heat applied to the reboiler serves to "boil-up" more vapor, akin to turning up the stove on a kettle of boiling water.

Reboilers can become an issue when the wine or beer fed to the column is rich in solids, as the tubes can readily foul and plug due to cooked solids. The alternative to this is direct steam injection—where steam is fed into the bottom of the column to provide heat for distillation. While the addition of water (as steam) may serve counter-intuitive when trying to purify ethanol, the energy provided from steam condensation drives the distillation without issues, so long as the steam is food-grade.

Some fraction of the liquid that falls down the column is boiled by the reboiler and returned as boil-up, and some fraction of the liquid is removed as bottoms. We define the "boil-up" ratio as the flow rate of vapor leaving the reboiler up the column divided by the rate of bottoms leaving the still. "Total boil-up" is when this ratio reaches infinity, as no liquid is withdrawn, and all liquid is boiled back up the column.

Boil-up ratio is one of the key ways of controlling a distillation column. More boil-up drives the more volatile components up the column. If ethanol is being lost in the bottoms, increasing the boil-up will reduce ethanol concentration in the bottoms. Increasing boil-up comes with associated energy costs, as it takes energy to heat and boil additional liquid. The ideal boil-up rate/ratio is enough boil-up to drive distillation and prevent losses in the bottoms, but no more. This boil-up rate changes with the feed composition and flow rate: high concentrations of ethanol in the feed and high flow rates of feed will require more boil-up than slow, dilute ethanol feeds.

8.3.3 CONDENSERS

The condenser has the opposite effect of the reboiler—it removes heat from the system. The condenser is typically a shell-and-tube heat exchanger, this time with the spirits condensing on the shell side and water flowing through the tube side.

In a pot still, condensing liquid is removed as soon as it has been condensed as product. In continuous distillation, vapor rising to the top of the column is condensed in the condenser. Some fraction of the condensate must be returned down the column. This returned liquid is known as "reflux." Without reflux, only vapor would exist in the column above the feed point. The mixing between the falling reflux and the rising vapor above the feed point is what enriches ethanol in the distillate. "Reflux ratio" is the ratio of liquid returned to the column as reflux versus the liquid removed as distillate. When all condensate is returned to the column, this is known as "total reflux."

Reflux is the other key way of controlling distillation operation. Just as boil-up "pushes up" more volatile components, reflux "pushes down" less volatile components. Increasing the reflux ratio will drive water (and to a lesser extent, fusels) down the column, enriching the ethanol concentration in the distillate. When making a vodka, a high reflux ratio is critical to produce concentrated ethanol, typically >90% v/v. When making brandy or whiskey, a distillate of ~75% ethanol is usually desired, which would require a lower reflux ratio. The required reflux flow rate will change with feed composition and flow rate, and the distiller will need to tune the reflux to hit the desired distillate quality.

8.4 MCCABE–THIELE METHOD

Analysis of continuous distillation columns is a full field unto itself, and worthy of dedicated textbooks. However, a simple graphical method for analysis of binary distillation (where two only two components are split, such as ethanol–water) was developed in 1925 by Warren McCabe and Ernest Thiele at MIT and is still used to this day, called the McCabe–Thiele Method. This method utilizes mass balances on the top "rectifying" section of a column, and on the bottom "stripping" section of column, along with an equilibrium relationship known as a "X-Y" diagram, to predict distillate and bottoms compositions, compositions at each tray, and the number of trays needed.

When designing a column via McCabe–Thiele, typically the Feed Rate (F, mol/hr), the feed composition z (mole fraction), the column pressure P, the distillate purity X_D (mole fraction), the bottoms purity X_B (mole fraction), and the ratio of actual to minimum reflux ratio (R/R_{min}) is specified. From this, the distillate flow rate D (mol/min), bottoms flow rate B (mol/min), minimum number of stages N_{min}, minimum reflux ratio R_{min}, true reflux ratio R, boil-up vapor flow rate V_B, and true stage number N, can be calculated.

First, we must introduce an X-Y diagram, seen in Figure 8.10.

The red curve is the X-Y curve. It shows the equilibrium between an ethanol–water liquid (X-axis) and an ethanol–water vapor (Y-axis) at their respective bubble and dew points. Reading off the curve, a liquid of 0.2 mole fraction (X_E) ethanol would be in equilibrium with a vapor of 0.54 mole fraction (Y_E) ethanol, while a liquid of mole fraction 0.6 mole fraction ethanol would be in equilibrium with a vapor of 0.67 mole fraction ethanol. Note that unlike a T-X-Y diagram, change in temperature is not shown explicitly, but does change along the curve.

The blue curve is titled the "45 degree line"—it draws a 1:1 equivalency between liquid and vapor mole fractions, the reason for which will be explained shortly.

To derive the McCabe–Thiele method, we first imaging a material balance around the top of a distillation column, above the feed point. We number the top tray "1," the one below it "2," and so on through stage "n." Liquid "L" (mol/min) leaves this material balance out the bottom (to the column below), as does distillate "D" (mol/min), while vapor "V" enters from the bottom (from the column below). We say the distillate has ethanol mole fraction X_D. At stage n, the liquid has ethanol mole fraction X_n, while the vapor entering stage n from below, stage n + 1, is Y_{n+1}. This, we can write an ethanol mole balance around the top of the column, assuming stead state:

$$V \times Y_{n+1} = L \times X_n + D \times X_D \tag{8.4}$$

If we want to solve for the composition of the vapor entering tray n, we need to solve for Y_{n+1}:

$$Y_{n+1} = (L / V) \times X_n + (D / V) \times X_D \tag{8.5}$$

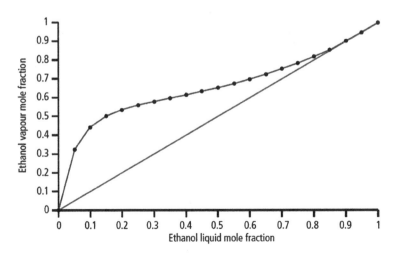

FIGURE 8.10 A X-Y diagram for and ethanol/water system, generated at 14.7 psia (atmospheric pressure) as predicted by the NRTL (non-random two-liquid) phase relationship.

We can extend this to any passing streams, with vapor leaving from tray n + 1 (they tray below tray n) and liquid standing on tray n, or:

$$Y = (L/V) \times X + (D/V) \times X_D \qquad (8.6)$$

We now define the "reflux ratio," R, equal to the ratio of liquid falling down from the top of the column divided by the liquid leaving as distillate:

$$R = L/D \qquad (8.7)$$

This is a critical parameter in distillation analysis. Since V = L + D, all liquid returned as reflux and distillate removed from the column must come from condensing vapor leaving the column, and returning some fraction as reflux while withdrawing some fraction as distillate. The reflux ratio varies between 0 (no liquid returned as reflux, all vapor removed as distillate) and infinity, or "total reflux," where no vapor is removed as distillate but is instead condensed and returned to the column. Typically, higher reflux ratios enhance separation, by providing more material for the rising vapor to contact along the length of the column.

Typically, both the reflux ratio and the actual number of stages used in a distillation column are free variables; but for a given separation target, the minimum reflux ratio and the minimum possible stages are not. At total reflux, the minimum number of stages are required, while at minimum reflux, an infinite number of stages are required. In practice, reflux ratios are often selected to be 1.3–2x the calculated R_{min}, with the actual number of stages is calculated from this true reflux ratio: see the example problems below for more detail.

With this new definition, we can write:

$$Y = x \times R/(R+1) + X_D/(R+1) \qquad (8.8)$$

This equation is an "operating line" and yields the relationship for the composition of **passing streams**—how the composition of a rising vapor stream relates to the composition of a falling liquid stream, as a function of (i) reflux ratio R and (ii) specified distillate purity X_D. This equation is the "rectifying operating line."

We can perform this same analysis on the bottom of the column. By defining the "boil-up ratio," V_B. This is exactly analogous to reflux ratio, with higher boil-up ratios enhancing separation.

$$Y = (V_B+1)/V_B - X_B/V_B \qquad (8.9)$$

This equation defines the compositional relationship between passing streams below the column feed point, as a function of bottoms purity and boil-up ratio. This equation is the "stripping operating line." Both the operating lines must intersect at least one point. At total reflux, they will overlap.

Equations (8.8) and (8.9), referred to as the "rectifying line" and "stripping line," respectively, can be plotted in a X-Y diagram. Since this is best demonstrated rather than explained, the next three examples will walk through common applications of McCabe–Thiele diagrams.

Example 8.4:

Consider a binary distillation of water and ethanol, with a minimum distillate purity of 0.65 mole fraction ethanol and a maximum bottoms mole fraction of 0.05 mole fraction ethanol. At total reflux, what are the number of stages required for this separation?

SOLUTION

Start by considering the rectifying section equation:

$$y = \left(\frac{R}{R+1}\right)x + \frac{x_D}{R+1}$$

At total reflux, R → ∞. This makes the slope ∞/(∞+1), or a slope of "1," and a y-intercept of X_D/(∞+1), or "0." That is, the rectifying section falls along the "45-Degree Line"!

Since the Rectifying and Stripping lines must intersect at least one point, they both must fall along the 45-degree line, as seen in Figure 8.11

We now "count of stages" by moving between the operating lines and the equilibrium X-Y curve: this is done under the assumption that the liquid and vapor leaving each tray are in equilibrium with each other, and so the X-Y curve allows us to "hop" between each stage! We can start at either X_B or X_D, but in this example we'll start at X_B and count "up," as seen in Figure 8.12.

This separation requires a minimum of three equilibrium stages at total reflux. Note that flow rates were not part of this analysis, and neither was feed composition! At total reflux, and for a feed between 0.05 and 0.65 mole fraction ethanol, a distillation column will always require three stages to achieve the desired spit.

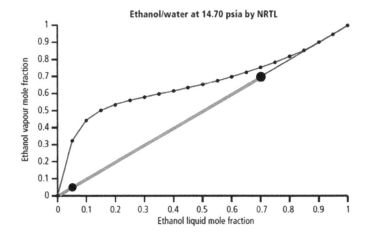

FIGURE 8.11   Example 8.4. The rectifying line falls along the 45-degree line.

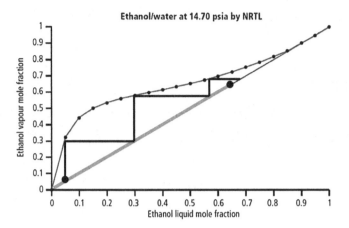

FIGURE 8.12 Example 8.4 continued. Counting off stages at total reflux.

It is important to note that total reflux is not and never can be a practical manufacturing condition—it means that no feed is entering the column, no distillate or bottoms is being withdrawn. This mode of operation is often done to equilibrate the column during a startup, but a practical operation requires a reflux ration < ∞ in order to remove product as distillate.

Example 8.5:

For a Feed of 0.15 Mole Fraction Ethanol, What Is the Minimum Number of Stages to Achieve the Separation from Example 8.4?

SOLUTION

First, we refer to the rectifying section operating line:

$$ y = \left(\frac{R}{R+1} \right) x + \frac{x_D}{R+1} $$

We know X_D, but R is unknown. However, consider that the rectifying line must stay below the equilibrium curve while to the right of the feed concentration (or else a separation superior to equilibrium is being effected, impossible!). With that, we can draw the curves seen in Figure 8.13.

Note an interesting development: we have had our operating line touch our equilibrium curve. During practical operation, this is not possible, but it is possible when there are an infinite number of stages, or N → ∞.

We can look at the y-intercept and see it is at 0.45. Looking at the rectifying operating line, we can look at how we define the intercept:

Y-intercept = 0.45 = X_D/(R+1)

With X_D known to be 0.65, we can rearrange and solve for a reflux ratio of R = 0.42. This is the minimum reflux ratio!

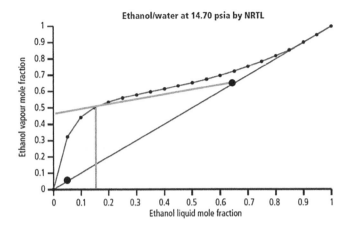

FIGURE 8.13 Example 8.5. Drawing the rectifying line at total reflux to find R.

Next, let's consider a different conundrum.

In this example problem, we ran into a "pinch"—where an operating line intersected with the equilibrium curve, and an infinite number of stages are required. While pinches always happen at minimum reflux, a pinch can also occur in azeotropic systems, like water–ethanol. Consider an attempt to refine this feed of 15% mole fraction ethanol to a 90% mole fraction distillate—when trying to draw the rectifying section (Figure 8.14), it ends up just touching (being tangent) to the equilibrium curve—a pinch point!

This pinch shows graphically why it is impossible to produce pure ethanol from distillation—you are always trapped by a pinch point!

Finally, let's use what we've learned in Examples 8.4 and 8.5 to specify a real distillation system:

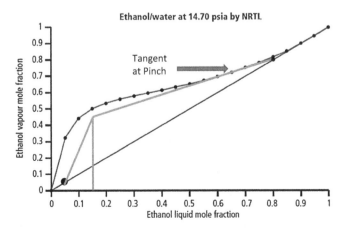

FIGURE 8.14 A pinch point is formed due to trying to distill beyond the water–ethanol azeotrope.

Example 8.6:

Design a Real Distillation System as Defined in Examples 8.4 and 8.5. This Time, You Want to Use an R/Rmin of 3.1.

SOLUTION

First, we can find our R by multiplying the Rmin from Example 8.5 by 3.1:
R = 3.1× 0.42 = 1.3

Now, we can use our rectifying line equation to plot the rectifying line (Figure 8.15). Plugging in our R and XD, we know our y-intercept is 0.28, and the rectifying line must touch the 45-degree line at the distillate purity of 0.65:

Now, how do we place our stripping section? We should have it (i) hit the 45-degree line at a bottoms purity of 0.05, and (ii) it should intersect with the rectifying section at the feed concentration of 0.15. With these two points known, we can just draw a line! Figure 8.16 shows the two operating curves on the X-Y diagram.

We can know count of stages between the operating lines and the equilibrium curve. Let's start at the bottom again, Figure 8.17.

Note that by starting at the bottom, we have exactly specified the bottoms purity at 0.05, while the distillate is only specified as XD ≥ 0.65, n this case it looks to be about 0.69. If we started counting at the distillate, the bottoms would be free to be less than or equal to 0.05.

Not only does McCabe–Thiele tell us the needed reflux ratio and ideal number of stages, it can also tell us the composition at each stage, by reading off where each stage intersects with the equilibrium curve:

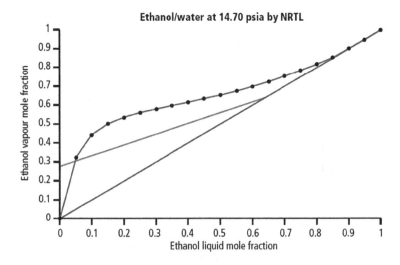

FIGURE 8.15   Example 8.6. Determining the rectifying line.

Stage	Liquid Mole Fraction EtOH	Vapor Mole Fraction EtOH
1	0.58	0.69
2	0.36	0.59
3	0.14	0.46
4	0.05	0.32

Note that the feed concentration is "inside" stage 3—this makes stage 3 the ideal stage to feed on!

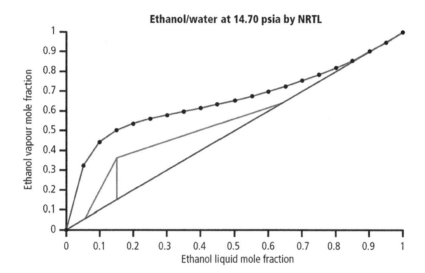

FIGURE 8.16   Example 8.6 continued. Drawing the stripping section line.

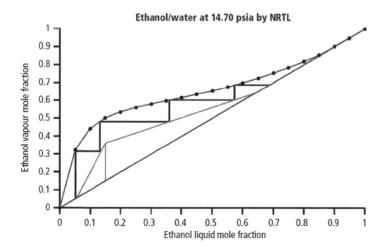

FIGURE 8.17   Example 8.6 continued. Counting off stages.

It is important to note that the configuration of the still will not change the number of stages needed, but may change the number of ideal plates needed. A partial condenser acts as an equilibrium stage between vapor and liquid—if one is employed, then the top most stage on the McCabe–Thiele diagram will be served by the partial condenser. Similarly, a reboiler acts as the mirror image of a partial condenser, so the bottom-most stage will be a reboiler. A total condenser, total reboiler, or direct steam injection enact no phase changes, so they do not serve as stages, and each separation stage must then be served by a tray.

REFERENCES

McCabe, W. L. and E. W. Thiele Graphical Design of Fractionating Columns. *Industrial & Engineering Chemistry* 1925 17 (6), 605–611. DOI: 10.1021/ie50186a023.
Seader, J.D., and Ernest J. Henley. *Separation Process Principles*. Wiley, 2006.
Kister, Henry Z. *Distillation Design*. McGraw Hill, 1992.

PROBLEMS

Consider a distillation column, operating at atmospheric pressure. This column is distilling a feed containing water ("W"), ethanol ("E"), and 1-propanol ("P"), and uses a side-draw ("S") to minimize 1-propanol in the distillate:

Stream	Flow Rate (kg/hr)	Ethanol Mass Fraction	1-Propanol Mass Fraction
Feed, "F"	1,000	0.05	0.0001
Distillate, "D"	?	0.85	?
Side Draw, "S"	?	0.3	0.1
Bottoms, "B"	?	?	0.0

The yield of ethanol, from the feed to the distillate, is 98%.
The yield of propanol, from the feed to the side-draw, is 50%.

1. Determine the following (the order presented may be helpful in solving!):
    (a) Distillate ethanol mass flow rate
    (b) Side draw propanol mass flow rate
    (c) Distillate propanol mass flow rate
    (d) Distillate total mass flow rate
    (e) Distillate water flow rate
    (f) Side stream total mass flow rate
    (g) Side stream ethanol flow rate
    (h) Side stream water flow rate
    (i) Bottoms total mass flow rate
    (j) Bottoms ethanol mass flow rate
    (k) Bottoms water mass flow rate

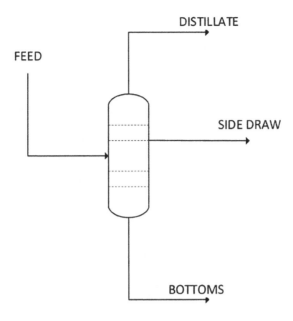

2. Determine the mole fractions of every stream. This can be done by hand, but an Excel file may be used too.

3. Assuming the feed and bottoms are saturated liquids and the distillate is a saturated vapor, determine the temperatures of the feed, bottoms, and distillate. Assume that the impact of propanol is negligible (i.e. use ethanol mole fraction when calculating).

4. Calculate the cooling rate required to condense the distillate if the latent heat of vaporization of the distillate is 1,200 kJ/kg. Give your solution in kW.

Consider the following process flow:

5. The first distillation column takes in a feed of $xE = 0.2$, and will create a distillate of $xE \geq 0.75$, and a bottoms of $xE \leq 0.025$. Assume this is a two-species system (ethanol and water). Note that reboilers are not shown. Ignore the impact of a side-draw on McCabe–Thiele analysis.

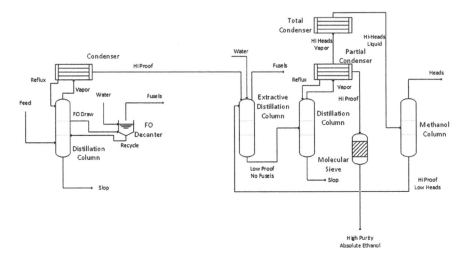

(a) Determine the ideal number of stages and ideal feed point if R/R_min = 3.
(b) For the ideal design in case, how many physical trays will we need to install if the tray efficiency is 70%?

6. The first distillation column will be fed 1,000 kg-mole/hr. Assuming this is a two-species system (ethanol and water), with a feed xE = 0.2, distillate xE = 0.75, and a bottoms of xE = 0.025 and that side-draws are negligible,

(a) calculate the total distillate and bottoms flow rates [in kg-mol/hr] if a 98% feed-to-distillate yield is specified.
(b) If the reflux ratio used is 3.0, calculate the total rate of liquid condensed in the first distillation column
(c) If we assume we can treat the vapor leaving the first column as all ethanol, where ethanol has a ΔH_vap of 43.5 kJ/mole, estimate the heat removal rate required by the condenser on the first distillation column. Give your answer in kW.

7. Explain the purpose of the

(a) Two condensers on the second distillation column
(b) The fusel oil draw and fusel oil decanter
(c) The extractive distillation column
(d) The molecular sieve
(e) The methanol column

8. McCabe–Thiele diagrams can also be used when separating one class of compounds from another, assuming the classes of compounds behave the same.

In the methanol column, we are trying to separate all heads from ethanol. Let us imagine that all heads, in aggregate, can be treated as a species that we wish to remove, with a relative volatility versus ethanol of 5. This can be plotted as:

For a feed of x_Heads = 0.4, a distillate of x_Heads ≥ 0.9, and a bottoms of x_Heads ≤ 0.1, determine:

Heads-Ethanol XY Diagram

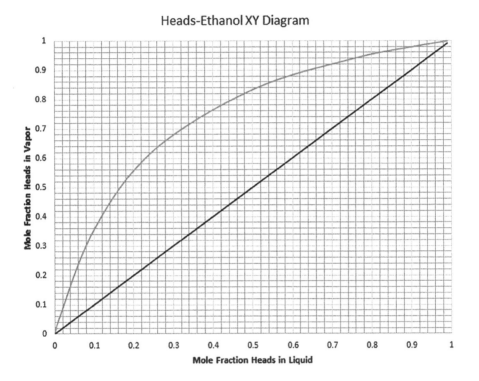

Mole Fraction Heads in Vapor

Mole Fraction Heads in Liquid

(a) Determine the minimum reflux ratio for this separation.
(b) Determine the minimum number of stages.
(c) Determine the ideal number of stages and ideal feed point if R/R_min = 1.5.

9 Wooden Cooperage and Oak Chemistry

9.1 BARRELS: A HISTORICAL PERSPECTIVE

Barrels have been used to store wine and spirits for thousands of years, introduced by the Romans and Gauls to replace clay pots and urns. Initial use of oak barrels led to the realization that wine quality actually improved during the storage, whether in a cave or during shipping. Therefore, barrel storage became a common integral practice in the process of producing wines in Europe and other parts of the world.

The shape of a standard barrel not only gives it strength, but also the ability to roll and move it with one person with relative ease, even given its weight when empty. The standard size for barrels (commonly around 225 L) likely evolved over time as a tradeoff between the number of individual barrels needed and the speed of extraction of oak-derived flavors into the wine or distilled spirit. The smaller the barrel used, the higher the surface area per volume and thus the faster the extraction. Therefore, if very large barrels are utilized, extraction of oak characteristics can become exceedingly low, especially after multiple uses.

It is worth noting that barrel ageing of beers is quite rare, for several reasons. First, oak flavor is not commonly thought to complement beer. Second, beer is substantially less microbially and oxidatively stable than wine (which has high alcohol, low pH, and antioxidants) or spirits (very high alcohol, nothing to oxidize or oxidation is desirable). Putting beer in a barrel is typically a fast way to spoil it! Finally, beer is typically carbonated. Just as sparkling wines are not barrel aged, all the dissolved carbon dioxide would be lost during ageing.

While the size and shape of barrels is somewhat uniform around the world, innovations do occasionally appear including spherical and egg-shaped barrels that come with their own interest and challenges.

9.2 USES OF BARRELS IN WINERIES AND DISTILLERIES

There are five major uses of barrels in the process of making alcoholic beverages. These are: (1) as a storage or fermentation vessel; (2) for oak extraction for addition of flavor; (3) to allow evaporation for manipulating the concentration of volatiles; (4) to create an environment that promotes oxidation; and (5) to create value through a perception of added value and quality. All five of these uses are described in more detail below.

One of the main uses of barrels is exactly its original historical use. That is, they are used as vessels for storage. Prior to stainless steel and concrete storage tanks, barrels would have been a main means of holding the wine during storage. This continues to be true, especially for premium wines. Barrels are also used as fermentation

DOI: 10.1201/9781003097495-9

vessels, though this is somewhat less common. Many producers have found that fermentation of white wines in barrel gives them desirable flavor profiles, as does subsequent storage of the wine on the resulting yeast lees (sur lees aging). This practice is usually preceded by an initial inoculation and nutrient adjustment of the juice in tank, followed by a barreling down after the yeast inoculum has begun to grow and is well distributed in the tank to assure uniform distribution. This practice can help avoid issues with non-homogeneity in barrels that can lead to sporadic stuck fermentations in barrels, especially as it is typical to only sample select barrels. Of course, other producers that are utilizing natural inoculation are content to barrel down without these precautions, though a higher degree of vigilance would be required in this case. While red wine barrel fermentations are not as common, due to the need to get skins and seeds into and later removed from the barrel, this practice has been used at the small scale. Doing this requires a large amount of hand labor, in addition to the need for careful and specific cleaning regimens between uses of the barrel. For some wines and distilled spirits globally, storage in oak is prescribed by law or by regional regulatory or quality control agencies. For instance, Bourbon requires a minimum of 3 years in charred new American oak barrels and American Brandy requires aging for a minimum of 2 years in oak.

The second use of barrels in a winery or distillery is to extract additional flavors from the barrel into the final product. As discussed in much greater detail below, these flavors can come from the wood itself or from the toasting of the wood, and winemakers often use these aromas and flavors as a chef would use a spice rack—to add interest and complexity to the profile of the finished wine. Winemakers using barrels primarily for the other reasons discussed here may choose to use "neutral" barrels, typically ones that have been used multiple times previously which generally decreases the extraction of volatiles into the wine or spirit. Extraction often follows saturation-type behavior in that it is quite rapid initially and becomes slower as aging progresses. This is likely due to a chemical equilibrium; however, this behavior is not well understood and warrants further research.

The third common use for barrels is to promote evaporation. Evaporation in barrels is highly dependent on the storage environment of the barrels and how they are sealed. However, evaporation can be the source of significant loss of product. On the other hand, depending on the humidity conditions in the barrel storage area, the ethanol concentration of the product can either increase or decrease during storage. This is also true of other volatile compounds that may impact product quality in either a positive or negative manner. Evaporation creates headspace in the barrels that is typically undesirable for most wines and leads to the need for "topping" of barrels to minimize this "ullage."

There are cases where this headspace is actually desirable, and this is the fourth use of barrels. That is, a product that relies on oxidation to create characteristic flavors will usually require storage in a barrel with ullage. Having oxygen in the headspace that slowly diffuses into the wine or spirit will induce oxidative reactions that result in oxidative characters usually associated with caramel, butterscotch, or nutty aromas. Examples of this would include Cognac, Sherry or Brandy de Jerez, as well as most whisk(e)ys. When these characteristics are not desired, regular topping becomes even more critical.

The final use of barrels is more for marketing than for technical or sensory reasons. Their use may increase the perception of quality for some consumers. That is, a notation on a label that the wine has been aged in small oak barrels may allow the wine to be sold successfully at a higher price point, even if barrels are all neutral and the storage time is quite brief.

Beer can and has been aged in barrel. However, as beers do not necessary improve with age, this is not a common practice. Beer is generally lower in ethanol and higher in pH and lacks the oxygen scavenging capacity of wine, which all tends to limit its stability during aging, especially in a barrel. Beers that are aged in barrel are generally high ethanol/gravity that are more stable microbially or lambic styles that already have complex microbial ecology.

9.3 ATTRIBUTES OF WHITE OAK AND THE ANATOMY OF A BARREL

Nearly all wood used for storing wine or spirits is white oak. White oak evolved as the chosen wood for this purpose because it is strong and durable, as well as watertight. This last attribute comes from tyloses or processes from parenchyma cells that enter the xylem of the tree and close off flow (see Figure 9.1). Interestingly, another key attribute of white oak is that it does not have an overpowering aroma. While one might question this attribute after we just stated above that extraction of aroma/flavor is a key use of barrels, it is important to think about other types of wood such as cedar or pine that would have considerably more aroma contribution to the point of overpowering the inherent characteristics of the beverage.

There are two types of white oak that are used for the vast majority of wine and spirits barrels. "American" oak comes from the species *Quercus alba*. It is basically found along the Mississippi River valley from the northern Midwest of the US down

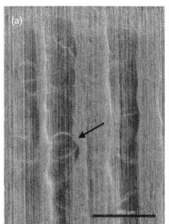 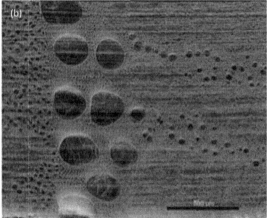

FIGURE 9.1 Tyloses visualized by micro-Computed Tomography (McElrone Group, USDA/UC Davis). (a) Cross section showing xylem with tyloses protruding into the vessel (arrow), (b) tyloses normal to vessel surface. The large openings are the xylem openings.

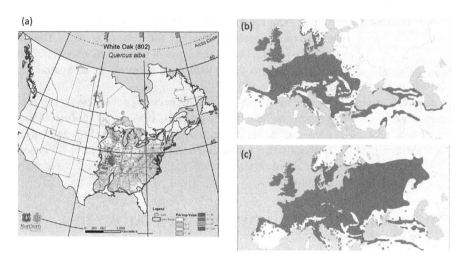

FIGURE 9.2 Geographic distribution of white oak species. (a) *Q. alba*, in the US (Prasad, A. M., L. R. Iverson., S. Matthews., M. Peters. 2007-ongoing. A Climate Change Atlas for 134 Forest Tree Species of the Eastern United States [database]. https://www.nrs.fs.fed.us/atlas/ tree, Northern Research Station, USDA Forest Service, Delaware, Ohio), and (b) *Q. robur* By Giovanni Caudullo - Caudullo, G., Welk, E., San-Miguel-Ayanz, J., 2017. Accessed in July 2016, CC BY 4.0, https://commons.wikimedia.org/w/index.php?curid=50449111 and (c) *Q. petraea* By Giovanni Caudullo - Caudullo, G., Welk, E., San-Miguel-Ayanz, J., 2017. Chorological maps for the main European woody species. Accessed in September 2016, CC BY 4.0, https://commons.wikimedia.org/w/index.php?curid=50388061

through Missouri (see Figure 9.2a). "French" oak, in reality, is composed of two species, *Q. robar* (sometimes referred to as *Q. pedunculata*) and *Q. petraea* (sometimes referred to *Q. sessilis*). In addition to being two species, French oak could actually be referred to "European" oak, as these two species grow throughout portions of France, as well as Hungary, the Baltics, and Russia (see Figure 9.2b). In France, trees are harvested from both privately owned forests, as well as state-managed forests called "futaie." Interestingly, these futaie date back to Napoleonic times when these trees were needed to maintain a secure source of wood for naval ships. Both French oak species have lighter tyloses than *Q. alba*. Practically, this affects the strength and water-fast characteristics of the wood, making it necessary for French oak logs to be split with a wedge so they break at their weak points prior to being sawed into individual staves keeping the vasculature in the right orientation to minimize leakage. On the contrary, American oak logs can be sawed from the beginning, allowing more staves per log and creating less waste in the barrel production process—one of the reasons that American oak barrels are typically less expensive, on average, than French oak barrels.

The general strategy for going from log to the staves that compose a barrel is the same, regardless of the oak. The log is cut into lengths dependent on the intended height of the barrel. The log cross section (see Figure 9.3) will have concentric growth rings with perpendicular rays. The log is then either split (French oak) or sawed (American oak) into quarters. Each quarter log is then sawed into staves of

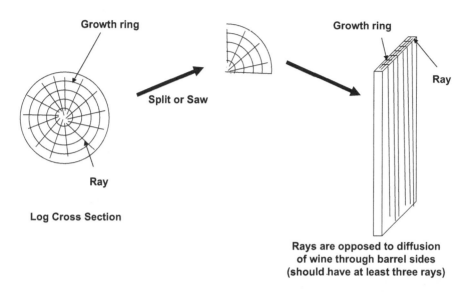

FIGURE 9.3   Diagram of the process from log to stave.

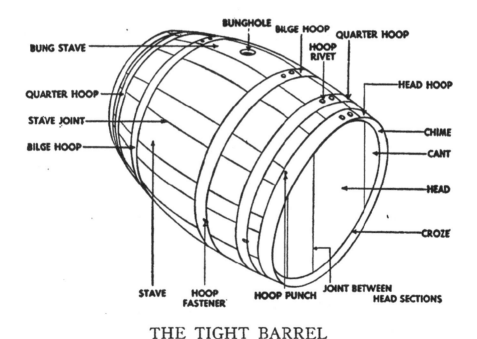

THE TIGHT BARREL

FIGURE 9.4   Schematic of a barrel. Credit: The Cooperage Handbook by Fred Putnam Hankerson.

successively smaller width in directions meant to keep the rays in a direction opposed to diffusion of wine across the stave as shown in Figure 9.3. When cut in this manner and readied for use in a barrel, the staves have an interesting anisotropy in terms of swelling and drying. As the staves dry, they will only shrink 0.2% in length, but 4% in width and 8% in thickness. The most important of these when thinking about the geometry of the barrel is the 4% reduction in width. This means that a 250 cm circumference barrel (at the widest part) will actually leave 10 cm of gap upon drying. Obviously, this will cause a huge functional problem, which is why barrels need constant humidity to maintain them.

The parts of a barrel are shown in Figure 9.4. The staves, already described, make up the sides of the barrel. One wider stave is used as the bung stave through which the bung hole is drilled as the main access to the inside of the barrel. The heads of the barrel make up the two ends and fit snugly into the croze, an indentation in the ends of the staves. The staves extend past the end of the heads slightly to form the chime. The barrel is held together by a series of metal hoops, generally six, that are named for where they sit, head, quarter, or bilge. The bilge is the point at which the barrel is widest, and the bung hole should be located at the bilge so that the barrel fully drains when turned with the bung hole pointed down toward the ground.

9.4 BARREL PRODUCTION

Many barrels are still made largely by hand by skilled coopers, though increasingly aspects of the process have been partially or entirely automated. Here, we discuss the general process for constructing a barrel.

Logs are often sold through auctions using accelerated methods such as a modified Dutch auction. That is, auctioneers start with a high price and then reduce the price until the first person bids. Alternatively, simultaneous single bidding on each lot (with high bidder winning) is utilized. Logs are then transported to the cooperage and stored until processed (Figure 9.5).

Initial processing begins by sawing the logs into appropriate lengths, followed by splitting (French oak, Figure 9.6) or sawing into quarters. Quarters are then sawed into rough staves keeping rays perpendicular to potential flow of gas or liquid. This requires alternating the cuts at roughly 90-degree angles and making progressively narrower staves. Keeping the rays at this angle is significantly more important for French oak in order to minimize barrel leakage because of the weaker tyloses. Once the rough staves are cut, they are layered into stacks to be placed outdoors for seasoning (Figure 9.7). The seasoning serves multiple purposes. First, it allows the wood to dry out overall to the point that it can be worked into the shape of a barrel. Second, precipitation will naturally leach phenolics and undesirable compounds out of the wood over time and allow microbial growth that can modify the surface of the wood. Wood can be seasoned for 1–3 years prior to forming the barrel, and the stacks are generally rebuilt periodically to allow for more uniform seasoning.

Following seasoning, staves are planed to remove the outer surface of the wood and shaped for barrel formation. This means that the ratio of stave width at the ends to stave width at the bilge needs to be the same as the ratio of the barrel end circumference to barrel bilge circumference. This part of the process can now be completely

FIGURE 9.5 Storage of oak logs prior to processing.

automated. Staves are then lined up to create the appropriate total width for a barrel. The cooper then "raises" the barrel using one end hoop and then secures the staves in place with additional temporary hoops (Figure 9.8) that are hammered into place. The next step is to bend all of the staves at the bottom of the barrel inward to be able to secure this end with hoops as well. In order to bend staves, coopers traditionally place the barrel over a fire (using scrap oak wood from the process) and add moisture. The staves are slowly winched closed and then secured with more temporary hoops. Because historically people found that they liked the taste of the toasting that ended up in their wine, further toasting past that needed to bend the staves is now the norm. The barrels are maintained over the fire and toasted to a specified level using a

FIGURE 9.6 Splitting (a) and sawing (b) French oak logs into rough staves.

(a)   (b)

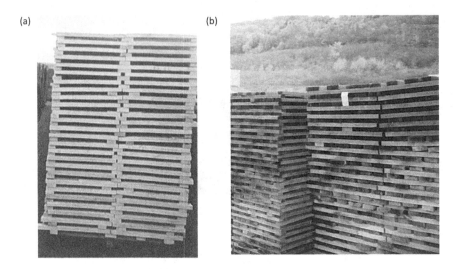

FIGURE 9.7   Staves stacked for seasoning outdoors just after stacking (a) and after some seasoning (b).

FIGURE 9.8   Preliminary barrel forming.

FIGURE 9.9   Toasting of oak barrels.

combination of time on flame and air flow to adjust temperature (Figure 9.9). For advanced cooperages, setting a time–temperature regime and monitoring/controlling closely with temperature sensors on the inside surface of the barrel helps them to achieve very uniform and reproducible toast levels. Traditional methods for toasting tend to be highly dependent on the particular cooper in the cooperage.

After toasting, heads need to be constructed. Short staves are used to build the heads. The staves are assembled into heads using dowels at the joints. River reed is traditionally used as a natural gasket between head staves (Figure 9.10). The rough heads are then cut into a circle and the edges planed to a point to fit securely in the croze (Figure 9.11). Then, the end hoops are removed, and the head is secured in place sometimes using a flour and water mixture as a food grade gasket or sealant material. The temporary hoops are then replaced. The heads can be branded to mark key attributes like toast level or cooperage, though more likely they are now being printed by computerized laser engraving into the head. This allows endless options to add winery logos and other information.

Next, the entire outside of the barrel is sanded down on a rotary sander (Figure 9.12). The final barrel hoops are fixed in place either by hand or using a hydraulic hoop press. The croze is then cut into the barrel ends to seat the heads (Figure 9.13). The bung hole is drilled in the wider bung stave (Figure 9.14), and then the barrels are leak tested with low pressure steam and/or hot water. Leaks are

FIGURE 9.10   Forming staves for barrel heads.

FIGURE 9.11   Barrel heads ready to be installed.

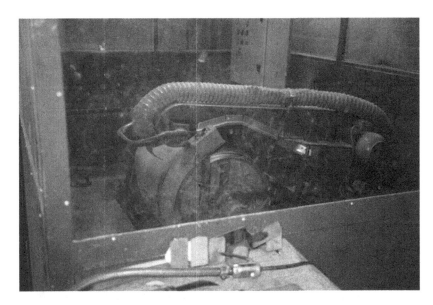

FIGURE 9.12   Sanding a barrel down after assembly.

FIGURE 9.13   Cutting a croze into a shaped barrel to seat a head. A barrel with croze cut into it, ready to accept a head, can be seen in the foreground.

repaired, usually by inserting tiny wood dowels where the leaks leave the barrel, effectively closing off open xylem-based vessels in the wood. Chestnut hoops (Figure 9.15) can be applied, mainly for decoration, though they have been suggested as providing easier rolling or as sacrificial material for oak wood boring bugs. Finally, the barrels are shrink wrapped for shipment to the customer.

FIGURE 9.14   Drilling a bung hole into a completed barrel.

FIGURE 9.15   Finished barrels wrapped with chestnut hoops.

Barrels used for distilled spirits are constructed using a similar process. However, specific types of spirits have come to use specific barrel attributes traditionally or legally. For instance, Bourbon barrels are actually charred to a finish called an "alligator" char because the surface looks like the skin or leather of an alligator.

9.5 OAK CHEMISTRY

The chemical composition of oak and its subsequent processing during the barrel making process determine the potential sensory impact of the resulting barrel. Here, we will discuss the general composition of white oak, its sensory impact molecules that can contribute to wine attributes, the kinetics of extraction of these compounds into wine, and factors affecting these extraction rates. More complete information can be found in the references at the end of this chapter.

Oak is primarily composed of four families of molecules (Figure 9.16). Cellulose makes up approximately 40–45% of the oak mass. Cellulose is made up of insoluble polymers of glucose monomers. Cellulose provides strength to the wood but does not generally contribute to the flavor extracted from the oak. Hemicellulose (20–25%) is a more heterogeneous polymer of five and six carbon sugars, including xylose, arabinose, mannose, glucose, and galactose. Hemicellulose is degraded by heat during toasting into compounds with caramel and associated aromas. Lignin is also a heterogeneous polymer composed of repeated phenylpropanoid subunits. It provides structural rigidity to the wood and is highly hydrophobic. It is totally insoluble and degrades into smoky and spicy aromas with elevated temperature during toasting.

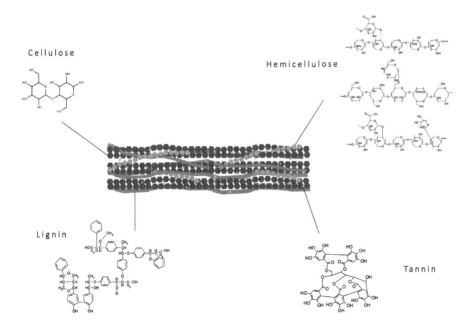

FIGURE 9.16 Structure of oak wood, a lignocellulosic material. Image credit: Dr. Jaime Moreno Garcia.

Lignin makes up 25–35% of the oak mass. Finally, tannins make up 5–10% of the oak. The tannins in oak are ellagitannins that hydrolyze readily to gallic acid and ellagic acid. While different oak sources seem to have different tannin concentrations, the sensory impact of these tannins on wine are not clear and are different than the impact of grape tannins coming from seeds and skins.

Figure 9.17 illustrates some of the key sensory compounds found in toasted oak barrels that are extracted into wine during fermentation or storage. First, cis-oak lactone and trans-oak lactone (sometimes referred to as whisky lactone or methyloctalactone) are two isomers of a molecule derived from the oak, itself (prior to toasting). Cis-oak lactone is the more important of the two from a sensory point of view, often associated with aromas of coconut, vanilla, and generic oakiness. Eugenol is formed through thermal degradation of lignin. It is generally associated with aromas of clove and spicy. Similarly, guaiacol (and its derivatives) and vanillin are also degradation products of lignin. Guaiacol has a smoky aroma and vanillin is associated with vanilla and similar aromas. Finally, furfural (and its derivatives) is associated with degradation of hemicellulose. The aroma impact of furfural is not clear, though it may act through affecting other compounds such as oak lactone. It is clear from data, such as that in Figure 9.18, that the matrix influences the aroma impact of oak-derived compounds heavily. While coconut aroma always seems to be associated with cis-oak lactone, the impact of other compounds seems to change depending on the matrix (e.g. red wine vs white wine vs model wine solution).

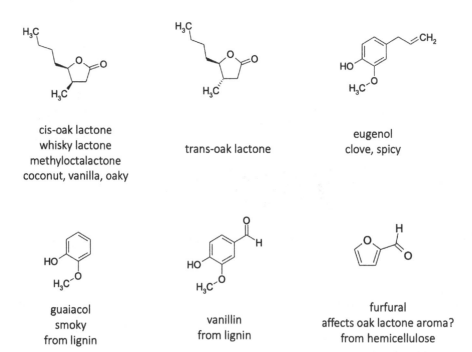

cis-oak lactone
whisky lactone
methyloctalactone
coconut, vanilla, oaky

trans-oak lactone

eugenol
clove, spicy

guaiacol
smoky
from lignin

vanillin
from lignin

furfural
affects oak lactone aroma?
from hemicellulose

FIGURE 9.17   Key sensory impact molecules from oak barrels.

24 Barrel-aged Chardonnay wines at 55 weeks

	cis	trans	eug	guaiac	4mg	van	malt	furf	eef	5mf	falc	5mfalc	fee	5mfee	vee	4vg	4eg	4vp
Coconut	0.744***	0.324	0.381	-0.432*	-0.327	0.021	-0.320	-0.095	-0.356	-0.158	-0.235	0.280	-0.101	-0.104	0.242	0.349	-0.355	0.232
Pencil	0.271	0.359	0.455*	0.303	0.612***	-0.032	-0.149	0.599**	0.158	0.549**	-0.417*	0.009	-0.034	-0.593**	0.289	0.383	0.283	0.325
Caramel	-0.446*	-0.544**	-0.433*	0.105	-0.345	0.095	0.553**	-0.332	0.210	-0.309	0.571**	-0.190	0.078	0.507*	-0.330	-0.556**	-0.046	-0.507*
Vanilla	0.094	0.137	-0.059	0.057	-0.047	0.365	-0.009	0.179	0.200	0.190	0.165		0.141	0.237	0.065	-0.085	0.066	
Butter	-0.161	-0.423*	-0.374	-0.349	-0.536**	0.099	0.092	-0.508*	0.048	-0.522**	0.434*	-0.170	0.064	0.235	-0.164	-0.233	-0.618***	-0.304
Allspice	0.196	0.029	-0.075	0.305	0.354	0.234	-0.058	0.436*	0.179	0.481*	-0.122	0.144	0.120	0.079	0.187	0.084	0.342	0.078
Smoky	-0.286	-0.058	0.013	0.762***	0.801***	0.438*	0.344	0.699***	0.717***	0.704***	0.162	-0.078	0.072	-0.100	0.020	-0.174	0.607**	-0.092
Cashew	0.329	0.140	0.306	0.036	0.335	0.348	-0.125	0.280	0.070	0.282	-0.091	0.311	0.203	0.032	0.329	0.217	0.206	0.171
Green apple	0.340	0.300	0.114	-0.664***	-0.486*	-0.277	-0.535**	-0.398	-0.627***	-0.459*	-0.422*	0.158	0.055	-0.181	0.291	0.349	-0.609***	0.152
Cinnamon	-0.224	0.017	0.032	0.300	0.340	0.561**	0.084	0.303	0.494*	0.296	0.374	0.010	0.368	0.199	0.273	-0.054	0.108	-0.130

24 Barrel-aged Cabernet Sauvignon wines at 93 weeks

	cis	trans	eug	guaiac	4mg	van	cyc	malt	furf	5mf	falc	5mfalc	fee	4vg	4eg	4vp	4ep
Coconut	0.861***	0.768***	0.865***	0.468*	0.337	0.427*	0.101	0.252	0.351	0.545**	0.288	-0.041	0.306	0.119	-0.020	0.057	-0.031
Pencil	0.045	0.113	0.123	0.057	0.340	-0.082	0.146	0.169	0.183	0.180	0.211	0.035	0.172	-0.035	0.012	-0.245	-0.101
Allspice	0.317	0.231	0.291	0.097	0.301	0.427*	-0.212	0.038	0.392	0.273	0.351	-0.378	0.209	-0.365	-0.513*	-0.630**	-0.534**
Berry	0.597**	0.434*	0.633**	0.087	0.038	0.493*	-0.170	-0.106	0.164	0.218	0.134	-0.166	0.083	0.112	-0.059	-0.031	-0.236
Smoky	-0.026	-0.027	-0.293	-0.007	0.196	0.214	0.187	-0.113	0.353	0.031	0.453*	0.120	0.348	-0.074	-0.005	-0.261	-0.262
Caramel	0.319	0.407*	0.390	0.478*	0.440*	0.546**	0.319	0.171	0.350	0.380	0.481*	-0.151	0.441*	-0.236	-0.128	-0.103	-0.196
Vanilla	0.776***	0.551**	0.713***	0.417*	0.330	0.482*	0.264	0.273	0.571**	0.545**	0.497*	-0.074	0.350	0.116	-0.107	-0.091	-0.197
Coffee	0.443*	0.440*	0.305	0.438*	0.522**	0.581**	0.199	0.141	0.565**	0.506*	0.738***	-0.136	0.566**	-0.212	-0.257	-0.429*	-0.480*
Dark choc	0.564**	0.514*	0.446*	0.356	0.348	0.482*	0.124	0.149	0.349	0.362	0.471*	-0.023	0.684***	0.025	-0.014	-0.120	-0.359
Band-aid	-0.247	-0.181	-0.414*	-0.054	0.024	-0.053	-0.056	-0.229	0.230	0.068	0.298	-0.073	0.055	-0.048	-0.056	-0.283	-0.096
Earthy	-0.504*	-0.226	-0.579**	-0.095	0.091	-0.377	-0.046	-0.079	-0.172	-0.231	-0.236	0.183	-0.006	-0.133	0.174	-0.187	0.083
Mint	0.196	0.227	0.135	-0.039	0.178	0.329	-0.086	0.066	0.385	-0.129	0.231	0.111	0.020	-0.331	0.026	-0.236	-0.287

*, **, *** denotes significant at 5%, 1% and 0.2% respectively. Cis = cis-oak lactone; trans = trans-oak lactone; eug = eugenol; guaiac = guaiacol; 4mg = 4-methylguaiacol; van = vanillin; cyc = cyclotene; malt = maltol; furf = furfural; eef = estimated extracted furfural (= furfural + furfuryl alcohol); 5-mf = 5-methylfurfural; falc = furfuryl alcohol; 5-mfalc = 5-methylfurfuryl alcohol; fee = furfuryl ethyl ether; 5mfee = 5-methylfurfuryl ethyl ether; vee = vanillyl ethyl ether; 4vg = 4-vinylguaiacol; 4eg = 4-ethylguaiacol; 4vp = 4-vinylphenol; 4ep= 4-ethylphenol.

FIGURE 9.18 Correlation between chemical composition and aromas related to oak in different matrices. Numbers represent Spearman's rank-order correlations. Spillman et al. https://onlinelibrary.wiley.com/doi/epdf/10.1111/j.1755-0238.2004.tb00026.x.

The extraction of volatiles into wine from a barrel seems to generally follow saturation kinetics. That is, extraction is initially rapid and then decreases in rate until a maximum concentration is reached. These kinetics seem to be compound-specific, with some compounds reaching their maximum sooner than others (e.g. in some studies, guaiacol may reach a maximum in 2–3 months, while other compounds do not seem to reach a plateau for 9 months or longer. For the most part, extraction rates are slower and concentrations less likely to reach a maximum in the second and subsequent uses of a barrel.

While the literature is filled with seemingly contradictory results on oak extraction of specific compounds, likely due to a large number of factors and barrel-to-barrel variability, there are several factors that affect extraction rates and extents. Oak variety, in most cases, has an effect on extraction. Ellagitannin extraction is typically higher in French oak species and cis-oak lactone is generally higher in American oak, but not throughout all of the literature. The ratio of cis-oak lactone to trans-oak lactone always seems to be higher for American oak species. For French oaks, *Q. robur* tends to be higher in ellagitannins and lower in oak lactones, compared with *Q. petraia*. Tree variability, within the same species and forest, is quite high, at least in terms of ellagitannin extractability, and height in the tree does not seem to affect extraction either. However, tannins are definitely greater in the outer parts of the log where growth is newer, and tannins have evolved to help protect the trees from invading insects. It is not clear what impact geography has on the final extractability of oak. While some geographic origins do seem to produce oak with specific characteristics, these results are often compounded by the fact that these regions are generally higher in certain species (like Limousin is higher in *Q. robur* trees). It seems likely that seasoning regimens and weather during seasoning would affect extractability.

However, the data in the literature are inconclusive, possibly due to large variation in seasoning based on stave positioning in the stack or the removal of the wood surface prior to barrel fabrication. Finally, toasting certainly affects extraction, as many of the compounds extracted into wine come from the thermal degradation of lignin during this process. However, historically toasting has been quite variable and highly dependent on the individual cooper. This is changing now as cooperages find ways to quantify and control temperature during the toasting process. This has allowed some cooperages to achieve very specific aroma profiles in their barrels that become a selling point for their product—as noted above, a veritable spice rack for the winemaker. The literature on aroma compound formation in the oak during barrel fabrication and the subsequent extraction would benefit from further studies, as many of the published studies were completed prior to the establishment of tighter controls on the variables briefly discussed here.

Because barrels can be quite expensive to purchase and require specialized facilities for optimal storage, a wide variety of oak alternatives have been developed to impart oak aromas into a wine without the barrel. These alternatives include oak chips, oak cubes, staves, or even sawdust. Alternatives are generally added during fermentation or storage in a stainless steel tank, and they are freely added, introduced in a large "tea bag," or fastened in place in the tank where they will be submerged. The rate of extraction from these alternatives can be significantly more rapid, as surface area per mass will be much higher than for traditional barrel extraction. Toasts and exposed wood surfaces may also be different than what can be achieved with a barrel. This can lead to results that are different than those achieved with barrels. While early on, oak alternative suppliers focused on recreating the barrel experience with their products, many now market their products as a means to achieve very specific aroma targets in the finished wines.

9.6   BARREL STORAGE SYSTEMS

Barrels need to be stored correctly in order to minimize wine loss, ease operations, and maintain a safe work environment. The first choice that the winemaker needs to make is whether they will be storing their barrels with the bung upright, called "loose bunged," or with the bung offset from upright, called "tight bunged and rolled" (Figure 9.19). The latter places the bung beneath the liquid level in the barrel, thus leading to reduced evaporation and SO_2 loss from the barrel, as the bung is the major route for evaporation. The advantage of loose bunged barrels is that they allow for barrel fermentation, either primary or malolactic fermentation without the fear of pressure buildup and subsequent wine loss. However, this advantage comes at the cost of increased topping due to more rapid evaporation.

Traditional barrel storage systems include barrels on the ground or rails chocked to avoid rolling. It is also quite common to see a second row of barrels resting on the first row (two high) as seen in Figure 9.20. The barrels are arranged so it is still possible to reach the bungs of the barrels on the bottom row to avoid the necessity of restacking barrels. Occasionally, wineries will build these traditional stacks more than two high. This arrangement is fraught with danger, as one loose chock could cause the barrel stack to fall. This could happen through operator error (e.g. a forklift

Loose Bunged

Tight Bunged And Rolled

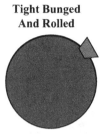

•Allows for ML
•More evaporation and SO_2 loss
•Requires periodic topping
•In situ sampling possible

•No ML possible
•Less evaporation and SO_2 loss

FIGURE 9.19 Different modes of wine storage in barrels. Orientation has less impact on spirits ageing, as malolactic fermentation and sulfur dioxide are not considerations, while evaporation is still a consideration.

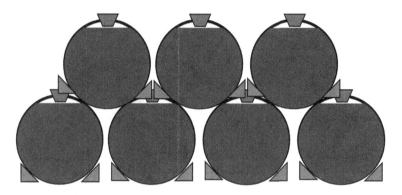

FIGURE 9.20 Traditional barrel stack (2-high) with chocks.

mishaps) or seismic activity, as is common in many wine producing regions. Empty barrels are already quite heavy, but filled ones are even more dangerous if the barrels are not properly secured and stored.

A more secure way of storing large numbers of barrels is using rectangular racks, often called Western Square racks after one of the common suppliers. These racks come in two- and four-barrel configurations and are shown in Figure 9.21. Barrels are placed on these racks, and then a rack with additional barrels is placed directly on the layer of barrels below. These racks (with barrels) can be stacked 6 barrels high, but it should be noted that all of the weight of the stack is on the bottom barrels since each rack sits on top of the barrels below. This can put an enormous amount of weight on the lower barrels, which begins to bend the barrel staves, resulting in leakage. Building these stacks requires a forklift. Four-barrel racks are significantly more seismically stable, but also require a specialized forklift with an appropriate counterweight and a larger space for maneuvering the forklift and racks. While these racks have traditionally been made out of metal, some newer versions have been

Portable Steel Barrel Racks
Western Square racks them up

All Western Square racks (WS29 Series) are designed to accommodate all oak barrels from 15 to 70 gallon capacity in both Bordeaux and Burgundy barrel shapes. Pictured here are the wine industry's most popular models. Specifications for these racks and other more specialized racks are listed on the following page.

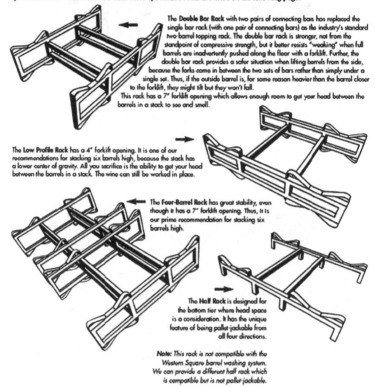

The **Double Bar Rack** with two pairs of connecting bars has replaced the single bar rack (with one pair of connecting bars) as the industry's standard two-barrel topping rack. The double bar rack is stronger, not from the standpoint of compressive strength, but it better resists "weaking" when full barrels are inadvertently pushed along the floor with a forklift. Further, the double bar rack provides a safer situation when lifting barrels from the side, because the forks come in between the two sets of bars rather than simply under a single set. Thus, if the outside barrel is, for some reason heavier than the barrel closer to the forklift, they might tilt but they won't fall.
This rack has a 7" forklift opening which allows enough room to get your head between the barrels in a stack to see and smell.

The **Low Profile Rack** has a 4" forklift opening. It is one of our recommendations for stacking six barrels high, because the stack has a lower center of gravity. All you sacrifice is the ability to get your head between the barrels in a stack. The wine can still be worked in place.

The **Four-Barrel Rack** has great stability, even though it has a 7" forklift opening. Thus, it is our prime recommendation for stacking six barrels high.

The **Half Rack** is designed for the bottom tier where head space is a consideration. It has the unique feature of being pallet-jackable from all four directions.

Note: This rack is not compatible with the Western Square barrel washing system. We can provide a different half rack which is compatible but is not pallet-jackable.

FIGURE 9.21 Barrel storage racks. Storage racks commonly come in two or four barrel capacity. Image courtesy Western Square.

made out of synthetic materials (like strong plastic) or steel reinforced synthetic materials. Although potentially not as strong or robust as the metal racks, the inherent give or flexibility in the material can actually help to avoid stack failure in a seismic event.

Other racking systems have been developed where the weight of the barrels is purely on an independent metal structure and not on the barrels themselves (Figure 9.22). Many of these systems have metal frameworks with built in casters that hold the barrels and allow for easy turning. This feature facilitates cleaning and topping, as well as storing the barrels tight-bunged and rolled. In addition, the casters allow facile battonage of the wine in barrels to get increased lees contact. The barrels

(a) 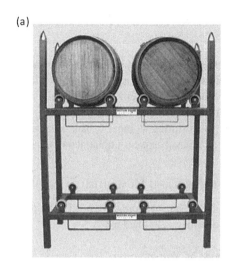 (b)

FIGURE 9.22   Alternative barrel rack system with castors. In this configuration, the weight of the racks are supported by the rack below and not by the barrel below. The castors allow for easy mixing. (a) A system that allows for forklift movement of barrels on the racks (Image courtesy Western Square). (b) A fixed-in-place system with more dense packing of barrels and a similar castor system for mixing, emptying, and cleaning in place.

are simply turned in place enough to disturb the layer of yeast and other sediment in the bottom of the barrel. So-called OXO systems have also been developed that work on the same principle, but with somewhat closer packing of barrels in the barrel room.

With any of these systems, seismic stability is a major concern. Based on experience with previous earthquakes in wine regions like California and Chile, it is clear that stack failures are common and catastrophic. Limited studies have been performed in California by placing barrel stacks on earthquake simulating shaker tables. Failure modes have been identified for these stacks: (i) Rack walking, where the entire stack of racks shifts, (ii) rack sliding, where the top rack displaces off, (iii) barrel ejection, where barrels from the top rack are thrown off of their mounting, (iv) longitudinal rocking, where the entire rack rocks as if it were a rigid body, (v) transverse walking, where the entire rack while "shuffle," and (vi) stack sliding, where the entire rack "hops" in a direction. For two-barrel racks, stacked six racks high, this is the order of importance (from most to least) of failure modes. For four-barrel racks, type (iii), top barrel ejection, is the most important.

Some of these failure modes can be minimized using straightforward means, such as using four-barrel racks, fastening the top barrel onto the rack to avoid ejection, being sure that racks sit on the wood of the barrel below and not on a hoop, and by using longer barrels (e.g. Bordeaux vs Burgundy style). Anecdotally, seismic failure of four-barrel racks has not been observed in California.

Similar systems for barrel storage are used for distilled spirits, though access to barrels for operations during storage may be far less frequent.

9.7 CONTROLLING TEMPERATURE AND HUMIDITY DURING BARREL STORAGE

Controlling the temperature and humidity of barrels being stored is critical to a successful barrel operation. These two parameters affect the evaporation rate, which in turn determines topping requirements and product loss. Temperature and humidity during barrel storage also determine the ratio of ethanol to water loss and whether the final wine ethanol concentration increases or decreases during storage. A lack of humidity can also lead to stave shrinkage, barrel leaking, and subsequent unplanned oxidation.

9.7.1 DEFINITIONS OF IMPORTANT TEMPERATURES AND HUMIDITIES

To begin the discussion of controlling humidity in barrel rooms, we need to begin by defining a series of terms. First, **absolute humidity** is defined as:

$$Absolute\ Humidity = \frac{Amount\ of\ Water\ Vapor}{Amount\ of\ Dry\ Air} \tag{9.1}$$

Absolute humidity can be expressed in units of moles of water per mole of dry air, for example, though other units including units of mass are common. Units of volume are also used. The next definition is for **relative humidity**:

$$Relative\ Humidity = \frac{Absolute\ Humidity}{Maximum\ Absolute\ Humidity\ at\ Saturation} \tag{9.2}$$

Relative humidity is usually reported as a percent from 0% to 100%, as opposed to the fraction form shown above. Air at 100% relative humidity is completely saturated with water (vapor).

A **dry bulb temperature** is the environmental temperature recorded using a normal thermometer. A **wet bulb temperature** is taken using a wet bulb thermometer. A simple wet bulb thermometer is shown in Figure 9.23. Essentially, a wet bulb thermometer is a normal thermometer equipped with a wettened wick over the temperature sensing bulb. When waved in the air, the water in the wick evaporates providing cooling that reduces the temperature read by the thermometer. The wet bulb temperature decreases as the air gets drier, as drier air promotes more evaporation. This is analogous to getting out of swimming pool on a dry day and feeling cold, even though the dry bulb temperature (or normal temperature) is quite warm. The **dew point** is the temperature at with the relative humidity reaches 100% for a given absolute humidity. It is important to note that warm air can hold more moisture, so as temperature decreases, relative humidity increases for a constant absolute humidity or moisture level in the air. It is also true that the maximum absolute humidity increases as temperature increases.

9.7.2 THE PSYCHROMETRIC CHART

With these definitions, we can begin to see the critical relationship between temperature and humidity. To understand this relationship more quantitatively, we often refer

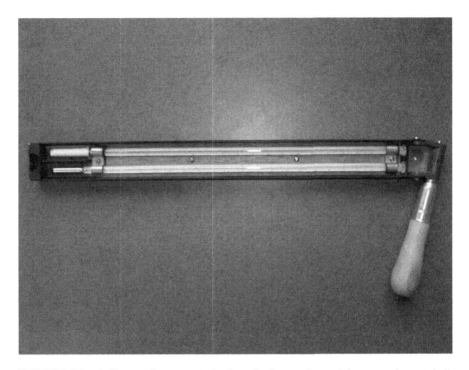

FIGURE 9.23 A sling psychrometer, a simple and robust analog tool for measuring wet-bulb temperature. Photo credit: Public Domain, https://commons.wikimedia.org/w/index. php?curid=247955

to a tool known as a **psychrometric chart**. An example of a psychrometric chart can be found in Figure 9.24. It is important to understand the components of this chart and how to use it as a tool to achieve the right combination of temperature and humidity for your barrel room. On the x-axis of this chart is the dry bulb temperature. The right y-axis is the absolute humidity (sometimes called the humidity ratio). Therefore, moving horizontally along this chart describes changing the temperature in a room without adding or removing water in any way. The curved lines represent constant relative humidity in percent from 0% to the left most curve which is 100% relative humidity or saturation. Any point to the left of this saturation curve represents an unstable situation in which water will be lost (in the form of fog and liquid water), dropping vertically on this chart until hitting the saturation curve again. Diagonal lines are called adiabatic saturation lines or wet bulb temperature lines. Moving from right to left diagonally along one of these lines describes cooling by addition of evaporating water without the addition of external energy/heat (therefore, the adiabatic label). If one follows these lines all of the way to the saturation curve, one reaches the wet bulb temperature for any point along that line. The psychrometric chart also allows calculation of the energy needed to get from an initial state of air (let's say outside air) to a final state (e.g. the desired conditions of the barrel room).

An example that illustrates how to use the psychrometric chart can be found in the example below for the Liquid Courage winery.

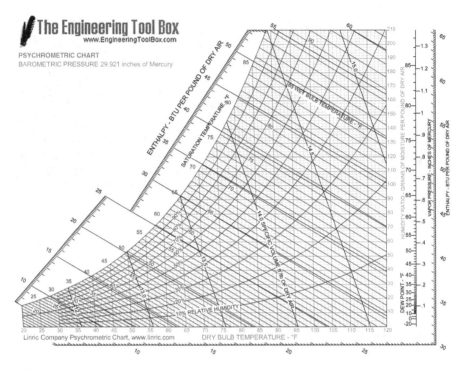

FIGURE 9.24 A psychrometric chart. Courtesy of: Engineering ToolBox, (2004). *Psychrometric Chart - Barometric Pressure 29.921 inches of Mercury.* [online] Available at: https://www.engineeringtoolbox.com/psychrometric-chart-d_816.html.

Example 9.1:   Designing a Barrel Room for the Liquid Courage Winery

THE LIQUID COURAGE CONCEPT (MACKENZIE DAVIS)

Liquid Courage is located in the hot southern California Temecula Valley. The estate vineyard of 300 acres is focused on producing varieties which thrive in this location. The varieties grown and sold offer a greater diversity beyond the scope of what the typical consumer is familiar. We hope to draw in consumers with the promise of delivering something unique and original. Liquid Courage provides a contrast to the surrounding area, showcased in the varieties produced as well as in the lively fun spirited atmosphere exuded at the winery. All of the crop is sourced from the estate vineyard. Everything is fermented in stainless steel tanks. For whites, the Torrontes is crafted in a way to exemplify its crisp characteristics and floral nose. While in contrast, the Vermentino goes through a malolactic fermentation to produce a richer and creamier white. The Tempranillo is aged in French oak for 12 months before bottling. The Aglianico blend is a unique rich chocolatey wine. All wines are distinct in character. The entire winery and vineyard are focused on sustainability. The label of a dauntless cowboy celebrates fearless pursuit of greatness. The wine boasts its quality without being pretentious or ostentatious. These moderately priced wines are for the average consumer. Anyone can take the reins and dive into the purchase of Liquid Courage. Often, wines suggest pairings of decadent expensive dishes which the average consumer would never prepare for themselves. Liquid Courage provides pairings that are for celebrating day to day experiences such as tailgating, picnicking, campfire circles, and gatherings around the dinner table.

PROBLEM

Barrel aging and storage will be an important part of several of the wines at Liquid Courage. This includes 30% of the Vermentino, Aglianco, Cabernet Sauvignon, Merlot, and Malbec, 70% of the Zinfandel, and 100% of the Tempranillo. Altogether, this represents 264,600 L of wine or 1,176 barrels (assuming 225 L barrels). All wines in barrels will be aged for 11 months or less—any times greater than this would necessitate double the barrel room space. Given these numbers and the expected weather in Temecula, we need to size the barrel room and understand the temperature and humidity control for this space.

SOLUTION

First, we can size the main barrel room. As this is a seismically active area, we will assume that only 4-barrel racks will be used, and these racks will be stacked a maximum of six barrels high. We need to leave room for forklifts to stack and unstack barrels, in addition to space for ladders or cherry pickers to sample barrels throughout the storage areas. Therefore, we are able to create the following scale diagram. While other configurations would certainly be possible, this one will fit all of the barrels projected to be needed immediately with some space for hospitality or expansion.

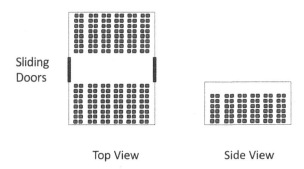

Sliding
Doors

Top View   Side View

Next, we can use the psychrometric chart in Figure 9.24 to understand the relationship between humidity and temperature in the barrel room. The wine-maker at Liquid Courage decides to maintain the barrel room at 60°F and 80% relative humidity (RH). What would the wet bulb temperature be at these conditions? This can be solved with the psychrometric chart. One can identify 60°F on the x-axis of the chart and move up until a RH of 80% is reached. Wet bulb temperature for this point can then be found in two ways: (1) move diagonally up and to the left from this point parallel to the adiabatic saturation lines on the chart until you reach 100% RH and read off the dry-bulb temperature or (2) identify the temperatures associated with the adiabatic saturation lines above and below the point identified and interpolate between the two of them. In either case, the wet bulb temperature can be found to be approximately 56.5°F. At the same time, the absolute humidity of the air in the room can be found by moving horizontally to the right on the graph from the original point. In this case, the absolute humidity can be seen to be 62 grains of moisture per pound dry air. This is equivalent to 0.014 lbmol water/lbmol air (where 1 lbmol = 454 mol). The dewpoint under these conditions can be found by moving horizontally to the left until reaching 100% RH, or in this case, 54°F. This means that, if the temperature further decreases to 52°F, we will get fog, and the total amount of water lost can be found by taking the difference between the initial absolute humidity and that found by dropping a line vertically down to the 100% RH line at 52°F.

Another important calculation that we can make with the psychrometric chart is to calculate the air conditioning load or energy needed to bring typical outside air to the desired conditions of the barrel room (60°F, 80% RH in this case). To do this, we can use the psychrometric chart to calculate the enthalpies of the beginning, or outside, air and the final, or barrel room, air. The difference is the amount of energy that must be applied to achieve this change. The typical worst-case scenario for this is the hottest air of the season in Temecula, which can be found to be July and August. Average conditions for Temecula in August (which is slightly warmer on average than July) are 91°F and 34% RH in the afternoon of a typical day. The winemaker decides to change her mind and target the air in the barrel room to be 60°F and 85% RH. Psychrometric charts handle enthalpy in different ways. For this chart, it is relatively straightforward. Enthalpies for a particular set of conditions are found by connecting the enthalpy lines on the far left and far right with a straight edge and finding the enthalpy that passes through your point of interest. Let's begin with the first point, 91°F and 34% RH. For this point, one can use the chart to find an enthalpy at the starting point of 32.8 Btu/lb$_m$ dry air. For the final point, 60°F and 85% RH, we find the enthalpy to be 24.6 Btu/lb$_m$ dry air. Therefore, the change in enthalpy needed to bring outside air to the desired conditions is 8.2 Btu/lb$_m$ dry air or 238 Btu/lb mol dry air. If we know the volume of the room, we can then calculate the total Btu's of energy needed for one room change of air.

Since this could represent a great deal of energy for a large space or one in which many room changes are needed (for instance, when barrel fermentations are actively releasing carbon dioxide into the environment), it is instructive to ask whether this change in air conditions could be accomplished by addition of misted water alone (i.e. evaporative cooling). To decide whether or not this is the case, one can use the psychrometric chart and follow a line parallel to the constant wet bulb or adiabatic saturation curves diagonally up to the left until the 100% RH curve. In this case, we can only reach 69°F and 100% RH (or 72°F if we want to maintain 85% RH). These temperatures are too high to be useful for our barrel storage purposes.

9.7.3 CHOOSING THE RIGHT HUMIDITY AND TEMPERATURE FOR YOUR BARREL ROOM

Now that we know the relationship between temperature and humidity, it is important to discuss optimal setpoints for these parameters in a barrel room. Of course, this depends on the key goals that the winemaker would like to accomplish during barrel storage. However, generally it is good practice to maintain relative humidity between 85% and 95% at the temperature desired for storage. If relative humidity is maintained higher, the probability of small deviations in control leading to condensation becomes unacceptably high. Condensation on the outside of barrels, floor, or other parts of the barrel room is likely to induce mold growth, which can negatively affect cleaning and wine quality, in addition to air quality. If the relative humidity is maintained at too low a setpoint, the loss of production due to evaporation is likely to be unacceptably high.

This point begs the question. How much evaporation will occur if the relative humidity is too low? Blazer (1991) developed correlations for water and ethanol loss from barrels as a function of temperature and relative humidity. While these are just correlations, these equations do approximate the relative losses of water and ethanol in units of mL per month per 225 L barrel:

$$water\ loss = 180 VP_W \left(1 - \frac{RH}{100} \right) \tag{9.3}$$

where RH is the relative humidity in percent and VP_W is calculated in this way:

$$VP_W = 10^{a_{H2O}} \tag{9.4}$$

and

$$a_{H2O} = 7.97 - \frac{1,668}{T + 228} \tag{9.5}$$

where T is in the units of °C. It should be clear from these relationships that vapor pressure, and therefore water loss, are a strong function of temperature.

Similarly, for ethanol loss:

$$ethanol\ loss = 5VP_{EtOH}\left(4.167e^{-\frac{707.77}{T+273.3}}\right) \tag{9.6}$$

where

$$VP_{EtOH} = 10^{a_{EtOH}} \tag{9.7}$$

and

$$a_{EtOH} = 8.04 - \frac{1,554}{T+223} \tag{9.8}$$

where T is again in units of °C. Example data calculated from these equations are shown in Table 9.1 to illustrate the absolute and relative magnitude of water and ethanol from a barrel at different common barrel storage temperatures and humidity levels.

It is instructive to think more broadly about the physical phenomena underlying these data. At high relative humidity in the barrel room, the driving force for evaporation (see equation (9.3)) will be zero or close to zero, as the relative humidity in the headspace of the barrel can be assumed to be at 100%. On the other hand, if the relative humidity is low in the barrel room, a large driving force will be present for evaporation of water from barrels. Since the concentration of ethanol vapor in the air is essentially zero outside the barrel, regardless of the concentration of water (i.e. humidity), the driving force for ethanol evaporation will be independent of barrel room humidity and solely a function of temperature. This is also consistent with the equations above.

Using specific examples, we can see how the humidity in the barrel room can also affect the ratio of ethanol to water loss and, therefore, the concentration of ethanol in the product during aging. For instance, if the relative humidity of the barrel room is 100%, no water will be lost from a barrel, but ethanol will be lost at a constant rate. Therefore, the percent ethanol in the wine or spirit will decrease over time. If the relative humidity is very low in the barrel room, water loss will be large, but the ethanol loss will be the same (assuming the same temperature), and the ethanol concentration will actually increase. This means that at some relative humidity, the concentration of wine, for instance, will stay the same (say around 14% alcohol by volume). This relative humidity is generally around 80 to 85%.

While control of relative humidity can affect the concentration of alcohol in the wine, it also dramatically affects the loss of product (and topping requirements). This can be a substantial financial loss. For example, if a barrel room is maintained at 60% relative humidity and a temperature of 15°C, a barrel of wine will lose approximately 5% of its volume over 12 months, a typical aging time for a barrel. For a 10,000 case annual production, this would amount to 500 cases of evaporative losses. If a case sells for $150 wholesale, this will amount to $75,000 per year, just in evaporative losses!

TABLE 9.1

Relative Humidity	(ml/225 L barrel/month)			Relative Loss
	Water Loss	Ethanol Loss	Total Volume Loss	% ABV Loss
Barrel Storage at 55°F				
100%	0.00	49.44	49.44	100.00
95%	99.39	49.44	148.83	33.22
90%	198.78	49.44	248.22	19.92
85%	298.16	49.44	347.60	14.22
80%	397.55	49.44	446.99	11.06
75%	496.94	49.44	546.38	9.05
70%	596.33	49.44	645.77	7.66
65%	695.71	49.44	745.15	6.63
60%	795.10	49.44	844.54	5.85
55%	894.49	49.44	943.93	5.24
50%	993.88	49.44	1043.32	4.74
Barrel Storage at 60°F				
100%	0.00	60.51	60.51	100.00
95%	119.39	60.51	179.90	33.64
90%	238.77	60.51	299.28	20.22
85%	358.16	60.51	418.67	14.45
80%	477.54	60.51	538.05	11.25
75%	596.93	60.51	657.44	9.20
70%	716.31	60.51	776.82	7.79
65%	835.70	60.51	896.21	6.75
60%	955.08	60.51	1015.60	5.96
55%	1074.47	60.51	1134.98	5.33
50%	1193.86	60.51	1254.37	4.82
Barrel Storage at 65°F				
100%	0.00	73.21	73.21	100.00
95%	141.91	73.21	215.12	34.03
90%	283.82	73.21	357.03	20.50
85%	425.73	73.21	498.94	14.67
80%	567.64	73.21	640.85	11.42
75%	709.55	73.21	782.76	9.35
70%	851.46	73.21	924.67	7.92
65%	993.37	73.21	1066.58	6.86
60%	1135.28	73.21	1208.49	6.06
55%	1277.19	73.21	1350.40	5.42
50%	1419.10	73.21	1492.31	4.91
Barrel Storage at 70°F				
100%	0.00	88.81	88.81	100.00
95%	169.10	88.81	257.91	34.43
90%	338.19	88.81	427.00	20.80
85%	507.29	88.81	596.10	14.90
80%	676.39	88.81	765.20	11.61
75%	845.49	88.81	934.30	9.51
70%	1014.58	88.81	1103.39	8.05
65%	1183.68	88.81	1272.49	6.98
60%	1352.78	88.81	1441.59	6.16
55%	1521.88	88.81	1610.68	5.51
50%	1690.97	88.81	1779.78	4.99

9.7.4 Controlling Humidity and Temperature at the Desired Setpoints

Finally, it is important to understand how humidity and temperature can be controlled in a barrel room. Most commonly, winery and distillery barrel room environments are controlled by different systems—humidity using a water misting system and temperature with a separate heating, ventilation, and air conditioning (HVAC) system. However, an alternative to these typical systems is having an HVAC system with built in steam injection into the air flow, thus using just one system to achieve both humidity and temperature control. There are advantages and disadvantages to both types of systems.

The more common approach in wineries involves an array of emitters, evenly placed across the ceiling, that spray a fine mist of water in tiny droplets (Figure 9.25). The spray continues until relative humidity sensors placed among the barrels reach their setpoint, and then the misting is turned off. This type of system is mechanically straightforward. However, there are common issues that need to be understood and dealt with during design and operation. First, nozzle design is critical. The water droplet size created by the emitter is inversely proportional to the evaporation rate. That is, small droplets (i.e. fine mist droplets) evaporate faster. At the same, large droplets will potentially fall with a higher velocity. If the nozzles clog due to hard water deposits or particulates, a problem will occur in that the resulting larger droplets will have less time to evaporate prior to hitting a surface below the emitter, even though they require more time to completely evaporate. This is a common problem in winery barrel rooms, as it will lead to standing water on barrels or floors that can initiate mold growth. Second, the nozzle placement is also critical. Placement of the array of emitters needs to assure even humidity distribution within the barrel room space. Care needs to be taken to consider air flow in the room, as well as the intended positioning of the barrels. Enough space should be left between the emitter and the highest barrel to assure complete evaporation of mist under normal operating conditions. Third, placement of relative humidity sensors is important. If only one sensor is utilized, placing it in the room at a point likely to give the average relative humidity is likely the best idea, although it would still be important to understand the range of humidity in all parts of the barrel room, so as to understand the danger of mold growth or excessive evaporation in other parts of the room. Placing a humidity sensor near an outside door that opens frequently can lead to poor humidity control, especially in a dry climate, as the emitters will be on throughout the workday as the sensor will remain dry, even as the remainder of the barrel room may develop a deep fog

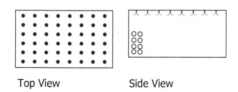

Top View   Side View

FIGURE 9.25 Typical relative humidity control for barrel rooms. Ceiling-mounted emitters creating a fine mist of water are commonly used in barrel rooms to control relative humidity. Temperature is then controlled by a separate system.

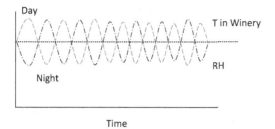

FIGURE 9.26   Temperature and relative humidity diurnal fluctuations at constant absolute humidity.

as you reach 100% relative humidity. Use of multiple sensors at different heights or throughout the barrel stacks may help to maintain the desired humidity. Finally, misting systems need to account for diurnal patterns in temperature. As shown in Figure 9.26, the temperature in a building can fluctuate dramatically throughout a 24-hr period, especially if the building is not well insulated or the external temperatures vary dramatically between night and day as is common in wine growing regions. If the absolute humidity remains constant in a barrel room that is experiencing

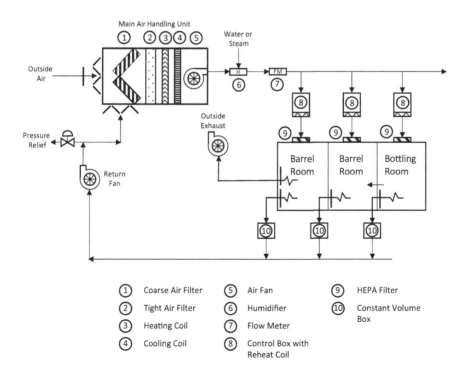

FIGURE 9.27   Schematic of a HVAC steam injection system, used to maintain constant humidity and temperature. In this case, the same system is used to control both temperature and humidity. After Lyderson et al.

diurnal changes in temperature, the relative humidity in the room will also experience diurnal changes. This can lead to condensation in the barrel room at night.

An alternative method for controlling humidity (and temperature) in barrel rooms is to use an HVAC unit with direct steam injection. This type of system is commonly used in hospitals and biopharmaceutical production facilities where control of humidity and temperature are mission critical. There are several installations of this type in the wine industry as well. The advantage of this type of system is that the temperature and humidity of the room are controlled by the same system, making control more accurate. This accuracy comes at the cost of a slightly higher level of complexity in the building systems. In this type of system (Figure 9.27), outside air is filtered and conditioned to a constant temperature. Then, humidity is introduced into the air stream by injecting pure steam, and this injection is controlled by a humidity sensor internal to the air handling unit of the HVAC. The temperature of the air is then adjusted to the desired temperature as it enters the room. While there are variations to this idea, the key here is that air is delivered to the room at the temperature and humidity desired.

While it would be easiest to control humidity in a room without ventilation and fresh air, regulations exist to protect workers against exposure to carbon dioxide and volatile organic carbon. Carbon dioxide evolution is of particular importance, as primary or malolactic fermentations in barrel can cause a lethal buildup of carbon dioxide in a short time without proper ventilation. All intended uses of barrel room spaces should therefore be considered in their design, as well as all safety issues.

REFERENCES

Blazer, R.M. Wine Evaporation from Barrels. *Practical Winery and Vineyard,* **Jan/Feb**, 20–22, 1991.

Boulton, R.B., Singleton, V.L., Bisson, L.F., and Kunkee, R.E. *Principles and Practices of Winemaking.* Chapman and Hall, New York, 1996.

Doussot, F., De Jeso, S. Quideau, B., and Pardon, P. Extractives content in cooperage oak wood during natural seasoning and toasting: Influence of tree species geographic location, and single-tree effects. *Journal of Agricultural and Food Chemistry, 50,* 5955–5961, 2002.

Gomez-Plaza, E., Perez-Prieto, L.J., Fernandez-Fernandez, J.I., and Lopez-Roca, J.M. The effect of successive uses of oak barrels on the extraction of oak-related volatile compounds from wine. *International Journal of Food Science & Technology, 39,* 1069–1078, 2004.

Lydersen, B.K., D'Elia, N.A., and Nelson, K.L. (Eds.). *Bioprocess Engineering: Systems, Equipment and Facilities.* John Wiley & Sons, Inc., New York, 1994.

Masson, G., Moutounet, M., and Puech, J. L. Ellagitannin content of oak wood as a function of species and of sampling position in the tree. *American Journal of Enology and Viticulture, 46,* 262–268, 1995.

Marrow, J.M. *Building a Better Barrel Stack.* Practical Winery & Vineyard Magazine, Jan/Feb 2003 Issue, pp. 71–79

Perez-Coello, M.S., Sanchez, M.A., Garcia, E., Gonzalez-Vinas, M.A., Sanz, J., and Cabezudo, M. D. Fermentation of white wines in the presence of wood chips of American and French oak. *Journal of Agricultural and Food Chemistry, 48,* 885–889, 2000.

Sefton, M.A. How does oak barrel maturation contribute to wine flavor? *Practical Winery and Vineyard,* November/December, 1991.

Spillman, P.J., Sefton, M.A., and Gawel, R. The effect of oak wood source, location of seasoning and coopering on the composition of volatile compounds in oak-matured wines.
Australian Journal of Grape and Wine Research, *10*, 216–226, 2004.

Spillman, P.J., Sefton, M.A., and Gawel, R. The contribution of volatile compounds derived
during oak barrel maturation to the aroma of a Chardonnay and Cabernet Sauvignon
wine. *Australian Journal of Grape and Wine Research*, **10**, 227–235, 2008.

Towey, J.P. and Waterhouse, A.L. The Extraction of Volatile Compounds From French and
American Oak Barrels in Chardonnay During Three Successive Vintages. *American
Journal of Enology and Viticulture*, *47*, 163–172, 1996.

Towey, J.P. and Waterhouse, A.L. Oak Lactone Isomer Ratio Distinguishes between Wines
Fermented in American and French Oak Barrels. *Journal of Agricultural and Food
Chemistry*, *42*, 1971–1974, 1994.

Zhang, B., Cai, J., Duan, C.-Q., Reeves, M.J., and He, F. A Review of Polyphenolics in Oak
Woods. *International Journal of Molecular Sciences*, *16*, 6978–7014, 2015.

PROBLEMS

1. The owner of Domaine de la Ingeniere tastes the Sauvignon blanc wine after
   cold stabilization and decides that he would like to add oak to half of the production to make more of a fume blanc style wine. Your job is to set the humidity and temperature levels in the barrel storage room. Unfortunately, the owner
   keeps changing his mind as to what the goals should be during barrel storage,
   and you need to accommodate his wishes.

   (a) At first, the owner wishes to minimize production losses due to evaporation.
       Qualitatively, how should the humidity be set in the room to achieve this
       goal? Explain. What are two potential consequences of making this choice?

   (b) Next, he changes his mind and decides that he wants to adjust the humidity
       to increase the alcohol content slightly to hit a "sweet" spot. How should
       the humidity be set? Explain.

   (c) Finally, the owner decides that he would just like to maintain the ethanol
       level at 13.5% during the aging process. How should the humidity be set?
       Explain.

   (d) You finally decide to set the initial conditions in the barrel room at a dry
       bulb temperature of 75°F and a wet bulb temperature of 70°F. What is the
       relative humidity at this point? The next morning, you get to work and find
       fog in your barrel room. How cold must it be in the room for this to hap-
       pen? Explain and show your work on the attached psychrometric chart.

   (e) Finally, the owner asks you to help out with the release party for the new
       Sauvignon blanc. Unfortunately, the weather is not cooperating. Even in the
       big tent outside, it is 105°F (and 25% relative humidity). You decide to rig up a
       water misting (i.e. evaporative cooling) system in the tent to cool off the guests.
       What is the minimum temperature that the tent will reach with this system?
       Explain and show your work on a psychrometric chart, such as Figure 9.24.

2. It is the beginning of spring when the weather suddenly gets hot at your Central
   Valley winery. Tomorrow is expected to reach 90°F with 35% relative humidity (RH).

(a) To what temperature should you cool your barrel room (with air conditioning) in order to maintain ethanol levels at their current level of 13% v/v? Explain your answer and show on a psychrometric chart.

(b) If the air conditioning does not work when you turn it on for the first time this season, what temperature could you get your barrel room down to using only evaporative cooling (i.e. misting)? Under these conditions, will your ethanol levels go up or down in the barrel relative to its original concentration? Explain.

(c) How much moisture (i.e. water) must be added to the room to get to this temperature from the original conditions? The barrel room is approximately 4,000 m³ (approximately 400 lb-moles of air).

(d) Initially, all of your barrels were American oak. Recently, however, you have decided that you would prefer to blend in some French oak into your barrel aging program. Your accountant wants to know why French oak barrels are so expensive compared with American oak barrels of the same size. Give one reason why this might be.

3. With the current design of the Wickson–Hall winery, we will only be able to fit about 1,200 barrels in the cellar. We will need to store about 800 barrels in an offsite barrel storage facility.

(a) What dimensions (H × W × L) should this facility be if you are willing to stack the barrels six high on four-barrel racks, but want access to all barrels during storage (without moving them around with a forklift once they are in place)? Assume American oak barrels in a Bordeaux Export-type shape. You can use information on barrel dimensions and racks from information you received in class or on the internet. Be sure to cite your sources.

(b) If you were to choose a misting system to maintain humidity, where would you place the relative humidity sensor in the room to assure adequate humidity control? How would your answer change if you could place four relative humidity sensors that all fed back information to the humidity controller?

4. **[Integrative Problem]** At Hetch Hetchy Vineyards, you are planning to barrel age your Zinfandel. At first, the barrels will be kept in the main large fermentation room, along with the fermentors.

(a) When you go to buy your first barrels, the first barrel salesman tells you that you should:

"...buy my Hungarian and Russian oak barrels as an alternative to the more expensive French oak barrels. They are made from the same varieties of oak as are French oak barrels, but are less expensive than the French just because of the "name." We are very careful when choosing the trees in these Eastern European locations, and then only use the bottom of the trees—the middle and top of the tree are not as good for barrels. Our wood is cut with saws—never split. We season the staves in our own yards for 18 months, restacking every 3 months for uniformity. If you don't do this, the seasoning is really uneven. Our toasting levels, light, medium, and heavy, are extremely uniform because we toast every barrel by hand over an open fire. We can do this because we have lots of coopers on staff. You know, the toasting process we use brings out lots of oak lactone and guaiacol…"

Name three things that are correct and three that are likely to be incorrect in what the salesman tells you. Explain each answer briefly.

(b) The outside air during the peak of harvest in this area is typically 85°F and 20% RH. If we want to maintain the relative humidity of the fermentation/barrel area at 80% RH (to maintain ethanol concentration) using just a misting system, what will the temperature be in the room? Use a psychrometric chart and show your work on the chart.

(c) If at the peak of harvest, we expect CO_2 to be released into the room at a rate of 164,430 L/hr, what flow rate of fresh air will be necessary if we want to keep the room below 3,000 ppm CO_2 for safety purposes? The fresh air in the Yosemite area typically has 0.05% CO_2 (= 500 ppm).

(d) What flow rate of water (e.g. L/min or gpm) will we need to mist into the air to maintain humidity and temperature at the desired levels while keeping a safe CO_2 concentration in the environment?

10,900 L air = 1 lb mole air

8.2 L water = 1 lb mole water

5. During the first harvest at Eagle's Pride, your Chardonnay turns out really well, but your marketing group feels that it would be even better if you decreased the alcohol from 13% to 12.5%.

(a) At first, you think about doing this during barrel storage. If the outside air at the winery is 98°F with a wet bulb temperature of 72.5°F, what (dry bulb) temperature would you have to chill the barrel room down to using air conditioning to decrease the alcohol? Explain your answer and show your work on a psychrometric chart.

(b) After establishing these conditions in the barrel room, your barrel room unexpectedly falls to 50°F at night (because of cold weather outside). How much water would you expect to find condensed on the floor and barrels the next morning in your 180,000 ft³ barrel room?

385.5 ft³ air = 1 lb mole air

8.2 L water = 1 lb mole water

(c) During the second vintage at Eagle's Pride, the marketing department still wants a lower ethanol level in the Chardonnay, but they decide that they want to lessen the influence of oak in the final product. Name two other unit operations (i.e. pieces of equipment) that you could use to lower the alcohol content of the finished wine and **briefly** explain how they work.

6. **[Integrative Problem]** At the Breakfast Winery, two separate rooms will be used for barrel storage. One will be used for Chardonnay barrel fermentation and subsequent aging, and the other will be used for aging the red wines.

   (a) In the Chardonnay barrel fermentation room, you are concerned that the air flow may not be sufficient to reduce the CO_2 concentration to below what you consider to be a safe level. The level that you choose is 3,500 ppm. The architectural company designing the winery tells you that they have designed the room for 5 room changes per hour (i.e. 5 room volumes of fresh outside air pass through the room each hour). If the barrel room holds 800 barrels and is 40 ft wide by 50 ft long by 22 ft high, is this a high enough air flow rate to maintain safe levels of CO_2? At the peak of fermentation, you can expect that each barrel will generate 65 L/hr of pure CO_2 and outside air contains 500 ppm of CO_2.

   $$1 \text{ ft}^3 = 28.3 \text{ L}$$

   (b) For the red wine barrel aging room, you want to take the outside air in Cloverdale (where the winery is located), which will be 90°F and 25% RH (Relative Humidity) and condition it to 60°F and 85% RH. How much energy will be needed to accomplish this conditioning (in Btu/lb mol air).
   (c) If you decide instead to keep your barrel room at 69°F and 85% RH, will you save energy? Show your work.
   (d) Will the ethanol level of the Cabernet and Petit Syrah go up or down in these conditions? Explain your answer briefly.
   (e) f your main goal in barrel aging your Cabernet is extraction of flavor components from the oak, should you plan to keep the wine in barrel for on the order of 6 weeks, 6 months, or 2 years? Explain your answer.

7. The barrel rooms for the new UC Davis Winery will have 6 room changes per hour as standard air flow. Humidity control will be accomplished by direct steam injection into the air flow.

   (a) What is the other, more typical means of controlling humidity in barrel rooms in wineries? Name two advantages of the direct steam injection system.
   (b) The average outside air in Davis during the summer is 95°F and 25% relative humidity. We want to maintain the temperature and humidity in our barrel room at 60°F and 90% relative humidity. If the room is 20 ft × 20 ft × 12 ft high, what flow of water is necessary to the steam generator to maintain these conditions.

$$385.5 \text{ ft}^3 \text{ air} = 1 \text{ lb mole air}$$

$$8.2 \text{ L water} = 1 \text{ lb mole water}$$

(c) Will the ethanol concentration go up or down in our barrels? Explain.

8. Initially, the Yi-Di Winery is planning to have a barrel program, and therefore, needs to plan for a barrel room. You get this assignment.

   (a) If the air outside in the He-Tao region of China is typically 80°F and 50% relative humidity during the hottest part of the year and you would like your barrel room to be at 62°F, what will the relative humidity be at this temperature if air conditioning is used to accomplish this cooling (i.e. no water added)? Show your work on a psychrometric chart.

   (b) How large will your air conditioning unit have to be? That is, what is the energy necessary to bring your barrel room down to its operating conditions?

   (c) In the first year of operation of the winery, there is a problem with a local power plant and you have rolling power outages. To maintain a cool and humid environment, you decide to cool the barrel room using only misters. If you want to maintain the same relative humidity as in Part (a), what will the temperature be in the room?

   (d) After operating for 2 years, the Yi-Di Winery decides to expand, but does not have space for a new barrel room. Therefore, they decide to explore oak alternatives to introduce oak aromas and flavors to their wines. You are asked to contact oak companies and explore the possibilities. The salesman from Stump Oak Alternatives comes to visit you with an array of products. First, he tells you that all of his products are better because they all come from the very bottom of the tree-within 4 ft of the bottom of the log. He tells you that he has staves, cubes, chips, and oak dust made from American oak and French oak (from various forests), seasoned 1 or 3 years. All products are toasted as staves and then cut. The salesman tells you that this gives extracted flavors that are more similar to those you get from barrels. The salesman tells you that they have a very rigorous quality control regimen for toasting that includes several temperature sensors reading the wood surface temperature during toasting—he tells you that this is the only effective way to reduce variability in toasting. Finally, he tells you, with a wink, that all of these products really impart the same flavors to the wine, even though the longer seasoning and French wood cost more. Evaluate his sales pitch by briefly discussing four points with which you agree and/or disagree.

9. **[Integrative Problem]** The Lakeshore Cellars project proves to be so successful in the first few years that you decide to expand your line and include a barrel-fermented Chardonnay. To do this, you have to expand and build a new barrel storage room just for this program. You decide to build this barrel room

underneath the crush pad, which has an area of 80 ft long by 40 ft wide by 32 ft high. This room should easily fit three hundred 225 L barrels with room to expand.

(a) We expect all of the Chardonnay to come in at about the same time (at 24 Brix) and for the maximum fermentation rate to be approximately 3 Brix/day. Total CO_2 evolution is expected to be 60 L CO_2 per L must over the course of the entire fermentation. How much fresh air would we need to bring in to keep the CO_2 levels below 4,000 ppm? How many room changes per hour does this correspond to (this is a common way of looking at HVAC)?

$$1 \text{ ft}^3 = 7.48 \text{ gal}$$

$$3.78 \text{ L} = 1 \text{ gal}$$

(b) If you are not worried about your alcohol levels lowering during fermentation and subsequent storage, what relative humidity should we maintain in this new barrel room? Explain your reasoning.

(c) As crush approaches, your new air conditioning unit for the barrel room has not arrived and it looks like it may not until several weeks into harvest. You call your neighbors and find that one of them has an old home air conditioner cooling coil that will provide 36,400 Btu/hr of cooling to the outside air you bring in. If the outside air during harvest in Tahoe City is a near record 85°F and 30% relative humidity, calculate the final enthalpy, H_{final}, you will achieve based on the air flow rate you found in Part (a) and the outside conditions. Assuming that you will not be adding water, estimate your final temperature and humidity level.

$$1 \text{ lb mole dry air} = 385.5 \text{ ft}^3 \text{ air}$$

(d) You are more concerned about the oak barrels you will use for the new Chardonnay line than for your previous wines at Lakeshore Cellars. Therefore, you have your assistant winemaker talk to as many cooperages as possible while at the Unified Grape and Wine Symposium in Sacramento in January. After talking to ten different cooperages, she tells you that she has found one that she trusts and has picked out barrels for the first year. She tells you that the salesman has told her that Hungarian oak and American oak are the same species of oak and both are inferior to French oak. For French oak barrels, the salesman has told your assistant winemaker that they have experienced coopers that know how much toast is on their barrels just by looking at them while they are on the fire, so they are very uniform. The salesman also said that as long as you stick with one of the "futaie" forests in France, the origin of the wood doesn't matter very much. Your assistant winemaker is excited because the salesman has helped her pick out special 100 L barrels that he says are better for longer aging than the more typical 225 L size. Help assess your assistant wine-maker's judgment by evaluating each of these four statements. Briefly explain why you agree or disagree with each one.

10. At Ekaya, they decide to build a new barrel room and ask for your help in designing it and helping to decide what barrels to use.

(a) You invite in three major cooperages to tell you about their barrels. The winemaker tells you that they want to have a quality wine product, but at their current price point, they need to be careful about how much they spend on barrels. Of the three cooperages that visit, the first one tries to sell you Baltic oak barrels. The salesman tells you that it is a cheaper alternative to French oak—he says that it is a good deal because it is the same wood anyway and higher in tannin than American oak. The second supplier is a long-time French producer that only makes barrels from French oak and specifically *Q. robur*. They pride themselves in doing everything, including toasting, in an old-fashioned way by hand. Every barrel from this company is unique, as each cooper at this cooperage has their own signature method for doing each step of the process. These are the most expensive barrels you are considering (by a factor of 2 over the next most expensive). The third supplier sells mainly American oak. She tells you that they have just set up a temperature control system for their barrel toasting which makes the process (and their barrels) very uniform. She also tells you that her barrels have extra large amounts of oak lactone, which will give the wine a lot of vanilla aroma. The price of the American oak barrel is in between the other two. Rank the order in which you would purchase barrels from these three vendors and explain your reasoning.

(b) The warmest time of year for the Stellenbosch (South Africa) is January, in which the average daily high temperature is 80°F with 50% Relative Humidity (RH). If you would like your barrel room to be 68°F and 90% RH, how much energy will be needed to supply your barrel room with fresh air at the right conditions? Back up your answer with calculations.

(c) Under these conditions, will the ethanol concentration go up or down while the wine is being stored in the barrel? Explain.

(d) How much water will you need to supply to the humidification system to make this work?

11. A new winery has just started up in the Bay Area town of Moraga. One of their wines that they have decided to barrel age is a Syrah. They have a small barrel room that will fit 40 barrels in a single level.

(a) In looking at the chemical data of the Syrah barrels from 6 months apart, they notice that the ethanol level has gone up slightly over time from 13.0% to 13.2%. They know that their fermentations were complete before placing the wine in barrel. You decide that this increase in ethanol could be due to the humidity setting in their barrel room. With your knowledge of humidity control and wine ethanol levels, what would you guess is the relative humidity in the barrel room? Explain.

(b) In August, the average daytime temperature in Moraga is 72°F and the average relative humidity is 60%. What is coolest that we can make this barrel room by just misting water, if we also want to avoid excessive mold growth on the barrel surfaces? Show your work on a chart.

(c) If instead, we decide to reach the same final humidity level as in (b) but using only cooling (without addition of water), what would the final temperature be? Show your work on a psychrometric chart.

(d) How much energy will be needed to bring the average August outside air to these final conditions (from Part (c))?

12. After years of using oak alternatives exclusively, Gunrock Winery decides to build a barrel room and start a barrel program for their premium products. Initially, they are planning for room for 3,000 barrels. The barrel room is approximately 500,000 ft³ (1,298 lb mol air).

(a) The average daily high temperature for the area where the winery is situated in the Central Valley is 96°F with 20% relative humidity during the summer months. If we wanted to maintain the barrel room at 60°F and 80% relative humidity, how much energy would this take for one air change (i.e. the volume of the barrel room)?

(b) Instead of having emitters for humidity control, you decide to use a more straightforward method that you've seen at another large winery. That is, you just water down the floor of the barrel room when the humidity is low and let the water evaporate off the floor. What volume of water would have to be placed on the floor to achieve the desired conditions (for each room change of air)?

$$8,172 \text{ g water} = 18 \text{ lb water} = 1 \text{ lb mol water}$$

(c) Name one advantage and one disadvantage of this approach to humidity control.

13. **[Integrative Problem]** After completing the design of Wickson–Hall Winery, other wineries start hiring you to consult on new winery projects. One of these wineries, Hey-Dude Winery, needs some help with the design of their barrel room. They will be making a Zinfandel and a barrel fermented Chardonnay as most of their production volume. As this winery is in the Lodi area, it is not uncommon for outside temperatures to get up to 105°F with 30% Relative Humidity.

(a) One of the partners in this winery (Mr. Hey) would like to reduce the alcohol level on the Zinfandel, which is around 15% v/v after the primary fermentation. What would be a good operating relative humidity to accomplish this alcohol reduction? Explain.

(b) What temperature should you bring the room down to with air conditioning in order to reach this humidity level without adding any extra moisture? Use a psychrometric chart.

(c) How much energy (per lb mole dry air) is needed for the air conditioning to get from conditions outside (in Part (a)) to the conditions inside (in Part (b))?

(d) Mr. Dude seems to be more interested in the barrel fermented Chardonnay than the Zinfandel. He would like to know how much fresh air flow is needed to maintain the carbon dioxide below 5,000 ppm. Assume that Mr. Dude has 500 barrels of Chardonnay in one of the rooms. On a typical day in the middle of the fermentation, all 500 barrels are fermenting at a rate of approximately 3 Brix/day, which corresponds to 70 L CO_2/hr barrel. What should the fresh air flow be?

10 Packaging and Bottling Lines

10.1 OVERVIEW OF PACKAGING LINES

There are many ways to package wine, beer, and spirits. The most common way to package wine and spirits is in a glass bottle with some type of closure such as a cork or screw cap, while glass bottles and aluminum cans are equally common for beer. However, there are many other common systems. These include bag-in-a-box, single-use plastic cups, TetraPaks, aluminum bottles, plastic bottles, and kegs. For now, we will focus on bottling lines, although goals for packaging are the same regardless of the format. That is, we need to get the product into the bottles or other packaging, avoid introducing chemical or microbial contaminants, fill the packaging evenly to avoid regulatory problems, and get the product information on the bottle that is required to attract consumers and fulfill governmental regulations.

10.1.1 REGULATORY ISSUES IN WINE PACKAGING

Several governmental regulations need to be considered for wine packaging. First, fill levels are governed by federal law. The applicable section is 27 CFR Part 24 Section 24.255b: "Allowable deviation from stated fill level." In this section, maximum deviations in fill levels are specified. For a standard 750 mL bottle, 2.0% variation in fill level is considered acceptable. This level is higher in 375 mL bottles (3%) and lower in large format bottles of 1–14.9 L (1.5%). This level of variation fluctuates with bottle size because the variation in fill on a bottling line is generally a fixed volume (i.e. number of mL), regardless of bottle size, which means there will be more fluctuation as bottle size gets smaller. It is not, however, acceptable to always fill a 750 mL bottle 15 mL low. An equal distribution of volumes above and below the target volume is mandated by this law as well. Finally, not more than ±0.5% aggregate deviation is allowable in six consecutive tax returns.

In addition to fill levels, label information must be approved by the appropriate authorities (like the federal TTB) prior to use. This includes an accurate description of grape variety, vintage, appellation, ethanol level, and the appropriate use warnings.

10.1.2 GENERAL CONSIDERATIONS FOR BOTTLING LINES

Bottling systems are made up of several components and these generally include a glass dumping station and conveyor system, glass rinsing units, fillers, corkers, capsule spinners, and labelers. Alternatives include screw caps instead of cork finish with a capsule and capsules that are heat sensitive instead of spun.

DOI: 10.1201/9781003097495-10

While bottling lines are becoming increasingly automated, there is still a range of automation existing in industry. On one end of the commercial spectrum (not counting completely hand bottled products), the unit operations are laid out in a U shape with glass bottles unpacked on one end and the empty cartons filled with packaged wine on the other. Unit operations are connected by conveyor belts, and each piece of equipment could be manufactured by a different supplier. Quality control functions of examining fill level and bottle integrity would be performed by hand, as would putting capsules on each bottle neck. With the unloading, loading, and QC functions, it would be typical to have five to six people working on this type of line to package hundreds of bottles per hour. The other end of the spectrum would be a fully integrated unit, optimized to have each operation coordinating with the other units. All of the QC functions and replenishment of raw materials would be automated. These units can process as much as 300 bottles/min or more, including automating the glass dumping, returning the finished product to the cartons, and palletizing. For small to medium size bottling lines, it is also possible to purchase a monobloc unit that puts all or many of the unit operations on one linear skid and may even allow easy crossover between closures.

For all of these units, it may be desirable to have the entire bottling line—or at least the fill and closure subunits when the product is open to the environment—in a positive pressure room with HEPA filtered air. By having a higher pressure around the filler, contaminants will tend to be pushed away from this sensitive environment.

Regardless of the type of unit or setup, the relative timing of the individual units will be critical, as a downstream unit operation that is too slow will cause backup of the bottles along the conveyor (i.e. the total bottling line speed is determined by the slowest step). Changes in bottle shape and size, as well as label type and size, may affect tooling of the whole line. Changeovers between products with different packaging can be painful and time consuming, not to mention expensive if using mobile bottling services. The impact of these changes, which may be suggested by a marketing group, should be discussed as early in the change process as possible.

10.2 GLASS DUMPING AND CONVEYERS

The first operation in the bottling line is the glass dumping station where the bottles are taken out of the cartons and placed on the conveyor. Care should be taken to make this table at a convenient height for workers. In addition, having a conveyor system that is at least as wide as a case of bottles will speed loading and reduce the manual operations needed to get the bottles onto the conveyor belt. Otherwise, bottles need to be manually pushed onto the conveyor belt 4 bottles at a time (Figure 10.1). Conveyance systems for empty cartons to be delivered to the bottle loading/palletizing area are also extremely useful, and may only require an inclined unpowered conveyor, though more complicated, powered systems are also common. Lubricants are often applied along the conveyor belt to facilitate bottles sliding onto and off of the belt to reduce bottles sticking and breaking during processing. Equipment that automates this part of the process is now starting to be implemented in industry. An example can be seen in Figure 10.2.

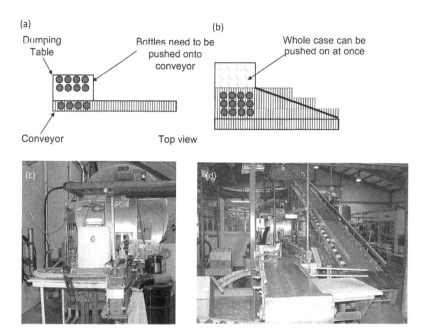

FIGURE 10.1 Comparison of glass dumping areas with single-bottle conveyor (a) or multiple-conveyor systems (b). The multiple conveyor system allows an entire case of glass to be dumped and moved onto the conveyors in a single movement. Examples of these systems are shown in (c) and (d).

FIGURE 10.2 Automated bottle dumping station. Glass is dumped out of cartons onto the conveyor belt and the empty carton is sent to the end of the line for packing filled and finished bottles.

10.3 GLASS RINSING

Bottles coming directly from a glass factory are generally clean and do not require any extensive treatment prior to use. However, these bottles can contain small dust particles from the cardboard carton in which they are packaged or elsewhere in the environment. The bottles are also obviously filled with air, and therefore a healthy dose of oxygen. For these reasons, it is common to "rinse" the bottles with an inert gas prior to filling them with product. There are various designs for accomplishing this, though all of them involve inverting the bottles over a gas inlet and injecting the gas for a time sufficient to displace the air present in the bottle. In a small bottling line, this can be done manually with a cylinder of the inert gas and a tip to direct the gas into the bottles efficiently. Alternatively, liquid nitrogen can be dosed into each bottle, with subsequent expansion filling the bottle. At a larger scale, several types of machines are widely used. One common type of rinser picks up the bottles in a rotating, inverting wheel. The rotational speed of the wheel determines both the throughput and the amount of gas delivered, as each bottle is engaged with the gas delivery from the "10 o'clock" to the "2 o'clock" positions. A more recent type of rinser picks up each bottle individually and inverts and sparges as the bottle travels around a horizontal carousel. At the end of its traverse, the bottles are righted and sent through a spacer to the filler. Both of these pieces of equipment can be seen in Figure 10.3.

Nitrogen is commonly used for rinsing/flushing. Argon may also be used. However, argon is considerably more expensive, and its use may not be warranted since air will largely be eliminated by a sweeping motion that will not be significantly affected by gas density. It is also important to note that after the bottle is filled with wine, residual inert gas will be limited to approximately 30 mL in the bottle headspace if using a screw cap, and likely about 15 mL if a cork closure is utilized. Studies that we have done comparing the use of nitrogen, argon, and carbon dioxide

(a)   (b)

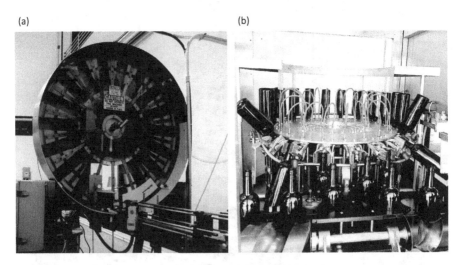

FIGURE 10.3 Rinsers. (a) is a vertical circular bottle rinsing station and (b) is a horizontal rinsing station.

gases during rinsing have shown no significant differences when evaluating the sensory characteristics of the packaged wine. Inadvertent introduction of oxygen at this point in processing, however, can significantly impact wine chemistry and aging.

10.4 FILLERS

Fillers are currently employed that use gravity, counter-pressure, or vacuum with counter-pressure. The choice of fillers used will depend on desired throughput and the properties of the product being packaged. Gravity fillers are the simplest of filler machines and fill based on a positive static head. Units used by home winemakers and brewers would fall into this category, but there are also units that are quite large and mechanized that use gravity as the motive force for filling (with anywhere from 8 to 40 spouts or more) (Figure 10.4). Increasing the number of spouts will increase the throughput as more bottles will be filled simultaneously. As the rinsed bottle enters the filler carousel, spaced to the correct spacing using helical or star wheel spacers, the bottle is lifted on a mechanical pedestal to engage with the filler head above the pedestal. These filler heads are spring-loaded, so the pressure of the pedestal pushing the bottle up against the head, opens the filler head and allows wine to begin flowing into the bottle. This filling continues as the bottle travels around the carousel. As the bottle arrives at the other side of the carousel, the pedestal lowers to disengage the bottle from the filler head. The filling is driven by the difference in static head or height between the wine in the filler bowl and the wine level in each bottle. Fill levels can be adjusted by changing this height difference, which will also change the timing of the engagement and disengagement of the filler head for each bottle.

It is important to note that while standard wine fillers can often be used to bottle slightly carbonated wine, beer and sparkling wine must be packaged in counter-pressurized fillers. As explained in Section 6.6, dissolved carbon dioxide concentration in wine and beer is an extremely strong function of the carbon dioxide pressure above the liquid. Pressurized fillers are connected to a carbon dioxide source, and actively

(a)   (b)   (c)

FIGURE 10.4   Multi-position filler. (a) The filler bowl supplies wine to multiple filler tips as the bottles travel around the filler on individual pedestals. (b) Star wheels space the bottles appropriately to move onto the pedestals and are bottle size and shape dependent. (c) Filler heads are spring loaded, opening and closing based on the position of the pedestal.

maintain the carbon dioxide partial pressure in the filler bowl headspace to ensure desired carbon dioxide concentrations immediately prior to bottling.

A pre-vacuum step can be added on the filler to remove air or gas from the bottle, though the rinser can serve a similar function. Electro-pneumatic filler heads, a newer technology, can be programmed to use different approaches for different products or packaging. Most fillers will accommodate changes in bottle width, bottle height, neck size, or other shape parameters, but each of these changes takes time and requires quality checking prior to starting up production runs. Canning lines will be sensitive to changes in packaging as well.

Many wineries will choose to filter the wine just prior to filling. This will likely stabilize the wine from a microbial point of view, as well as further clarify the wine. This is usually accomplished with a filter train including a 0.5 or 1 μm nominal pore size, depth prefilter and a 0.45 μm final membrane filter. Final membrane filters should be integrity tested prior to and after use for quality control using procedures described in an earlier chapter. Also prior to filling, the entire transfer line including the filters and filler (all the way to the filling tips) must be sanitized. As will be discussed later, this can be completed effectively using hot water or steam, or possibly chemical sanitizers, but in the latter case, care must be taken to assure compatibility with bottling line gaskets and other materials.

10.5   CORKERS AND CAPSULERS

If packaging is to have a cork finish, the next two steps are to insert the cork into the neck of the bottle and then cover the top of the bottle with a decorative capsule that covers the cork. For the corker, the goal is to get the cork into the bottle with a neutral or slightly negative pressure in the bottle. This will help to avoid corks "pushing" especially in warm or hot weather or during transit. To do this, the corker takes the filled bottle, compresses a cork, and then pushes it rapidly into the neck of the bottle in a single movement (Figure 10.5). While manual corkers do not have a means of pulling a vacuum, prior to this action, nearly all commercial corkers will have this

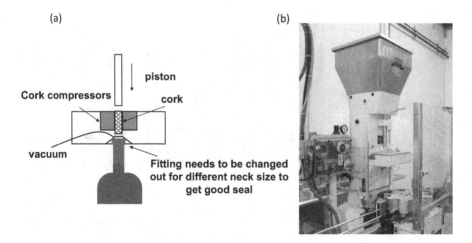

FIGURE 10.5   Corker. (a) Schematic of key components of a corker. (b) A single-head corker.

feature. A filled 750 mL bottle of wine will have approximately 30 mL of headspace prior to cork insertion. If the cork itself has a volume of 15 mL, this means that, in the absence of a vacuum draw, the headspace will be compressed to half of its volume, thus resulting in a final pressure of 2 atm. If a vacuum can be applied to the bottle just prior to cork insertion, the hope is that the final pressure will be 1 atm or less. This is actually a parameter that can be easily measured on site with a simple pressure gauge attached to a needle that can be inserted through the cork, and therefore can be used as quality control during preparation of the bottling run. In order for this vacuum draw to occur and occur with the right timing, the corker head will be equipped with a fitting specifically for the neck of the bottle being used to assure a good seal. Timing on a multi-head corker is accomplished by holes on the carousel placed in exactly the right place so they engage with the vacuum line just prior to cork insertion. If the bottle neck changes, this fitting also needs to be changed, again to assure a tight seal. After the corker, the product is sealed and there is no longer danger of introducing microbial contaminants.

The next operation is to place a metal foil or capsule over the neck of the bottle. The placement can be performed manually, usually by an operator that is also serving a quality control function on the fill level and closure. In larger operations, a register is filled with a stack of foils, and a flywheel picks up the foils one at a time with suction and places it over the top of the bottle. At this point, the foil is loose and will fall off easily if perturbed. In order to secure it in place, the foil will be spun down in the capsuler (Figure 10.6). This equipment uses a series of spinning rollers that surround the neck of the bottle to spin the foil down. By elongating the metal foil, the capsuler reshapes the foil and tightens it around the neck of the bottle. The goal is to do this reproducibly without creases, which often means adjustment of the rollers for each

(a)   (b)

FIGURE 10.6   Multi-position capsule spinner (a) with a close up of a spinner head (b).

(a)   (b)   (c)

FIGURE 10.7   Screw cap closures used with glass bottles. To use a screw cap finish, bottles need to have threads as part of the mold (a). The screw caps come without threads (b), and the threads are applied by rollers during application, along with a crimping at the bottom to assure a successful seal (c). Caps with internal threads require different equipment for the bottling line and create a flatter surface on the finished product.

new bottle shape. Like the other unit operations on the bottling line, throughput is increased by having a carousel with multiple heads that is spinning multiple bottles simultaneously.

An alternative to metal foils are plastic capsules that can be applied to the bottle and then shrink wrapped to the bottle by passing the neck of the bottle through a heater.

10.6   SCREW CAPS

Screw caps (Figure 10.7) are an alternative to a cork and capsule finish for bottles. With this type of finish, glass bottles must have integral threads. In the most common type of screw cap, closures do not come with threads, the bottles do. After the cap is placed on the neck of the bottle, a roller is used to seal the skirt of the cap and imprint the threads on the cap to match the glass. An alternative type of screw cap comes with internal threads and is torqued onto the thread of the glass. This type of cap has a different smooth look on the outside of the cap. All of these caps can be purchased with different liners including tin and special polymers that allow specific amounts of oxygen ingress depending on the wine chemistry and intended aging.

10.7   LABELERS

The goal for the labeler is to get all labels on straight and flat and evenly spaced (e.g. front, back, and neck, when used). Bottles that have large, sloped surfaces such as some sparkling wine bottles and Burgundy-style bottles present more of a challenge. This challenge increases as labels increase in size.

Two types of labelers are commonly used, ones using glued paper stock and ones using pressure-sensitive labels, which are essentially stickers that come on a roll with a paper backing. There are advantages and disadvantages to these systems. Glue-type labelers generally have higher throughput and are less sensitive to moisture on the outside of the bottle. With pressure sensitive labels, there is no need to worry about glue changing in consistency over time and the equipment is not as label specific. Up until the early 2000's, glue labelers were considerably more common, mostly because there were more options for paper stock, finishes, and inks. As the pressure sensitive label technology matured, options have grown to the point where they are similar or even more extensive than those of glue-style labels and are now used on the vast majority of bottles in California wineries.

In a pressure sensitive unit (Figure 10.8), labels are spooled on a paper backing. Like the filler and corker, labels are applied while the bottle is carried in a circle past the labeling head. A "knife" separates the label from its backing and applies the label to the bottle. Brushes assure that the applied label is flat against the bottle. The bottle is then turned 180 degrees if there is a back label and the label applied in an identical fashion to the front label. If bottles are embossed and the label needs to be applied directly beneath the embossing, a notch is included in the bottom of the glass mold so that the bottle can be oriented in the correct direction prior to label application. Inverse tapered bottles (where the shoulder of the bottle is wider than the bottom of the bottle) are fairly common and can be an extra challenge as they may need to be tipped slightly to apply the label correctly. The labels in these cases also need to be tapered or they cannot be applied straight across the bottle. For glue-type units, a stack of paper labels is fit into a register. Labels are lifted individually by the labeller brushed with glue, picked up by two metal "fingers" and pressed again the bottle

(a)   (b)

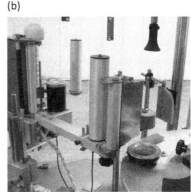

FIGURE 10.8 Multi-position self-adhesive labeler (a) with detail of front label application arm (b). In (b), from the back, the cylinder that holds the labels on their backing can be seen, followed by the "knife" that pulls the label off its backing as the bottle transits the correct position, a brush to flatten the label on the bottle and cylinders to pick up and spool the spent backing after label application.

using a form exactly the shape of the label in order to assure that the label will stay affixed. If the label size or shape changes slightly, this form needs to be replaced.

10.8 PALLETIZING

After labeling and all other packaging is completed, bottles are placed back into their cardboard cartons, either manually or using a vacuum lift device. Case size can vary based on winery or intended market. For example, cases are typically 6 bottles in Australia, but 12 bottles in the US. The boxes are then sealed and printed with filling data/SKU information. Cases are stacked in pallets either manually or using palletizing robots and stored until shipped.

10.9 WAREHOUSING

Pallets are stored in controlled temperature space until needed. For large wineries, this warehousing can be completely automated using robotics. In this case, pallets are stored on shelving units in high bay space and robots on rail systems or free driving are tasked with retrieving the correct pallets to complete a shipment.

PROBLEMS

1. The bottling line at Domaine D'Kaj is typical of many wineries in that it has a rinser/purger, filler, corker, capsule spinner, and labeler. The maximum capacities of the various units are shown below.

Unit	Maximum Capacity (bottles/min)
Rinser/Purger	300
Filler	80
Corker	40
Capsule	120
Labeler	60

   (a) What is the maximum speed of the entire line? Explain.
   (b) At what speed should the labeler be set? Explain.
   (c) At what speed should the filler be set? Explain.
   (d) If you could only buy one new piece of bottling line equipment in the next 10 years, which piece would it be and what capacity would you be looking for when you were ordering it? How much would your throughput increase with this new equipment?

2. You have been the winemaker at Chateau Ordinaire for 5 years. Your Chardonnay has traditionally been fairly oaky (made in 100% new medium-toast French oak barrels) and buttery (from 100% malolactic fermentation) and packaged in a traditional Burgundy-style bottle with a non-descript paper label. However, your sales are dropping in this very competitive market. Your marketing department decides to try a new concept. Your Chardonnay will now be

a more fruity wine with very little phenolic extraction. The packaging will also be updated to reflect this style change. The wine will now be packaged in sleek, new bottles with elongated, narrow necks. The labels will now be clear plastic. From the grapes arriving at the winery to the product leaving, list all winery **equipment** that might need to be changed or replaced to accomplish this makeover. Be as specific as possible and explain your reasoning for each change.

3. **[Integrative Problem]** Five years after leaving UC Davis, you get a job as an assistant winemaker at Commonwealth winery in Sonoma which is now up and running successfully. Up until this point, Commonwealth has used a mobile bottling line. However, after several years of not being able to bottle at the optimal time, the winemaker decides to buy her own used bottling line. She gives you this job. You find a rinser, filler, corker, foiler, and labeler—all from different sources. Their speeds are as follows in the table.

Unit	Speed (bottles/hr)
Criveller Poggio Rinser, 20 clamp	18,000
Bertolaso Epica Filler	6,000
GAI 6-head Corker	9,000
Criveller Nortan Capsule Installer/Spinner	30,000
GAI Rotative Labeler	3,000

(a) If Commonwealth is planning to bottle 45,000 cases per year, what is the fastest this can be accomplished using this pieced together bottling line? Explain. Note: Each case has 12 bottles.

(b) After hearing your calculation, the winemaker finds some extra money to upgrade two of the units. Which two units would you upgrade in capacity? Explain. How much time will the bottling take now?

(c) Commonwealth winery will be bottling their Prieto Field Blend and Zinfandel in inverse taper Bordeaux style bottles. Their Chardonnay will be in a slightly shorter and wider Burgundy style bottle. Name four parts of the bottling line that will need to be changed out to change bottles.

(d) Swabbing and liquid sampling of the filler shows that a previous owner had had some issues with Brettanomyces. You decide that you would like to make sure that this will not be an issue for you by sterilizing the filler with hot water at 80°C (= 353.15 K) for 30 min. If your testing indicates that the initial number of live cells is 80,000 for the entire filler, what will the probability of contamination be after this treatment? You can assume that Brettanomyces has an Arrhenius constant of 1.29×10^{13} min^{-1} and an activation energy of death of 21,350 cal/mol.

$$R = 1.987 \, \text{cal} / \text{mol} \, K$$

(e) After this initial sterilization with hot water, you decide to start a sanitization regimen using peracetic acid (PAA) at 5% w/v. What BOD can you expect going into your wastewater treatment system from this waste stream? Is this reasonable for an aeration pond to handle? Explain. PAA has a molecular formula of CH_3CO_3H.

$$1\%w/v = 10\,g/L$$

4. **[Integrative Problem]** In a third large room at Chateau de Pommes, you find a left over building system that creates a positive pressure in a small cordoned off booth in this room. You decide to put a used bottling line that you purchase and case goods storage in this large room.

   (a) The Chateau de Pommes Chardonnays and Pinot noirs will all have a cork finish with a capsule and will be bottled in identical Burgundy style bottles. They have a front and a back label that are pressure sensitive. You expect all of the wine to be sold in the US. Draw a block diagram of the unit operations that will be needed for this bottling line. Point out which one of the unit operations needs to operate the fastest and which the slowest.

   (b) If you can only fit two of the bottling unit operations in the positive pressure booth in this large room, which two would they be? Why is this? What if it had room for three unit operations? Explain.

   (c) After you get your bottling line up and running for the first time, you decide to sterilize the filters and filler heads to assure that you will not have a problem with contamination in your first vintage. You have heard that the previous owner of the bottling line had a huge issue with lactic acid bacteria, so you decide to target this organism for killing. You can assume that lactic acid bacteria have an Arrhenius constant of 3.8×10^4 min^{-1} and an activation energy of death of 8,300 cal/mol. Your swabbing of filter housings, filler bowl, and filler heads gives you an idea that you are starting with approximately 1,000,000 live cells. On this premier bottling, you would like to sterilize down to a 1/10,000 chance of contamination. If you are using 82°C water for sterilization, how much time do you need to achieve this level of kill?

$$R = 1.987\,cal/mol\,K$$

   (d) Right next to the positive pressure booth, you find an old steam generator. You fire it up and find that it works well. If you were to use steam at 100°C, how much more rapidly could you accomplish this same level of kill?

5. **[Integrative Problem]** After doing an excellent job figuring out the temperature and humidity control for the barrel rooms at Vests and Flannels Winery, you are hired by Casa Carnevale in Sonoita, Arizona to help with setting up a new bottling line. The average high temperatures in Sonoita are 92°F in July when you bottle with a relative humidity of 20%. You know that you will want a cool environment for the bottling line to protect your wine and make it more comfortable for the workers. Since the air is so dry, you decide you can potentially

use evaporative cooling (a swamp cooler), but you have had issues when the humidity is too high (>60% RH) as the adhesive on the labels does not work consistently.

(a) How much can you lower the temperature with evaporative cooling without jeopardizing the bottling? Show your work on the attached psychrometric chart.
(b) What volume of water would have to be added to each room change of air to achieve this temperature decrease if the bottling room is 50 ft long × 20 ft wide × 15 ft high.

$$1\,\mathrm{lb\,mol\,air} = 385\,\mathrm{ft}^3$$

$$8,172\,\mathrm{g\,H_2O} = 1\,\mathrm{lb\,mol\,H_2O}$$

(c) Casa Carnevale will produce four wines: (1) a Malvasia bianca to be bottled in a Burgundy style bottle with a screwcap; (2) a Malbec to be bottled in a Bordeaux style bottle with a cork and capsule finish; (3) a reserve Malbec to be bottled in an inverse taper Bordeaux style bottle with a cork and capsule finish; and (4) a GSM blend in a Burgundy style bottle with a screw cap. If you wanted to bottle all of your wines in a 3-week period, what order would you bottle your wines to minimize downtime? Between bottling runs, what equipment will need to be changed out?

11 Utilities

While the process equipment discussed in the previous chapters is critical to the production of wine, beer, and spirits, the utilities that support this production are just as important to operations. These utilities include equipment for cleaning and sanitization, wastewater treatment, refrigeration, and heating, ventilation, and air conditioning (HVAC). All are discussed in this chapter.

11.1 STERILIZATION, SANITIZATION, AND CLEANING

In the spectrum of cleanliness in a winery or brewery setting, three general levels can be defined. These levels are cleaning, sanitizing, and sterilizing. **Cleaning** is defined as the process of removing dirt and nutrients for potential growth of contaminants. **Sanitizing** is the process of reducing the microbial population to some level (e.g. kill/remove 99% of all viable microbes). **Sterilization** is the process of killing or removing *all* microbes to some probability. In a commercial winery, sanitization and cleaning are more important than strict sterilization, while commercial breweries will often target sterilization or sanitization of at least 99.999% of viable microbes. In a commercial distillery, cleaning is of main import, as fermented material will be rapidly rendered microbially stable via distillation.

11.1.1 STERILIZATION AND SANITIZATION

Three main methods are used for sterilization or sanitization. These are thermal inactivation, chemical inactivation, and filtration. Here, we will focus on the first two methods, as we have already described filtration in detail in an earlier chapter. We will begin with thermal inactivation, as it is the best studied and characterized quantitatively and then discuss chemical inactivation qualitatively.

Thermal inactivation of microorganisms can be described using first-order death kinetics. Mathematically, this means that the change in the number of viable organisms can be described as:

$$\frac{dN}{dt} = -kN \tag{11.1}$$

where N is the number of live organisms (NOT concentration) and k is the "death constant." Separating this differential equation and integrating, we get:

$$\int_{N_0}^{N} \frac{dN}{N} = -\int_{0}^{t} k\,dt \tag{11.2}$$

DOI: 10.1201/9781003097495-11

or

$$\ln \frac{N}{N_0} = -kt \qquad (11.3)$$

where N_0 is the initial number of viable microorganisms. Therefore, if the $\ln N/N_0$ is plotted versus time, the slope is equal to $-k$. Expanding on this, we can look at Figure 11.1 of a semilog plot of N versus time. It is clear at first that the viable cell concentration is being reduced from 10^4 to 10^3 and so on. When we reach 10^0, only one viable cell remains. Below this point, a fractional viable cell number exists. Physically, this fractional number corresponds to a probability of contamination. For example, $N = 10^{-3}$ corresponds to a 1/1,000 probability of contamination. The explanation for this stems from the fact that N is a number of cells and not a concentration. If we were to do this same sterilization starting with the same number of live cells 1,000 times for instance, overall, we would have collectively started the sterilizations with 10^7 cells (i.e. 10^4 cells $\times$ 1,000 sterilizations). We can see from this figure, assuming the slope of the line (k) stays the same throughout, that we would end up with $N = 1$ or one live cell in 1,000 sterilizations for that given time. From these arguments, we can derive our definitions of sanitization and sterilization. Any final N equal to or greater than 1 is sanitization and any N less than one will be sterilization.

If we look closer at the death constant, k, we will find that this constant is a function of the organism we are trying to kill and the temperature. The temperature dependence can be described using an Arrhenius dependence as in Equation (11.4):

$$k = \alpha e^{-E/RT} \qquad (11.4)$$

where α is the Arrhenius constant, E is the activation energy for death, R is the gas constant in appropriate units, and T is the temperature in absolute scale (i.e. °K or °R). The activation energy for death is highly dependent on the organism to be killed and typically ranges from 50 to 100 kcal/mol. This value is higher for spores and lower for vegetative cells. While the amount of desired kill will depend on the setting,

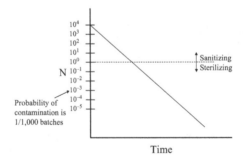

FIGURE 11.1 Thermal deactivation kinetics. N, the number of live cells, is plotted on a semi-log plot versus time. Achieving $N = 1$ is the cut off between sanitization and sterilization.

hot water at 82°C (180°F) for 20 min has been given as a guideline in wineries and breweries.

Example 11.1:   Bottling Line Sterilization for Traverse Winery (Trevor Grace)

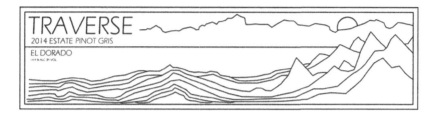

Traverse is a medium-sized winery located in El Dorado County in the Sierra Foothills. The region's prolific microclimates allow for a number of different varietals to be grown here. The 110 acre estate will mainly focus on Spanish and Rhone varietals in addition to well-known marketable varietals. Other varietals will be grown in small lots for the wine club and for blending use. Additionally, quality fruit will be sourced from neighboring areas such as Amador County, Calaveras County, and San Joaquin County to pad volumes or to make catchy lower-cost blends. Fermentations will be inoculated to ensure predictability and control. Whites will be made in stainless tanks and reds will be put in French and American oak barrels after malolactic fermentation. Some premium barrels will be purchased down in Napa after one vintage from reputable outfits to save on cost. 33% of the barrels will be new oak. The winery will target populations from the Sacramento, CA and Reno, NV regions. A satellite tasting room will be placed in Truckee, CA to capture affluent individuals, typically from the SF Bay Area, that often recreate at Lake Tahoe throughout the year.

Traverse means to travel across or through. The label depicts abstract topographic elements of elevation gain from flatland, to the Sierra Foothills, to the large Sierra Nevada Mountains that surround Lake Tahoe. The target consumer will be professionals with supplementary income ranging from 35 to 65 years old.

PROBLEM

Because Traverse is making so many different wines, they have chosen to have their own bottling line. They prepare their wines and hold them in a bottling hold tank. From there, they pump their wine through a 1 μm nominal pore size depth filter cartridge followed by a 0.45 μm membrane cartridge filter directly into the filler bowl of the bottling line (see figure below). To assure no microbial issues during bottling, the system is sanitized or sterilized with hot water from the inlet of the filters through the filler tips. Assuming the hot water temperature is 76.5°C and the organisms you are trying to kill have an activation energy of death of 42,500 cal/mol and an Arrhenius constant, α, of 1.22 × 1026 min-1, how long will we need to flow hot water through the filters and filler to get 99.5% kill (i.e. sanitization)?

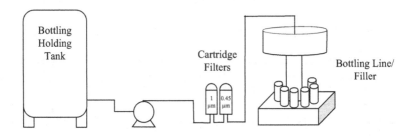

SOLUTION

To solve this problem, we start with the integrated equation for death kinetics:

$$\ln\frac{N}{N_0} = -kt$$

where

$$k = \alpha e^{-E/RT}$$

Substituting, we get:

$$k = 1.22 \times 10^{26}\,min^{-1} \times e^{\frac{-42,500\,cal/mol}{\left(1.987\frac{cal}{mol\,K}\right)(349.65\,K)}}$$

or

$$k = 0.331\,min^{-1}$$

For 99.5% kill, N/N_0 = 0.005. Rearranging the first equation

$$t = -\frac{1}{k}\ln\frac{N}{N_0} = \frac{-1}{0.331\,min^{-1}}\ln(0.005)$$

$$t = 16\,min$$

So, we need to run the hot water through the filters and filler bowl/filler tips for 16 min.

For 1/1,000 chance of contamination (i.e. sterilization), how much time would this take if we have 10^3 bacteria per filler tip (and an 8 tip filler)?

In this case, N/N_0 = 0.001/(8 × 10^3) = 1.25 × 10^{-7}. This is because N = 0.001 for a 1/1,000 chance of contamination and there are 8,000 live cells at the beginning of the process.

We can now use the same equations as above:

$$\ln\frac{N}{N_0} = -kt$$

$$t = -\frac{1}{k} \ln\frac{N}{N_0} = \frac{-1}{0.331\,min^{-1}} \ln\left(1.25 \times 10^{-7}\right)$$

$$t = 48\,min$$

Finally, what if we want to achieve this level of sterilization, but in 20 min. What temperature is needed? We again start with the same equation:

$$\ln\frac{N}{N_0} = -kt$$

or substituting:

$$\ln\left(1.25 \times 10^{-7}\right) = -k\left(20\,min\right)$$

$$k = 0.79\,min^{-1}$$

But,

$$k = \alpha e^{-E/_{RT}}$$

Taking the ln of both sides of this equation, we get:

$$\ln(k) = \ln(\alpha) - \frac{E}{RT}$$

or

$$\ln(k) - \ln(\alpha) = -\frac{E}{RT}$$

Rearranging and substituting, we get:

$$T = \frac{-E}{R(\ln k - \ln \alpha)} = \frac{-42,500\,cal\,/\,mol}{\left(1.987\,\dfrac{cal}{mol\,K}\right)\left(\ln 0.79 - \ln 1.22 \times 10^{26}\right)}$$

$$T = 354.7\,K = 81.5°C$$

It should be noted here that thermal inactivation generally happens in the presence of moisture (e.g. hot water or steam). In this scenario, the microbial death is caused by denaturation of proteins. However, if we were to heat a surface in the absence of moisture, the same amount of cell death would take on the order of 10 times longer.

This is because the predominant mechanism of microbial death would become oxidation. While heat inactivation purposely in the absence of moisture would be rare, this point is still important. One can imagine sterilizing a surface of a barrel or tank after use. If there is, for instance, a clump of dirt on the side of the vessel during sterilization, the outer part of the clump may get wet, but the inside of the clump could remain dry. Even if the clump of dirt reached the desired inactivation temperature, it is likely that microbial inactivation would not be complete. If the clump were later to fall into the process stream, this could cause contamination issues. Therefore, it is imperative to clean thoroughly prior to sanitization or sterilization to assure the intended results.

Finally, this analysis assumes non-porous surfaces such that organisms on the surface of a piece of equipment instantaneously reach the temperature of the surface that they are on. The use of oak barrels complicates this analysis, as organisms can penetrate the wood to a depth of almost 1 cm. Steam introduced into the barrel does not instantly heat the inner part of the stave. In modeling the heat transfer and cell death in this scenario (Yao, et al., 2021), we found that it can take nearly 90 min to reach 85°C at a 0.8 cm depth into the stave. However, for common wine microbes like *Saccharomyces* and *Brettanomyces*, steaming for 12 min was sufficient to achieve a 5-log reduction in live cells, even at 0.8 cm into the stave. Of course, longer steaming times will achieve a greater kill, and organisms do exist in a winery environment that will take longer than this to kill with steam.

11.1.2 CHEMICAL SANITIZATION

For a given chemical sanitizer, effectiveness depends on concentration, temperature, and contact time. Unlike thermal inactivation, killing does not always increase with temperature. For instance, above 50°C, chlorine-based sanitizers may decrease in activity as the chlorine gas volatilizes. These effects will be specific to the sanitizing agent. Several common classes of chemical sanitizing agents are available for wineries. These include chlorine-based compounds, iodophores, quaternary ammonia compounds, peracetic acid, and ozone. None of these sanitization methods (including thermal inactivation) work well on "dirty" equipment as noted above. While we give a brief overview of these sanitizing agents here, one can find a much more in-depth assessment of these agents in texts such as Marriott et al. (2018).

One of the most common types of sanitizing agents is bleach or other chlorine containing solutions. Bleach is a sodium hypochlorite solution, but the active form of this sanitizing agent is undissociated hypochlorous acid, which predominates at a pH of around 4. At lower pH, degradation to chlorine gas occurs, and at higher pH, the anionic form predominates. It works by a mechanism of oxidation but loses activity rapidly above 50°C as the solubility of chlorine gas decreases. As mentioned in an earlier chapter, long-term use of chlorine containing compounds is not compatible with stainless steel. These sanitizing compounds have dipped in popularity as residual chlorine has been linked to the incidence of TCA (trichloroanisole) issues.

Iodophores are made up of molecular iodine (I_2) and non-ionic wetting agents. The I_2 is the active ingredient. While they are more compatible with stainless steel than chlorine-based sanitizers, they are also more expensive and still sensitive to pH and high temperatures (>50°C) like chlorine-based sanitizers.

Quaternary ammonias or Quats are ammonia molecules with various organic molecules substituted for the hydrogen atoms. While the activity is not quite as broad as chlorine-based sanitizers, they are active against most of the microbes critical to wine and beer production. They work by disrupting the cell membrane of microbes. They tend to be good at penetrating porous surfaces and are stable when heated.

Another sanitizer very commonly used in wineries is peracetic (or peroxyacetic) acid. Peracetic acid is a combination of acetic acid and hydrogen peroxide. It is temperature stable, effective against biofilms, and has low toxicity—it breaks down to water, oxygen, and acetic acid. Depending on the concentration used, it may not require rinsing after treatment.

Finally, ozone (O_3) is another common sanitizing agent. Because ozone is unstable, it must be generated on site. While ozone is a gas, when used in a winery, it is mixed with water to make an ozonated water solution that is used for sanitization. It works by oxidation. While ozone has been used safely in municipal wastewater treatment facilities for decades, its use in a winery can be more problematic if proper training and monitoring are not in place, as prolonged exposure to the ozone gas can be quite dangerous.

11.1.3 CLEANING

There are five main factors affecting cleaning in a winery. The first four are the chemistry (and concentration) of the cleaner, cleaning temperature, contact time, and turbulence (or physical action). The fifth factor is the surface characteristics of the equipment to be cleaned. However, only the first four factors can be varied to clean the equipment successfully while accounting for the fifth factor (covered at the beginning of this text).

11.1.3.1 Chemistry

Several types of cleaners exist and are used in an industrial setting. These are alkalines/bases, complex phosphates, surfactants, acids, and chelating agents. Each of these categories of cleaners has its advantages in terms of specific types of cleaning (i.e. sequestering, wetting, emulsifying/suspending, dissolving, saponifying, peptizing, dispersion, and rinsing). A summary is given in Figure 11.2 (after UC Davis extension publication). Many proprietary cleaning agents are mixtures of several of these types of cleaners. A choice of cleaner must be made based on knowledge of the dirt to be encountered in your facility and the equipment and surfaces to be cleaned. It should be noted that your choices may also affect your wastewater treatment facility by adding BOD (e.g. citric acid) or potential growth inhibitors (e.g. HOCl).

11.1.3.2 Temperature

Temperature choice may also be critical in the cleaning of winery or brewery equipment. While higher temperatures will generally enhance the cleaning action of many chemicals, high-temperature cleaning of a protein-rich solution will tend to "bake" this residue onto the surface as the proteins denature. In this case, ambient temperature rinses are utilized first, followed by subsequent steps at elevated temperatures.

Property/Function	Strong Alkalies	Mild Alkalies	Poly- Phosphates	Mild Acids	Strong Acids	Surfactants
Sequestering	0	+	++++	0	0	0
Wetting	+	++	+	+	0	++++
Emulsifying/suspensing	+	+++	++	0	0	++++
Dissolving	++++	+++	++	+++	++++	+
Saponifying	++++	+++	0	0	0	+
Peptizing	++++	+++	+	++	+++	0
Dispersion	++	+++	+	+	0	+++
Rinsing	+++	+++	++	+	0	++++
Corrosion	++++	+++	0	++	++++	0
Effectiveness; Extreme, ++++ ; high, +++; moderate, ++; low, +; none, 0;						

FIGURE 11.2 Cleaning agent comparison. All cleaning agents (across the top row) are able to clean some type of dirt. No one type of cleaner is effective against all types of dirt. Therefore, commercial cleaning agents are general combinations of several types. (After UC Davis Enology Briefs)

11.1.3.3 Current Methods for Varying Contact Time and Turbulence

Given a certain chemistry and temperature, methods must be devised to vary contact time and turbulence. Several general methods employed in the industry include "fill and flush" (basically soaking a tank or piece of equipment in/with a static liquid), high-pressure water/chemical pumps with manual spraying wands, portable recirculation loops, and hand scrubbing. Any of these methods can be effective (with the possible exception of soaking) as long as the equipment design allows all areas to be reached. Precautions must be taken to minimize worker contact with cleaning agents.

11.2 CLEAN-IN-PLACE SYSTEMS

Clean-in-Place or CIP systems are extremely common in brewery settings, and while rare, are increasingly popular in wineries as well, especially in packaging facilities. This technology began in the dairy industry and is widely used in the pharmaceutical industry and other food/beverage sectors. The general idea for this type of system is that one (or a small number depending on the size of the facility) cleaning skid is located in a fixed, centralized location in the facility. All cleaning agents and water used for cleaning are piped through this system to the equipment that needs to be cleaned. After hooking up a piece of equipment, all cleaning is automated. The skid, as described below, is able to vary chemistry, temperature, contact time, and turbulence, all of the factors that can be used to optimize cleaning for a particular piece of equipment or "target."

11.2.1   JUSTIFICATION FOR CIP

Several reasons exist for implementation of CIP systems over the other methods discussed previously. First, the process of CIP cleaning is significantly less labor-intensive with at least the same, if not better, results in most cases. Second, the cleaning is more uniform—the cleaning will be performed exactly the same way every time. Third, less waste is generated because less water and cleaners are used, especially when recycling is used. Finally, CIP systems are generally safer as there is less contact with the cleaning agents, no dismantling of large equipment, and no climbing inside tanks to scrub. The main downside to this type of cleaning is the extra capital outlay for the unit initially.

11.2.2   SKID DESIGN AND TYPICAL CLEANING CYCLES

Various designs have been employed in CIP systems. However, all of them share some common features (Figure 11.3). Almost all CIP systems have a hold tank (some have multiple ones for each solution). This tank is used as a reservoir for cleaning water, as well as a mixing tank for the automated preparation of cleaning solutions such as bases and acids. From this hold tank, a pump controls the flow of cleaner at an appropriate flow rate for the "target" through a heat exchanger in order to supply cleaner at the correct temperature. After passing through a strainer, the stream is then sent out to the target where it enters sprayballs or nozzles that direct the cleaner toward the top and sides of the tank. This provides the necessary turbulent sheeting action. Water returns through the target drain to the CIP skid via a CIP return pump or eductor system (or both). At this point, the cleaner can be recycled to the hold tank or sent to drain depending on the level of dirt in the return.

Any CIP design would have to provide for effective cleaning cycles that control all four cleaning factors. A typical cleaning cycle for a CIP unit would be:

(1) Pre-Rinse: Water at room temperature
(2) Alkaline Wash: Hot cleaner at 50–80°C
(3) Post-Rinse: Water
(4) Acid Rinse: Citric acid to neutralize base, remove mineral deposits, and perform a mini-passivation
(5) Final Water Rinse: Can be hot to promote more rapid drying.

Different cleaning targets in your production facility may require different cleaning cycles, so it is best to have a unit that allows each cleaning cycle to be targeted toward the dirt expected at each process step.

11.2.3   SIZING FLOWS FOR CLEANING

In order to clean pipes in the facility using CIP, flow rates will generally be maintained at greater than 5 ft/s to assure turbulent flow. For tanks, it has been found experimentally that good cleaning action can be achieved when a flow rate is used that is 2–2.5 gpm/ft of tank circumference for each cleaning solution. For instance, a

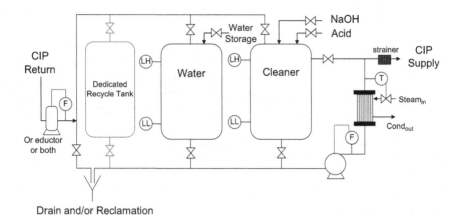

FIGURE 11.3 Schematic of a typical CIP skid. CIP systems have different numbers of tanks depending on their design. However, all should be able to control temperature, concentration, time, and turbulence—the four factors that affect cleaning.

6 ft diameter tank would require a flow rate of 38–47 gpm. The discharge pressure at the spray device in the target should be about 25–30 psig (with a higher pressure at the CIP pump discharge to account for head loss—e.g. 50–80 psig). In order to maintain efficient pumping, the flow rates for the smallest tanks are usually increased to maintain a ratio of highest flow rate/lowest flow rate of less than 2. If this is done, however, it is important to assure that drains are sized for this increased flow rate. Otherwise, puddling will occur during cleaning and the dirt will tend to rise to the surface of the liquid puddle, making cleaning significantly more difficult.

11.2.4 TESTING FOR ADEQUATE CLEANING

Several methods exist for verifying cleaning. Liquid samples can be taken from the equipment during cleaning and then assayed for pH, conductivity, juice/wort/wine/beer residuals, or cleaner residuals. Swabbing of a fixed equipment wall area using a cotton swab can also be performed. Another interesting method now widely used in the pharmaceutical industry is Riboflavin testing. In this test method, a dilute solution of Riboflavin (a non-toxic vitamin) is sprayed over the entire surface to be cleaned. Riboflavin fluoresces when a UV light is shined on it, so complete coverage can be assured. Then, a cleaning cycle is performed using the standard method (whatever that method might be). Finally, the UV light is used again to see if the cleaning method has reached all the necessary surfaces. It is important at this point to search for evidence of "shadowing." This is the term that describes the phenomenon where a baffle, for instance, might keep a sprayball from uniformly covering a tank wall. In wine and beer tanks, manways and sensors often cause shadowing that needs to be identified and remedied, often by the addition of new, specifically targeted spray devices. While the Riboflavin method tests for coverage, it does not assure cleaning efficacy as do the other assays for potential contaminants or residues.

11.3 WASTEWATER TREATMENT

The use and availability of water for processing is rapidly becoming a limiting factor in the operation of many wineries and breweries and in the development of new facilities. Current industry standard practice, when measured, is to use approximately 4–6 volumes of water to make one volume of wine. Companies that have really focused on water use have been able to reduce this to as little as 2–2.5. Breweries, which are less location dependent, can be built in areas with substantial water reserves, and often operate at a ratio of 2–3 volumes of water per one volume of beer (this includes the water used to actually constitute the beer!). However, this remains a large amount of water. Nearly all of this water (outside of the water that ends up in a final beer) is from cleaning and therefore contains spent cleaning and sanitizing agents, in addition to dirt from processing. In order to recover this water, or even to release it safely back into the environment, it needs to be processed. This processing could be as straightforward (for the producer) as releasing it to a municipal wastewater treatment facility, though in these cases, there are typically strict limits on the composition and characteristics of water entering these facilities (e.g. pH, organics, salts, temperature, etc.). For facilities located in rural areas, municipal wastewater treatment is likely not an option. Therefore, production facilities will need to install and operate their own facilities, even if the final treated water will just be released in the environment (e.g. in a leach field), in order to minimize impact on the environment and follow all local laws and guidelines.

As will be discussed in this chapter, it is critical to remove compounds from the water that lead to consumption of oxygen, either biologically (biological oxygen demand or BOD) or chemically (chemical oxygen demand or COD), as reduction or elimination of oxygen in a stream, for instance, can have disastrous implications on the native ecology. It is also important to adjust pH to be compatible with the surrounding environment in which the water is used or released, and remove or deal with all other solids, dissolved or not, that will affect the intended use of the treated water. In designing water treatment facilities, it is critical to understand all local, state, and federal regulations associated with water treatment and discharge. Because these regulations will vary with location, we will focus here on the technical issues of treatment.

11.3.1 SOURCE AND NATURE OF PLANT PROCESSING WASTEWATER

Some form of waste comes from every step of wine processing. Figure 11.4 (after Storm, 1997) gives an overview of the sources and nature of wine processing waste streams.

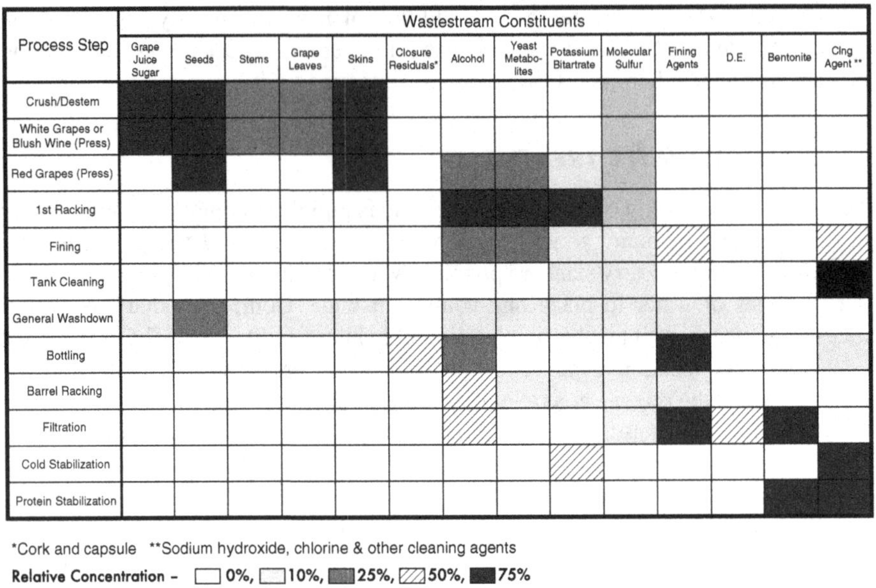

Process Step	Wastestream Constituents													
	Grape Juice Sugar	Seeds	Stems	Grape Leaves	Skins	Closure Residuals*	Alcohol	Yeast Metabolites	Potassium Bitartrate	Molecular Sulfur	Fining Agents	D.E.	Bentonite	Clng Agent**
Crush/Destem														
White Grapes or Blush Wine (Press)														
Red Grapes (Press)														
1st Racking														
Fining														
Tank Cleaning														
General Washdown														
Bottling														
Barrel Racking														
Filtration														
Cold Stabilization														
Protein Stabilization														

*Cork and capsule **Sodium hydroxide, chlorine & other cleaning agents

Relative Concentration – ☐ 0%, ☐ 10%, ■ 25%, ▨ 50%, ■ 75%

FIGURE 11.4 Sources and nature of wastewater in winemaking. Components of winemaking are listed across from the top of the table and winery operations and listed down from the side. M. Ogawa after Storm.

A summary of a typical composition of wastewater during harvest is given below:

TABLE 11.1
Typical Harvest Winery Wastewater Composition*

Composition Parameter	Level
BOD or COD	3,000–7,000 mg/L
Settleable Solids	5–20 mL/L
Suspended or Dissolved Solids	400–500 mg/L
PH	3–5

* From Storm (2001).

Though this is a typical harvest composition, the variation in composition throughout the year can be extreme. The wastewater system, therefore, needs to be able to handle this type of fluctuation. The ultimate goal for the treatment system is obviously to remove as much of the waste as possible and leave pure water. In reality, wastewater systems can be expected to remove 75–95% of BOD and COD.

11.3.2 BIOLOGICAL OXYGEN DEMAND (BOD)

Most organic components can be aerobically degraded almost completely into water, carbon dioxide, nitrate, phosphate, and sulfate. The quantity of oxygen consumed in this process, the "biological oxygen demand" or BOD (sometimes also called the biochemical oxygen demand), is usually expressed as mg O_2/L and was traditionally measured over a 5 day period. Today more rapid assays can be used, but it is also useful to be able to calculate the theoretical BOD associated with a waste stream in order to guide design or operation of the wastewater treatment facility.

The theoretical BOD of an organic compound with the generic empirical formula, $C_cH_hO_oN_nP_pS_s$ is estimated using the following equation:

$$\frac{moles\ O_2}{mole\ compound\ i} = c + \frac{h}{4} - \frac{o}{2} + 1.25n + 1.25p + 1.5s \qquad (11.5)$$

which is derived from the stoichiometric relationship for the oxidation of a given compound to CO_2, H_2O, H_2SO_4, HNO_3, and H_3PO_4. That is, we are simply solving for the stoichiometric coefficient a in this relationship:

$$C_cH_hO_oN_nP_pS_s + aO_2 \rightarrow cCO_2 + bH_2O + nHNO_3 + pH_3PO_4 + sH_2SO_4$$

using elemental balances on oxygen and hydrogen. The units of this form of BOD are not particularly useful from a practical point of view. Therefore, this number can be converted to more commonly used units using these expressions for BOD:

$$\frac{mg\ O_2}{g\ compound\ i} = \frac{moles\ O_2 \times \left(32g\ O_2/mol\ O_2\right) \times \left(1000\ mg/g\right)}{mole\ compound\ i \times \left(MW_i\right)}$$

$$\frac{mg\ O_2}{L} = \frac{moles\ O_2 \times \left(32g\ O_2/mol\ O_2\right) \times \left(1000\ mg/g\right)}{mole\ compound\ i \times \left(MW_i\right)} \times \frac{g\ compound\ i}{L} \qquad (11.6)$$

where MW is the molecular weight in g compound/mole. In order to find the BOD of a waste stream with more than one compound, the BOD is found for each compound and all of the contributions are summed to give the total BOD for the waste stream.

This provides a maximum, conservative estimate, of the BOD and in some systems, it may not be able to be degraded completely. Table 11.2 shows the oxygen demands for several juice and wine components, together with those of some cleaning chemicals. It is important to note here that juice and wine are usually diluted as they enter the wastewater stream (from cleaning), but clcaning solutions will enter the wastewater system at nearly full strength, making them a key factor in the design and sizing of the treatment system.

11.3.3 Chemical Oxygen Demand (COD)

In addition to the BOD, there will be some components, such as phenols and sulfur dioxide in juice and wine, that will undergo chemical oxidation more rapidly than they will be broken down microbially. They will still use up oxygen in the process and this is called the "chemical oxygen demand," or COD. This measure is quite pH sensitive since most ionizable components have only one oxidizable form. As wastewater pH may vary widely over time, COD may also vary widely over time.

As a note of clarification, some in the waste treatment field have used a Chemical Oxygen Demand test, which takes 3 hr to complete (as opposed to 5 days for BOD) to estimate BOD for a sample. For wineries, BOD is generally on the order of 40–80% of the measured COD. Lower percentages indicate a stream that is difficult to breakdown with microbial populations, while higher percentages represent a stream that is highly biodegradable.

TABLE 11.2

Table of BODs for typical winery wastewater components.

BOD Calculations For Common Juice, Wine, and Cleaning Constituents

Class	Name	Chemical Formula	C	H	O	N	P	S	BOD (moles of O_2/mole compound)
Juice/Wine	Glucose	$C_6H_{12}O_6$	6	12	6	0	0	0	6
Component	Fructose	$C_6H_{12}O_6$	6	12	6	0	0	0	6
	Tartaric Acid	$C_4H_6O_6$	4	6	6	0	0	0	2.5
	Ethanol	C_2H_5OH	2	6	1	0	0	0	3
	Acetic Acid	$C_2H_4O_2$	2	4	2	0	0	0	2
Cleaning/	Peracetic	$C_2H_4O_3$	2	4	3	0	0	0	1.5
Sanitizing	Acid								
	Citric Acid	$C_6H_8O_7$	6	8	7	0	0	0	4.5
Inorganic	Phosphoric	H_3PO_4	0	3	4	0	1	0	0
Acids	Acid								
	Nitric Acid	HNO_3	0	1	3	1	0	0	0
	Sulfuric	H_2SO_4	0	2	4	0	0	1	0
	Acid								
Inorganic	Sodium	NaOH							0
Bases*	Hydroxide								
	Potassium	KOH							0
	Hydroxide								

* inorganic bases will not be broken down further during microbial waste water treatment, so BOD is zero.

11.3.4 SETTLEABLE SOLIDS

Settleable solids are solids that are large enough and dense enough to settle out over time in the waste treatment system simply by gravity as described in the earlier chapter on clarification. These solids may include material such as live or dead microbes (biomass), hop cones or hop solids, trub, spent grains, grape solids, or inorganic precipitates. It is possible, and sometimes desirable, to increase the amount of settleable solids in the waste stream using flocculating agents. In this manner, particles that might otherwise remain suspended, perhaps because of their size or density, will be large enough or dense enough to settle out with gravity in a reasonable amount of time.

11.3.5 SUSPENDED, DISSOLVED, OR SOLUBLE SOLIDS

Once all BOD is broken down and settleable solids have settled out, what is left over is soluble or suspended solids. The soluble solids are essentially the inorganic ions left from BOD breakdown, along with inorganic chemicals added during processing (e.g. sodium from cleaning with NaOH). In most cases, these will still be in solution in the water leaving the treatment system. Suspended solids would be solids that are too small or with too low a density to settle out in a reasonable amount of time (e.g. the retention time of the aeration and/or settling ponds). Some of this latter group can be removed by the addition of flocculants to convert them into settleable solids.

11.3.6 pH and Buffering

The pH of the treated water stream will also have to be within certain limits, typically between 6.0 and 8.0. As wine and juice are significantly lower than this in pH, it is clear that there will be a need to adjust pH during processing. Similarly, acidic and basic cleaning and sanitizing chemistries will also cause fluctuations in wastewater pH. Overall chemical use, as well as the nature of typical waste from cleaning, will also determine the buffer capacity of the waste stream which will have to be considered as pH is automatically adjusted en route to the wastewater pond.

11.3.7 Amount of Waste Generated

Storm (2001) gives some typical numbers for winery waste flows in his book. The figure in this book for waste streams is approximately 0.06 gal/day for each case of wine produced per year. That is, a 25,000-case winery will have a peak daily process wastewater discharge on the order of 1,500 gal/day. Using these numbers, a "small" winery can be defined as one producing less than 42,000 cases/year (2,500 gal/day). Below this amount, subsurface septic tank/leach field systems can be installed. Above this level, above ground aeration ponds must be used.

Brewery wastewater production is about equal to the amount of beer produced in a year, and wastewater production rates can be calculated based on peak production rates, or by total production days in a year. For example, if a 15,000 barrel per year brewery operates 200 days in a year, the wastewater production rate will be 15,000 barrels/year, or 2,300 gal/day (1 beer barrel = 31 gal).

Distilleries can either be part of a winery/brewery (in which case the fermentation facility must also be included), or stand-alone. A stand-alone distillery will generate roughly 3–4 gallons of stillage (ethanol-free still bottoms) per gallon of 80 proof spirit produced. Note that this is an absolute volume rather than a rate; such that the rate of wastewater production is tied to production rate.

11.3.8 Wastewater Treatment for Small Facilities

In many small wineries and breweries, the treatment of wastewater has often been performed by septic tanks and/or leach fields. These options are not suitable for even moderate wastewater flows. A typical treatment system for a small facility is shown in Figure 11.5 from Storm (2001). The septic tanks in this case are preceded by some device for gross solids entrapment (Figure 11.6) so that the septic tanks do not get clogged with large solids. Typical devices of this sort include rotary drum screens, inclined screens, or self-cleaning strainers. Of these, inclined screens are perhaps the simplest, as they do not have moving parts. However, they do require a minimum static head for effective use (a driving force for the solids to roll off the screen while the liquid filters down through the mesh). This may require an extra pump to provide the head unless the production facility is sited on a hill that allows for gravity to provide the necessary head. Rotary drum screens are also an excellent option as they consist of a simple cylindrical rotating screen (similar in some ways to a destemmer basket) to separate gross solids from liquid. Actually, removable basket strainers in winery floor

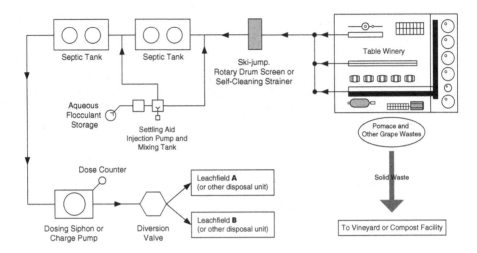

*Optional disposal by means of commercial septic tank pumping service
**Supernatant liquid to septic tank and leachfield system

FIGURE 11.5 Wastewater treatment facility for a small (<42,000 cases/year) winery. M. Ogawa after Storm (1997).

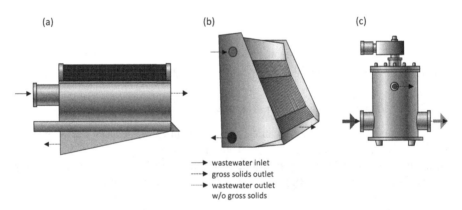

→ wastewater inlet
---→ gross solids outlet
......→ wastewater outlet
        w/o gross solids

FIGURE 11.6 Gross solids entrapment for winery wastewater systems. Three common gross solids entrapment methods are shown: (a) a rotary drum screen, in which wastewater enter in the center of a rotating cylindrical drum, and the solids exit the cylindrical screen at the opposite end; (b) inclined screen, in which wastewater enters at the top of the "ski jump" screen, water falls through the screen and the solids tumble down the screen to a catch basin; and (c) a self-cleaning strainer, in which wastewater enters through the center bottom of cylindrical screen and passes through clarified, while solids accumulate on the inside of the screen and are periodically cleaned off the screen automatically (M. Ogawa).

drains may be the best and simplest alternative as long as they are frequently emptied into solid waste disposal on a regular basis (and not just emptied into the floor drain that leads to the wastewater system!). Figure 11.5 shows two subsurface septic tanks in series. These systems are usually designed so that most of the BOD breakdown and settling will occur in the first of these septic tanks and the second tank is mostly there to protect against surges and as a backup. This way, regular maintenance and emptying can be focused on the first of the septic tanks. The septic tanks are followed by duplicate leach fields. Leach fields allow the water to percolate back down into ground water, thus making the soil effectively a very large depth filter for the water. The effectiveness of leach fields may be limited by soil permeability and the extent of soil saturation. Therefore, it is important to have experienced professional help site the wastewater system, but especially the leach fields.

11.3.9 WASTEWATER TREATMENT FOR LARGER FACILITIES

For wineries, breweries, and distilleries producing greater than 40,000 cases, an aeration pond system is a necessity. An overview of one of these systems is shown in Figure 11.7. In this type of system, a gross solids entrapment system is again needed as with the smaller systems above. The options for these larger systems are the same as for the smaller systems. The aeration pond, itself, can be thought of in much the same way as a fermentor as it basically is designed to provide an optimal environment for microbial degradation of BOD. Therefore, for efficient operation, the system needs to have a way to correct pH, add enough nutrients, maintain an acceptable

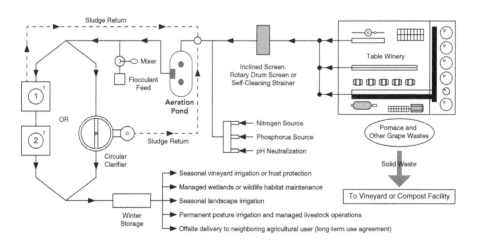

*Optional disposal by means of commercial septic tank pumping service
† Primary ① and secondary ② clarifier

FIGURE 11.7   Wastewater treatment facility for a large (>42,000 cases/year) winery. This type of system utilizes an aeration pond and secondary clarifier before sending the water to a storage pond. M. Ogawa after Storm (1997).

temperature, and supply oxygen and mixing. Some of these actions can be taken as pretreatments, and some are accomplished in the pond itself.

11.3.10 POND AND CLARIFIER DESIGN

Usually, ponds are rectangular in shape with rounded ends. The length to width ratio is generally 2. Operating height is about 4 ft with another 2 ft for freeboard (to account for unexpected fluctuations, waves, etc.). Deeper ponds will be difficult to aerate effectively. Sides should be sloped not steeper than a 2:1 (horizontal:vertical drop). Two or more aerator/mixers will be used to provide oxygen and mixing. The total working volume of the pond should be sized for a residence time of 5–8 days (with the lower end of the range used if sludge will be returned to the aeration pond from the clarifiers). For example, if the wasterwater output is 5,000 gal/day, the pond should be about 25,000 gal. A weir in the side of the pond regulates flow into the next stage clarifiers. Details of the design of an aeration pond can be found in Figure 11.8. An aeration pond should be thought of like any other fermentor and given the appropriate nutrients (BOD 5,000 mg/L, Nitrogen 250 mg/L, Phosphorus 50 mg/L, dissolved oxygen 3–5 mg/L), temperature (65–75°F), and pH (6.5–7.5) for a successful fermentation. The types of organisms that grow in this type of aeration pond are shown in Figure 11.9, though they will certainly change over time and with fluctuating nutrient levels and composition. Commercial inoculum can be added to the aeration pond (similar to the idea of buying active dry yeast to inoculate your beer or wine fermentations) or organisms in the environment of your aeration ponds can be relied on for natural inoculum. As these ponds are open to the environment, you will likely end up with some of the latter organisms anyway.

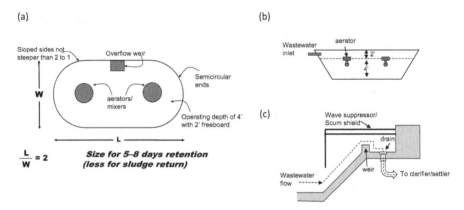

FIGURE 11.8 Typical aeration pond design. (a) General design parameters for an aeration pond (top view). (b) Side cutaway of an aeration pond with positioning of the aerators and inlet. (c) Detail of overflow weir and wastewater discharge to clarifier. The scum shield/wave suppressor hinders microbes/BOD on the pond surface from discharging directly to drain prior to appropriate residence time and degradation.

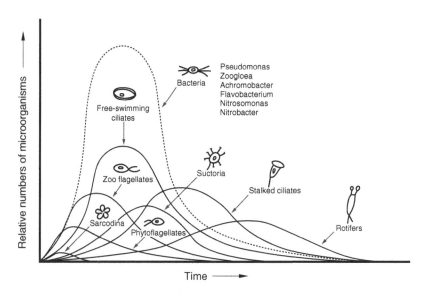

FIGURE 11.9 Relative abundance of microbes in a wastewater pond over time. M. Ogawa after Metcalf and Eddy (2014).

Clarifiers are used downstream of the pond(s) to separate the settleable solids from the clarified water. Settled "sludge" can be recycled to the pond to maintain an active inoculum. As noted above, this will reduce the required pond residence time. Figure 11.10 illustrates such a clarifier. From the clarifiers, the water will go to a holding pond until it is used.

It should be emphasized here that wastewater flow is highly seasonal. If the aeration ponds are working really well during harvest and then wastewater flow slows in the winter, the residence time in the aeration ponds will lengthen. If these times get too long, organisms in the pond will use up all available BOD and then start to die off or settle. If a bolus of BOD then arrives at the pond, the inoculum in the pond will likely be too low to get effective BOD degradation. This is one of the reasons that wineries have seasonal problems with their wastewater systems. This problem can be solved by having multiple smaller ponds in parallel. All of the ponds can be active during harvest and then fewer can be operated during times when less waste is generated in order to keep the ponds operating efficiently.

Because we are essentially trying to optimize a fermentation process in our wastewater system, people operating these facilities also need to be vigilant of waste streams that can disrupt the optimal conditions for cell growth in a pond. Excessive use of cleaners or sanitizers that inhibit cell growth is one example. Another interesting example is the generation of very dilute waste streams. Leak testing new barrels with water, for instance, can generate waste with essentially no BOD, as can rainwater entering process drains in outside facilities. Both of these situations should be avoided when possible.

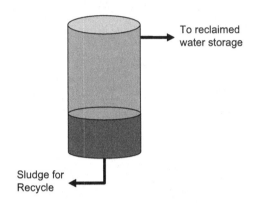

To reclaimed
water storage

Sludge for
Recycle

FIGURE 11.10   Schematic of a typical clarifier design.

If expansion is a future possibility for a winery, these plans should be considered when designing the original waste treatment facility. Be sure that space exists for the necessary size of waste treatment to handle the expansion plans on the production side, especially close to the original wastewater systems.

11.3.11   ANAEROBIC PONDS

The anaerobic degradation of organic components is far less common in wineries and breweries, even though it has the potential to be a more compact treatment process. The requirement for deep ponds, the formation of unpleasant smells, and the seasonal nature of the treatment requirements, have limited the acceptance of this approach in winemaking. However, many large commercial breweries and distilleries (including industrial ethanol plants) have figured out that an enclosed anaerobic digester can not only reduce wastewater BOD but can also generate methane (from the anaerobic microbial digestion of BOD). This methane can be used to fire boilers or generate electricity. There are some larger wineries that are now taking this approach.

11.3.12   USES FOR RECLAIMED WATER

Reclaimed water from the wastewater system can be used in combination with winter runoff for several purposes. These include seasonal vineyard irrigation and frost protection, managed wetlands, seasonal landscape irrigation, pasture irrigation, fire protection, and offsite delivery to agricultural users (using long-term contracts). The best uses are not always obvious, as winery water generation does not always coincide temporally with potential uses.

Example 11.2:   Wastewater Treatment for the Arboreal Winery (Rian Strong)

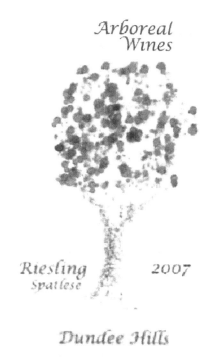

Arboreal Wines

Riesling
Spätlese   2007

Dundee Hills

Arboreal is a winery focused on locally based, sustainable practices from vineyard to bottling. The goal for this winery is to be negative: water-negative, carbon-negative, and electricity-negative. The name is derived from the wineries support of reforestation groups and a commitment to reversing the damage done to the environment by non-sustainable practices.

To become electricity-negative alternative energy sources should be utilized including solar panels or wind harvested electricity. If sufficient quantities of either are installed the initial capital investment can be recouped within a finite time-frame during and after which excess electricity can be sold back to energy provid-ing companies (PG&E; municipal utilities, etc.). This revenue stream could possibly generate enough income to support the higher costs of other, more eco-friendly practices such as use of alternative fuels (discussed later).

In order to minimize transportation costs and energy outputs fruit for the wines produced should be sourced from vineyards as close as possible without sacrificing fruit quality. Surrounding the winery are 75 acres of sustainably grown vines. Wherever possible, vineyard practices should include the use of biocontrols to pro-vide maximum vine fruitfulness and minimize pests' effects on fruit quality, that is, cover cropping, beneficial insects, etc. Use of eco-friendly chemicals should be used only when absolutely necessary. Any fruit not sourced from the adjacent vine-yards should be shipped by renewable alternative, cleaner-burning fuel-based vehicles, that is, biodiesel and electric (solar generated). Vehicles used on-site should be similarly run to reduce greenhouse gas (GHG) emissions and also reduce the winery's carbon footprint. Electric ATVs should be acquired for vineyard management. To further reduce the company's carbon footprint a portion of proceeds should be donated to reforestation groups. Use of alternative fuels and reforestation contributions should offset GHG emissions to make the winery carbon negative.

To increase water efficiency, vineyards should be dry farmed and the winery should be constructed to allow for minimal water use. Winery procedures should include processes that mitigate water waste such as sweeping up debris before spraying, etc. Rainfall capture systems should be installed to further reduce dependency on outside water resources.

Gravity flow should be implemented to reduce unnecessary pump use. Winery location should be chosen to provide a tiered system by which the product flow through successive unit operations occurs descending latitudes. In descending order (altitude): grape receiving/triage, crusher/destemmer, stainless steel fermentors, wine processing tanks, and finally barrel storage.

Consumption of non-local products is also a concern for maintaining ecologically sound procedures. Therefore, fermentations will be wild (grapes naturally have yeast present so why pay for yeast and contribute to our C-footprint with its transportation?); the wines will be unfined and unfiltered (again, minimizing use of products sourced far away and reducing equipment costs); Barrels used will be new, American oak (sourced closer than France) or 1- or 2-year-old barrels purchased from neighboring wineries.

PROBLEM

We need to size the wastewater treatment system for this 40,000 case winery. In order to allow for some planned expansion, however, we will actually size the system for 50,000 cases.

SOLUTION

We will design for a wastewater treatment system that uses aeration ponds for BOD degradation. Generally, this system will look like this:

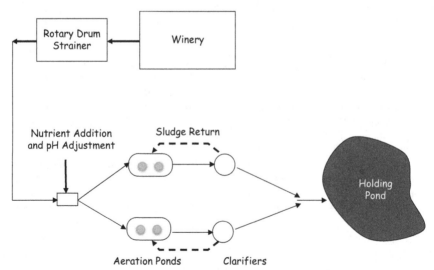

We will use two parallel aeration ponds during harvest and then use only one of the ponds for the rest of the year.

The wastewater output during harvest can be estimated using the rule of thumb from the Storm book:

$$Q_{wastewater} = \frac{0.06 \frac{gal}{day}}{case} \times 50,000\,cases = 3,000\,gal\,\frac{wastewater}{day}$$

With this flow rate, we can now size the aeration ponds. We can assume that the residence time needs to be 5 days (with sludge return).

$$V_{pond} = Q_{wastewater} \times t_{residence} = 3,000\,\frac{gal}{day} \times 5\,days = 15,000\,gal$$

Therefore, we should use 2 ponds, each 7,500 gal or 1,003 ft³. With this volume, we can get the approximate dimensions of each pond. If the pond looks like this:

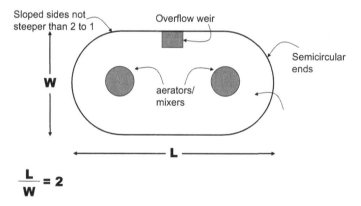

we can simplify to a rectangular shape with straight sides.

$$V_{pond} = W \times L \times D$$

We will choose to use a standard depth, D, of 4ft. If we want to keep L/W = 2 for good mixing, then L = 2W. Substituting,

$$V_{pond} = W \times 2W \times 4\,ft = 1003\,ft^3$$

Rearranging, we get

$$W^2 = \frac{1003\, ft^3}{8\, ft} = 125.4\, ft^2$$

Therefore, W = 11 ft, so L = 22 ft, and the depth is fixed at 4 ft. We add 2 ft of freeboard to account for waves or surges, so the total depth will be 6 ft. For the actual pond, we would likely use sloped sides as this will be cheaper to line with concrete. The volume calculations would be somewhat more complicated, however.

As we are thinking about the wastewater system, it is important to think about the BOD that might come out of the winery during harvest. Let's look at what the BOD would be if there was actually a juice spill. To do this, let's assume a simplified juice composition of sugar and tartaric acid:

Chemical	Formula	Concentration (g/L)
Sugar	$C_6H_{12}O_6$	240
Tartaric Acid	$C_4H_6O_6$	6

Now, let's figure out the BOD for the sugar:

$$BOD_{sugar} = c - \frac{o}{2} + \frac{h}{4} = 6 - \frac{6}{2} + \frac{12}{4} = \frac{6\, moles\, O_2}{mol\, sugar}$$

$$BOD_{sugar} = \frac{6\, moles\, O_2 \times \dfrac{32\, g\, O_2}{mole\, O_2} \times \dfrac{1000\, mg}{g}}{mole\, sugar \times \dfrac{180\, g\, sugar}{mole\, sugar}} \times 240\, \frac{g\, sugar}{L}$$

$$BOD_{sugar} = 256,000\, mg\, BOD\, /\, L$$

which is huge! Now, we can do the same calculations for the tartaric acid:

$$BOD_{tartaric} = c - \frac{o}{2} + \frac{h}{4} = 4 - \frac{6}{2} + \frac{6}{4} = \frac{2.5\, moles\, O_2}{mol\, tartaric}$$

$$BOD_{tartaric} = \frac{2.5\, moles\, O_2 \times \dfrac{32\, g\, O_2}{mole\, O_2} \times \dfrac{1000\, mg}{g}}{mole\, tartaric \times \dfrac{150.1\, g\, tartaric}{mole\, tartaric}} \times 6\, \frac{g\, tartaric}{L}$$

$$BOD_{tartaric} = 3,198\, mg\, BOD\, /\, L$$

and

$$BOD_{Total} = BOD_{sugar} + BOD_{tartaric} = 259,198\,mg\,BOD\,/\,L$$

This means that we will need to dilute the juice 52x to get to 5,000 mg BOD/L for the aeration pond.

Another scenario worth considering is the BOD of the cleaning/sanitizing solutions that we might use in the Arboreal Winery. Using the formulae above, we can find that the BOD of nitric acid (HNO_3) is 0 moles O_2/mol HNO_3, citric acid ($C_6H_8O_7$) is 4.5 moles O_2/mol citric acid, and peracetic acid ($C_2H_4O_3$) is 1.5 moles O_2/mol peracetic acid. Nitric acid has no BOD because it is already in its mineral form. Since the aeration pond is used for converting organic material to minerals, carbon dioxide, and water, an inorganic acid will not be further broken down by the microbes. Citric acid and peracetic acid will be an important source of BOD, because they will not be diluted prior to going down the drain (as juice or wine would likely be).

11.4   REFRIGERATION SYSTEMS

Refrigeration is critical to the operation of wineries and breweries. In wineries, it is used for grape chilling, juice or must chilling, temperature control in fermentors, temperature control in barrel rooms, and for cold stabilization of wines. In breweries, it used for chilling wort and beer and controlling fermentation temperature, in addition to conditioning room space (sometimes used as an alternative to cooling jackets on fermentors or lagering tanks). Refrigeration is not used as much in distilleries. Decisions on operations, for example, the use of must chilling, can have a huge impact on the choice and sizing of refrigeration. While there may be conditions and geographic areas that would lessen the need for mechanical refrigeration, most wineries and breweries will utilize these systems for critical cooling.

11.4.1   THE REFRIGERATION CYCLE

To understand refrigeration, it is important to begin with the refrigeration cycle. This is common to all refrigeration systems. To describe the refrigeration cycle (Figure 11.11), we will start with refrigerant flowing through the expansion valve. The expansion valve takes liquid refrigerant at a high pressure and expands it to a liquid at a lower temperature and pressure. The liquid refrigerant then flows through the evaporator where it is evaporated into a low temperature gas. The heat used to evaporate the refrigerant is the heat that you are removing from your process (i.e. this is the cooling coming from your refrigeration system). This low-temperature and low-pressure gas is then compressed in the compressor to a high-pressure, high-temperature gas. Finally, this high pressure gas is condensed in the condenser using air or possibly cooling water depending on the needed load and environmental

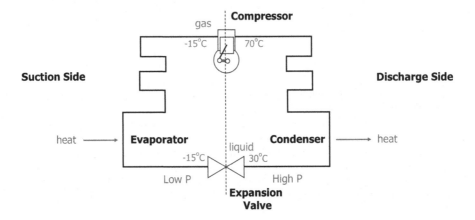

FIGURE 11.11   Schematic of a refrigeration cycle.

conditions. The cycle then repeats. In this way, heat is taken up by the refrigeration system at low temperature and released from the system at higher temperature (more or less ambient). The low-pressure, left side of the refrigeration system is called the suction side, as this is the suction side of the compressor, while the right side of the cycle is called the discharge side. The temperatures and pressures are a function of the refrigerant chosen for the application and the refrigeration requirements of the production facility (and, of course, the design of the refrigeration system).

Two kinds of refrigeration systems are possible in a winery or brewery setting and both are used. These are direct and indirect refrigeration systems (see Figure 11.12). In a direct refrigeration system, the evaporator is the process equipment. In other words, low-pressure, low-temperature refrigerant is passed directly through tank jackets and heat exchangers, evaporating to supply the chilling capacity needed. As the low temperature gas returns to the refrigeration, it enters the compressor to complete the cycle. In an indirect system, an extra coolant loop containing a heat transfer fluid like water, glycol, or brine is used to carry the cooling capacity of the refrigeration to the process equipment. That is, this coolant is used to evaporate the refrigerant in the evaporator of the refrigeration system and then is used to provide the cooling in process jackets and other heat exchangers. Because there always needs to be approximately 5°C between the streams in a heat exchanger to provide a sufficient driving force, an indirect system will need to operate about 5°C cooler than a direct one. Figure 11.12 illustrates this principle, using cold stabilization in a fermentation tank as an example. If you want to maintain the fermentor at 5°C, the evaporator in the direct refrigeration system needs to operate at a maximum of 0°C, whereas the evaporator in the indirect system will have to operate at −5°C to accommodate the two heat exchangers necessary. For this reason, the direct refrigeration system will be more efficient, as the cycle can be run at a higher temperature (generally 5–10°C) as discussed below in more detail. On the other hand, the indirect system negates the necessity of having potentially hazardous refrigerants piped throughout the production facility.

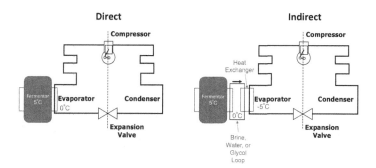

FIGURE 11.12 Two types of refrigeration systems. In a direct refrigeration system, the evaporating refrigeration fluid directly cools the target tank as it expands. In the indirect system, the evaporating refrigerant cools a secondary heat transfer fluid, such as glycol, which then chills the tank. The indirect refrigeration system needs to operate colder, which will reduce efficiency, but process piping will not be as critical as the refrigerant will not be piped through the entire production facility.

11.4.2 REFRIGERANTS

A number of refrigerants are available. All refrigerants share some common characteristics—they need to have temperatures for evaporation and condensation that are reasonable and useful at moderate pressures and they need to have high latent heats of vaporization. The latter point is important as less refrigerant will be needed to provide the same refrigeration, thus making the system more compact. Each refrigerant is given an "R" number, even if it has other common uses. For instance, ammonia is R717. Other common refrigerants are Freons R12 (CCl_2F_2) and R22 ($CHClF_2$). While still used as a refrigerant, ammonia is actually poisonous, so that refrigeration system leaks can be extremely dangerous. Freon is odorless and colorless. Freons are less toxic to humans. However, they have been shown to deplete the ozone layer in the atmosphere and have, therefore, been banned (e.g. R12) or are being phased out (R22) in favor of other alternatives (such as R-134a, CH_2FCF_3). The pressure-temperature relationship of a wide variety of refrigerants can be found in Figure 11.13.

11.4.3 SIZING A REFRIGERATION UNIT

The most common units used to describe refrigeration are "tons of refrigeration." A ton of refrigeration is equal to 12,000 Btu/hr. The term is defined as the rate of heat transfer required to melt a ton of ice in 24 hr. A ton of refrigeration, used as a unit, is abbreviated tonR, so 1 tonR = 12,000 Btu/hr.

There are various ways to estimate the size of a refrigeration system needed for a new winery or brewery, though a detailed analysis of the intended processing will always be important. One simple rule of thumb is that 1 tonR is needed for each 1,000 cases of wine or beer to be produced. For instance, a 40,000 case winery will require a 40 tonR refrigeration unit. Another way to estimate refrigeration needs for wineries is to use a prediction of grape delivery during the course of harvest to

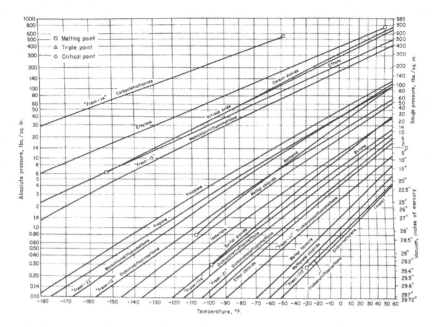

FIGURE 11.13 Saturation pressure versus temperature curves for common refrigerants. Perry's Chemical Engineers' Handbook (1997).

calculate peak fermentor refrigeration loads on each harvest day based on expected fermentation kinetics. Anticipated must chilling needs can also be added into this estimate. The refrigeration system can then be sized based on the peak anticipated load (plus some additional capacity to handle unexpected surges in refrigeration needs). Because refrigeration loads will be seasonal for wineries, it is possible to have multiple refrigeration units on site and only bring them online as needed. That is, at peak harvest, all of the units will be operating and at other times of year, a fraction of the units will be in operation. If the peak refrigeration period is particularly short, it may even be worthwhile to rent excess refrigeration capacity for that short peak, though having the capacity in house if a peak comes at an unexpected time may be worth the piece of mind.

11.4.4 THE EFFECT OF SUCTION OR EVAPORATOR TEMPERATURE ON SYSTEM CAPACITY

The temperature at which the refrigeration system operates can have a large effect on system capacity. If we look at Figure 11.14a (after Rankine), we can see that the same refrigeration system can remove twice as much process heat at a slightly higher temperature (e.g. 0°C instead of −15°C). Of course, if we need our process to be at 0°C or below, we cannot operate at 0°C, and we have no choice but to run our refrigeration at the lower temperature and therefore accept lower capacity.

Given this, it is important to understand why lower refrigeration suction temperatures give lower system capacity. To do this, let's picture the piston active in the

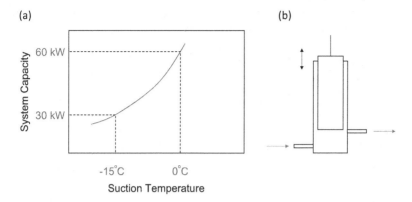

FIGURE 11.14 (a) Diagram of how a system's total cooling capacity greatly increases with small increases in suction temperature. (b) A piston in a compressor.

compressor (and pictured in Figure 11.14b). Every time the piston rises, a fixed volume, V, of refrigerant gas is drawn in. This volume has a number of moles of refrigerant, n. Therefore, we can examine the concentration n/V and see that the higher this concentration, the higher the molar or mass flow rate of refrigerant through the refrigeration cycle. A higher mass flow rate of refrigerant will give us a higher system capacity, given a constant latent heat of vaporization for the refrigerant. So, the question is then, how does n/V vary with temperature? To find this answer, we can use the ideal gas law:

$$\frac{n}{V} = \frac{P}{RT} \tag{11.7}$$

where P is system pressure and T is gas (in this case, suction side) temperature. At first, the result may seem counter intuitive, as it seems like a higher temperature will give a lower n/V and therefore capacity. However, the key here is that P and T are not independent. The T is the boiling point of the refrigerant at a specific P. As can be seen in Figure 11.13, this is not a linear relationship for any refrigerant. For every increase in T, the P associated with that boiling point increases exponentially, thus dramatically increasing system capacity.

11.4.5 Assessing Refrigeration System Efficiency

When examining the refrigeration cycle, one can see that the energy applied to the refrigeration system is through the compressor and the cooling capacity of the system is through the evaporator. Therefore, it makes sense to look at the efficiency of how the compressor work is converted to chilling capacity. The engineering term for this ratio is the coefficient of performance (COP):

$$COP = \frac{useful\ heat\ extracted}{power\ consumed} \tag{11.8}$$

In order to further examine the COP, we can perform an energy balance on the refrigeration system as shown diagrammatically in Figure 11.15. Here, the work supplied by the compressor is W_{in} and the heat removed (chilling capacity) is Q_L, while the heat discharged at higher temperature is Q_H.

An energy balance over the system gives us:

$$Q_L + W_{in} = Q_H$$

or rearranging:

$$W_{in} = Q_H - Q_L$$

Using the same nomenclature, we can define COP as:

$$COP = \frac{Q_L}{W_{in}}$$

or substituting:

$$COP = \frac{Q_L}{Q_H - Q_L}$$

If the system is working at its maximum theoretical efficiency, it can be shown that this relationship is equivalent to:

$$COP = \frac{T_L}{T_H - T_L} \qquad (11.9)$$

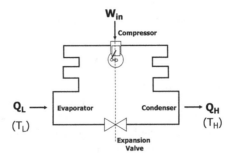

FIGURE 11.15 Energy balance around a refrigeration cycle to calculate the coefficient of performance (COP). Energy is drawn into the cycle during evaporation and compression, and energy is removed from the cycle during condensation.

where T_L and T_H are the temperatures of the evaporator and condenser, respectively, in absolute temperature. It can be seen from this equation that the COP will decrease with decreasing evaporator or suction temperature or increasing condenser temperature. While Equation (11.8) will give the maximum theoretical COP for a system, given the system temperatures, the actual COP will be somewhat lower given heat losses due to friction, inefficient heat exchange, and inefficiencies related to converting electrical energy to mechanical energy in the compressor.

11.5   HEATING, VENTILATION, AND AIR CONDITIONING (HVAC)

Wineries tend to be a mix of indoor and outdoor operations, while breweries are almost exclusively indoors. Large distilleries are entirely outdoors with the exception of offices and control rooms, while smaller distilleries can often be located indoors.

For indoor spaces, it will be necessary to have some type of HVAC system. HVAC systems control the temperature and humidity in rooms, as well as potentially the pressure if positive pressure is required (e.g. for a bottling line). The temperature and humidity are especially important for barrel rooms as discussed in an earlier chapter. Here, we focus on perhaps the most important aspect of HVAC in a winery or brewery and that is control of carbon dioxide concentrations at levels safe for operators. Buildup of carbon dioxide from fermentations in a closed workspace is extremely dangerous, as exposure to high levels of this gas and displacement of oxygen can cause a person to lose consciousness or even die. Alarms should be in place in a winery environment that alert workers to low oxygen and/or high carbon dioxide levels, but this should be a backup to the HVAC working properly and keeping a safe environment.

11.5.1   WORKPLACE CARBON DIOXIDE LEVELS

Carbon dioxide comes from yeast converting sugar to alcohol and carbon dioxide. Two moles of carbon dioxide are released for each mole of sugar used. This will amount to approximately 60 L of carbon dioxide to be produced per liter of must for a 240 g sugar/L wine fermentation. While this amount is liberated over the entire course of fermentation, at the peak of fermentation it is conceivable that 20% of this carbon dioxide will be produced in a single day. The building HVAC must bring in enough fresh air to dilute the carbon dioxide down to acceptable levels. This level is understood to be less than 5,000 ppm. As will be seen in the next section, this will require a large amount of fresh air and special air flow patterns to sweep out carbon dioxide, as it is heavier than air and will settle to the ground where operators will be most impacted.

In climates such as California and many more globally, outside temperatures can be quite high during harvest. Thus, bringing in large amounts of fresh air to maintain safe carbon dioxide levels will either result in an uncomfortably warm work environment or an extremely high energy load to air condition incoming air at a high rate. Simply venting tanks directly outside would alleviate this problem. This is more common in breweries than in wineries. Of course, as a safety backup, the building HVAC should still be able to evacuate carbon dioxide if the vent manifold fails for some reason.

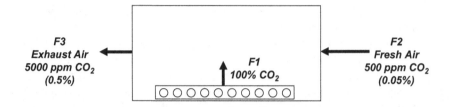

Total Flow Balance: F3 = F1 + F2
CO2 Balance: 0.0005 F2 + 1.0 F1 = 0.005 F3
F2/F1 = 220 = flow of fresh air/flow of CO₂ from tanks

FIGURE 11.16 Material balance around a winery for CO_2 evolving and being removed by sweep air.

11.5.2 Necessary Fresh Air Flow Rates

Fresh air flow rates needed to maintain safe work environments can be extremely high. To calculate just how high, we can use a mass balance approach. Figure 11.16 illustrates the mass balance for gas flow over an entire winery (or brewery).

In this case, tanks are producing gas flow (F1) that is essentially pure carbon dioxide, fresh air is being introduced with a flow rate F2 at 500 ppm carbon dioxide (0.05%), and air is being exhausted at a flow rate of F3. The maximum allowable carbon dioxide concentration in the winery is 5,000 ppm (0.5%). We would like to know how much fresh air must be brought in to keep the work environment safe, given the carbon dioxide evolved during fermentation.

We start with an overall mole balance (equivalent to a volume balance given constant temperature and pressure with no reaction).

$$F1 + F2 = F3$$

We can also perform a balance for carbon dioxide specifically.

$$0.0005F2 + 1.0F1 = 0.005F3$$

Substituting for F3, we get:

$$0.0005F2 + 1.0F1 = 0.005(F1 + F2)$$

Rearranging,

$$0.995F1 = 0.0045F2$$

Therefore,

$$\frac{F2}{F1} = 220 \qquad (11.10)$$

This means that for every volume of carbon dioxide formed in the fermentors, we need to bring in 220 volumes of fresh air to keep a safe work environment. Given the amount of carbon dioxide produced in fermentation, this will be a substantial flow that may require conditioning to keep a comfortable work environment. It will also represent a substantial energy cost.

REFERENCES

Marriott, N.G., Schilling, M.W., and Gravani, R.B. *Principles of Food Sanitation* (6th Edition). Springer, Cham, Switzerland, 2018.
Metcalf & Eddy. Wastewater Engineering: *Treatment and Resource Recovery* (5th Edition). McGraw-Hill Education, New York, 2014.
Green, D.W. (ed.). *Perry's Chemical Engineers' Handbook* (*7th* Edition). McGraw-Hill, New York, 1997.
Storm, D.R. *Winery Utilities: Planning, Design and Operation.* Springer, New York, 1997.
White, R., Adamson, B., and Rankine, B. *Refrigeration for Winemakers.* Winetitles, Underdale, 1989.
Yao, R., Kwong, G., Miller, K.V., and Block, D.E. Prediction of Effective Steam Sterilization Times for Wine Barrels Using a Mathematical Modeling Approach. *American Journal of Enology and Viticulture*, 72, 101–105, 2021.

PROBLEMS

STERILIZATION, SANITIZATION, AND CLEANING

1. Just prior to beer canning, you decide to sanitize the filler unit on the bottling line. To accomplish this you have both steam at 110°C and hot water at 85°C. You have done some swabbing and culturing around the bottling line, and have found that the organisms that are most difficult to kill thermally have an Arrhenius constant of 6.9×10^{22} min^{-1} and an activation energy of death of 38,400 cal/mol.

   (a) If the goal is to reduce the microbial population in the filler by 98.5%, how long should steam be applied? How long should the hot water be applied to get the same amount of kill?
   (b) When the steam generator in your brewery breaks down, you are left with only the hot water as an option. You decide that it makes more sense to sterilize the cartridge filters and filler simultaneously by running hot water through the whole system after it is connected. However, in doing this, you notice that by the time the water gets to the tip of the filler, it is significantly cooler than 85°C. Since your goal has actually changed to **sterilizing** the equipment to a contamination probability of 1/10 and each filling tip generally contains 10^3 viable cells initially, what is the minimum temperature in the filling tip that is necessary to achieve this kill in 60 min?
   (c) Name three ways in which you can verify that the 0.45 µm sterile filter has integrity (i.e. no holes or defects) prior to filtration into the filler.

2. You have been having some problems with sanitizing your bottling line. An unidentified spoilage organism is repeatedly showing up in your bottled wine that is ruining the clarity and causing an undesirable aroma that some have characterized as "stale subway station." You isolate this organism and do some thermal killing studies to find the Arrhenius constant and activation energy of death for the organism. These are 2.4×10^{31} min^{-1} and 52,000 cal/mol, respectively.

(a) If you sanitize the transfer line and filler for 20 min at 82°C (your normal conditions), what percentage of this organism should be left?
(b) Upon further investigation, you find that your hot water for sanitization is not always 82°C. Sometimes it is less. What is the minimum temperature for this water that will still give you 95% kill in 20 min?

3. **[Integrative Problem]** After 2 years of production, you decide to retrofit your winery with a CIP system.

(a) To clean the tank in the previous problem (the largest piece of equipment in the winery), what is the minimum flow rate out of the sprayballs that is likely to clean the tank? This tank is 7.9 ft in diameter and 16 ft tall. Total volume is 6,000 gal with a working volume of 5,000 gal.
(b) What size pipe should be used in CIP transfer lines to accommodate that flow? Verify that you have turbulent flow and a "reasonable" pressure drop. Assume the density of the cleaning solution is 62.5 lb$_m$/ft^3 and the viscosity is 1.2 cp. You can also assume that the transfer pipes are "smooth." (Note: 1 cp = 0.000672 lb$_m$/ft s)
(c) In between the CIP skid distribution pump and the tank there is a heat exchanger (10 psi pressure drop), strainer (5 psi pressure drop), 60 ft of straight pipe, 6 standard elbows, and the sprayball (25 psi pressure drop). The top of the tank to be cleaned is at 16 ft above the CIP pump. Size the centrifugal pump given a pump curve at the Waukesha Cherry Burrell website.
(d) Finally, you will use the same tank as a holding tank just prior to bottling. Therefore, you are programming the CIP skid to add a hot water sanitization step after the cleaning. If you want to reduce the number of microbes by 99.5%, what temperature water should be used to accomplish this process in 14 min? Assume that the most difficult microbes to kill have an Arrhenius constant of 7.1×10^{22} min^{-1} and an activation energy of death of 35,640 cal/mol.

4. **[Integrative Problem]** You get a job with a large winery and are put in charge of their custom crush business. In your first harvest, you get several lots of machine-harvested Chardonnay with a high degree of mold. This mold secretes a polysaccharide that causes a gel-like substance to form during fermentation. To avoid this problem in future years, you decide to install a pasteurization

system to kill the mold. As shown below, this system consists of a tube-in-tube heat exchanger for heating up the juice to 170°F from an initial 70°F, a tube-in-tube heat exchanger that maintains the juice at 170°F for the pasteurization (killing) step, and a third heat exchanger for chilling the juice back down to fermentation temperature.

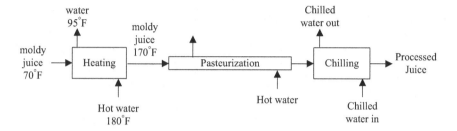

(a) To minimize the effects of this pasteurization on scheduling, you decide that you need the pasteurization unit to handle 10,000 gal (= 1,337 ft³) in 2.5 hr. The density of the juice is approximately 70 lb$_m$/ft³ and the viscosity is 2.6 cp. What size should the pipe be that runs through the pasteurizer in order to assure turbulent flow and a reasonable pressure drop. The inner surface of the pasteurizer can be assumed to be "smooth tube."

$$(1 \text{ cp} = 0.000672 \text{ lb}_m/\text{ft s})$$

(b) In the initial "heating" heat exchanger used for heating the juice, what is the necessary volumetric flow rate (i.e. gpm) for hot water? How long should the exchanger be (assuming a simple tube-in-tube design)? The heat capacity of the juice is 0.94 Btu/lb$_m$ °F and the heat capacity of the water is 1 Btu/lb$_m$ °F. The overall heat transfer coefficient for the exchanger is 500 Btu/hr ft² °F. The densities of the juice and hot water are 70 lb$_m$/ft³ and 62.4 lb$_m$/ft³, respectively.

(c) Your goal for the pasteurization step is to inactivate 98% of the polysaccharide-producing mold that enters the pasteurizer. If the temperature in the pasteurization step is kept constant at 170°F (= 76.7°C), how long does the pasteurizer have to be? Your search of the literature reveals that the Arrhenius constant for killing this mold is 2.25×10^{27} min^{-1} and the activation energy of death is 42,000 cal/mol.

5. The new UC Davis winery is being designed for CIP. A preliminary drawing of the fermentation tanks is shown in the diagram below. It has not yet been adapted for a CIP system. The tank is fairly simple. It includes a manway at the top of the fermentor and a manway in the side to remove grape skins after the fermentation. It also has an automated pumpover system to remove juice from near the bottom of the tank during fermentation (using a dedicated pump) and spray it downward over the top of the grape skins.

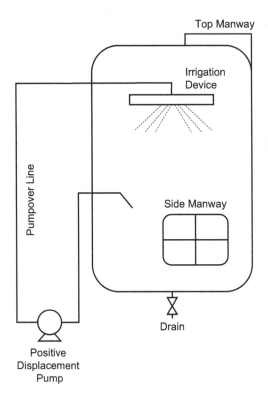

(a) Name three parts of this tank that you feel will be difficult to clean using an automated (i.e. CIP) system. Explain.

(b) How would you add to or modify this design to make it CIPable without reducing its functionality? This would include adding in CIP supply and return line(s) and any spray devices needed. Draw these changes on the figure above and briefly explain your changes here.

6. **[Integrative Problem]** Just prior to your Chardonnay entering the filler on your bottling line at Yi-Di, it is filtered using two cartridge filters, a 3 μm depth prefilter and a final PVDF 0.45 μm sterile filter. You think the sizing of the prefilter is going to be more critical, so you start with this.

(a) To size the prefilter, you use a 4 cm (= 0.04 m) diameter filter disk and filter some of your wine. Specifically, you take 3 L of your Chardonnay from the tank to be filtered and filter it using a pressure drop of 8 psi (= 55,200 Kg/m s^2). Your initial solids concentration is 5 Kg/m^3 and the viscosity of the filtrate is 0.0016 Kg/m s. In your experiment, you get the data shown in the following graph. Use these data to find the filter resistance, R_m, and the specific cake resistance, α. The equation for the regression line is shown at the top of the graph.

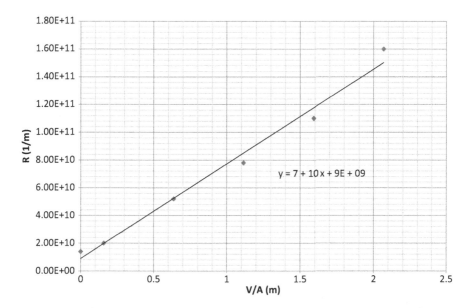

(b) Based on these data, how much time will it take to filter the 10,000 gal (= 37,800 L = 37.8 m³) of Chardonnay, if the cartridge filter has an area of 7.2 m². Assume that a pressure across the filter of 10 psi (= 69,000 Kg/m s²) is maintained.

(c) What would the average bottling speed be (i.e. bottles/min) with this filter? Would this be considered a low-, medium-, or high-speed line for wine packaging?

(d) If you can insert a centrifuge in the line prior to the filters, you think that you can get the solids concentration down to 1 Kg/m³ prior to entering the filters. What effect would this have on the average bottling speed possible? Explain and show your calculations.

(e) Prior to bottling, it will be necessary to sterilize the two filters and the bottling line. If you have done some swabbing and plating to find that there are typically 2×10^8 live cells left in this equipment after cleaning, how long would you have to sterilize with steam at 100°C to reach a 1/10,000 probability of contamination. Some lab testing indicates that likely contaminants would have an Arrhenius constant for death, α, of 1.5×10^{23} min⁻¹ and an activation energy of death, E, of 38,200 cal/mol.

(f) If the steam source fails and you need to use hot water at 80°C, how much longer would it take to achieve the same degree of sterilization?

7. **[Integrative Problem]** Instead of using an aerobic wastewater treatment pond at Lakeshore Brewery, you decide to filter your wastewater in a three-step process to reuse it. In the first step, you are sending your wastewater through a process centrifuge to reduce solids. This is followed by a lenticular stack made of cellulose with DE that is meant to remove the remaining larger particulates. Finally, the water will be sent through a crossflow filter using a nano-filter membrane cartridge.

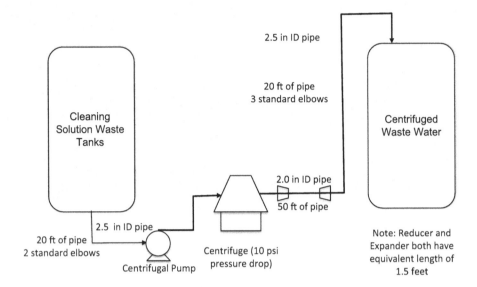

(a) To figure out the sizing of this system, we first have to calculate how much wastewater is being generated during cleaning. We will start with cleaning the largest tank, a 5,000 gal fermentor. This tank has a diameter of 7.5 ft. If the cleaning cycle is a total of 60 min long, how much wastewater is generated in this cleaning cycle?

(b) All water generated in this cleaning cycle is collected in an equalization tank (where the acid and base components of the cleaning can neutralize each other). From the equalization tank, it is sent through a centrifuge (to take out particulates) and sent into another hold tank of the same size. The diagram for the system is shown below. Unfortunately, the company constructing the wastewater treatment system is not used to their new computer-aided design (CAD) software and makes a mistake in purchasing stainless steel pipe. They are short 50 ft of the pipe required for this part of the system. In order to fix this problem, they decide to splice in 50 ft of smaller diameter pipe in the center of the transfer line. The original line was supposed to be 2.5″ ID pipe. For an intended flow rate of 50 gpm (= 6.7 ft³/min), is 2.0″ ID pipe going to be feasible here? Assume a density of 62.4 lb_m/ft^3 and a fanning friction factor (f_f) of 0.006.

8. **[Integrative Problem]** At Weeping Mountain Vineyards and Winery, you decide to filter all of your wines through a 0.45 μm pore size membrane cartridge filter just prior to bottling to make them more microbiologically stable. You have done extensive racking on your Pinot noir, so you feel that rough filtration will not be necessary prior to the sterile filtration.

(a) To evaluate the filtration characteristics of the Pinot noir, you use a small 0.45 μm pore size PVDF filter disk with a 0.045 m diameter. An initial filtration indicates that you have 4 g solids/L (= 4 Kg solids/m³). The viscosity of the filtrate is 1.4 cp ($= 0.0014$ Kg/m s). Your filtration apparatus controls the pressure upstream of the filter at 8 psi ($= 55{,}143$ Kg/m s^2). You measure the volume filtered over time and get the following data. Calculate the resistance of the filter media, R_m, and the specific cake resistance, α, using the first three data points.

Time (s)	Volume of Filtrate (mL)	Volume (m³)	V/A (m)	t/V/A (s/m)
16	32	3.20E-05	2.01E-02	7.95E+02
48	64	6.40E-05	4.02E-02	1.19E+03
120	100	1.00E-04	6.29E-02	1.91E+03
500	140	1.40E-04	8.80E-02	5.68E+03
1,000	154	1.54E-04	9.68E-02	1.03E+04

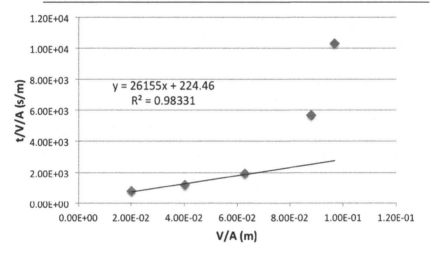

(b) You notice that the last two data points deviate considerably from the trend of the first two. What is your interpretation of what is happening during this part of the filtration?

(c) Given the parameters that you calculated in part (a), you decide to size the cartridge filters for the commercial scale filtration. In this filtration, you plan to run with a pressure drop of 30 psi ($= 206{,}785$ Kg/m s^2) and would like to filter 4,000 L ($= 4$ m^3) of wine in 4.5 hr or less. You find one filtration company that sells four sizes of cartridge, 10″ (0.78 m^2 area), 20″ (1.56 m^2), 30″ (2.34 m^2), and 40″ (3.12 m^2). Will any of these be sufficient? Show your calculations.

(d) After getting the filter train set up, you decide to sterilize from the filter through the pipe to the filler heads of the bottling line prior to filtration and bottling. Microbial screening tells you that the most difficult to kill bacteria in your filler heads has an Arrhenius constant of 6.44×10^{26} min^{-1} and an activation energy of death of 44,000 cal/mol. If your screening indicates that each of the 16 filler heads on your filler is likely to have 200 live cells, how long should you run your 80°C water through the system to have a 1/10,000 chance of contamination?

WASTEWATER TREATMENT

9. After completing the cold stabilization described above, you decide to clean the tank used for this process with a CIP system. The CIP cycle consists of an initial water rinse that removes much of the tartaric acid left in the tank (10 g/L in rinse stream leaving the tank), a NaOH cleaning step, another water rinse, a citric acid washing step, and a final water rinse. All cleaning solutions are delivered to the tank at 90 gpm. The tank is 10,000 gal total volume with a 12 ft diameter.

   (a) Name two purposes for the citric acid step in the cleaning cycle.

   (b) Is 90 gpm an appropriate flow rate for cleaning this tank? Explain.

   (c) For the water rinse stream that is rich in tartaric acid ($C_4H_6O_6$), what will the BOD be entering the waste water treatment system in mg O_2/L?

   (d) What will the BOD be for the citric acid ($C_6H_8O_7$) stream that is 5 g citric acid/L?

   (e) What kind of adjustments might you have to make to the waste prior to the aeration pond given that at this time of year, tartaric and citric acids are the main components of the waste?

10. Your winery, Chateau R'Efuse, is about to put in a new wastewater treatment system. Chateau R'Efuse has an annual production of approximately 100,000 cases. Harvest wastewater output is approximately twice the output during the remainder of the year.

   (a) To allow the aerobic ponds to work more efficiently year-round, you decide to put in two parallel ponds—each will treat half the peak output. In your experience, a residence time of five days will be sufficient for BOD reduction. Size the pond (i.e. find volume, length, width, and depth—assume a rectangular pond without sloped sides).

   (b) You also want to get an idea of the BOD in the waste streams from the winery. In the past, Chateau R'Efuse has inadvertently dumped nearly finished wine to the waste treatment system. This fermenting wine has had a composition of 120 g/L ethanol, 40 g/L fructose, 3 g/L tartaric acid, and 1.5 g/L malic acid. What is the theoretical BOD of this stream?

   (c) For cleaning, Chateau R'Efuse has always used a NaOH wash followed by a rinse with a 0.75% w/v solution of citric acid. What is the BOD of this citric acid stream? What would the BOD of the acid rinse be if the winery switched to dilute nitric acid or sulfuric acid for the acid rinse?

11. You are planning a new brewery in Colorado. Initially (over the first 5 years) you plan to produce about 3,4,00 barrels annually. However, between the fifth year and the tenth year, you plan to increase this production to 68,000 barrels annually if everything goes well. Assume 250 working days per year.

   (a) Draw a diagram of the major parts of the wastewater treatment system you would design for efficient treatment (do **not** size any units).

(b) Would recycling solids from late stage to earlier stage (septic tanks or aeration pond) treatment increase or decrease the size of the septic tank or aeration pond? Explain.

(c) Explain why dumping spent grains into the wastewater treatment system (as opposed to selling them for cattle feed) is problematic and may disrupt normal operation of the system.

12. Eagle's Pride has come up with a new wine-containing product to appeal to current beer drinkers. This beverage, called ZEAL, is an off-dry, low-alcohol, carbonated wine to be packaged in 12 oz. crown cap bottles. After 2 years of phenomenal success, the winery decides to make **only** this product—about 200,000 cases annually (assume the volume of a case is equivalent to the volume for a standard 12 × 750 mL wine case).

(a) At the end of harvest, you are cleaning out the product tanks. The final Eagle's Pride product is mainly 90 g EtOH/L (C_2H_5OH), 50 g sugar/L ($C_6H_{12}O_6$), and 8 g Tartaric Acid/L ($C_4H_6O_6$). What is the theoretical BOD of this stream in mg BOD/L if the product is diluted 50 times in the cleaning process?

(b) Assuming an aeration pond will be part of the wastewater facility for this winery, calculate the appropriate volume for this pond and the approximate dimensions (i.e. L × W × D). You can assume vertical walls instead of sloped walls to facilitate the calculation.

13. The property purchased for the Breakfast Winery has an old dry pond. Just from a quick measurement of its dimensions, you think you might be able to line this pond with concrete and use it for your aeration pond for your wastewater treatment (for your 80,000 case winery). This will save on some excavation costs. The dimensions of the existing dry pond are shown in the diagram below.

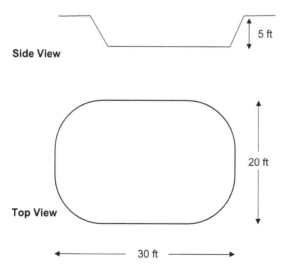

Side View — 5 ft

Top View — 20 ft — 30 ft

(a) Do you see any potential problems in using this existing pond? Explain.

(b) What else will you need to install between the winery and the final holding pond to make this work? Draw a diagram of the wastewater treatment facility and label it.

(c) If you decide to use this pond for aeration, what will the maximum capacity of the brewery be (in annual cases produced)? Assume that you will have sludge recycle and ignore the sloped sides of the pond.

$$7.48 \text{ gal} = 1 \text{ ft}^3$$

REFRIGERATION

14. **[Integrative Problem]** You are put in charge of sizing a new refrigeration system for Chateau Polar. The main loads on the refrigeration system will be 30 fermentors (each with a working volume of 16,000 L) and a must chiller.

(a) It is anticipated that at the peak of crush, 15 of the fermentors will be fermenting at a rate of 6 Brix/day, while another 10 will be fermenting at a rate of 3 Brix/day. The remaining 5 fermentors will be idle or very slowly fermenting. What is the total heat load that must be removed from the fermentors (in BTU/hr) in order to control fermentation temperature? (Note: 3.41Btu/hr = 1 W)

(a) At this same peak time in crush, it is also anticipated that a must chiller will be used. A typical load on the must chiller would include reducing the temperature of white juice from 80°F to 60°F at a rate of 45 gpm. If the heat capacity of the juice is 0.94 Btu/lb$_m$ °F and the density of the juice is 70 lb$_m$/ft^3, what is the load that must be removed by the refrigeration system?

(c) What is the minimum size of the refrigeration system that you will purchase?

(d) Draw a diagram of the refrigeration cycle that will be an integral part of the refrigeration system. Label all of the major components and the phase of the refrigerant at each part of the cycle.

(e) If we later decide that we want to use the refrigeration system at a lower temperature for cold stabilization during an off-peak period in the winery, will this change increase or decrease the performance or efficiency of the refrigeration system? Explain your answer.

15. **[Integrative Problem]** Since Golden Gate Winery has turned out to be a huge success, you decide to open your own winery overlooking a tourist attraction, this time near Yosemite National Park, called Hetch Hetchy Vineyards. You will produce about 60,000 cases comprised of both red (mostly Zinfandel) and white (mostly Sauvignon blanc and Pinot grigio) varieties. Initially you purchase thirty 4,000 gal (= 15,120 L) working volume fermentors, 15 for red wines and 15 for whites.

(a) What is the maximum rate of heat generation for each type of fermentor if the maximum expected sugar utilization rates for reds and whites are 4.5 Brix/day and 2.75 Brix/day, respectively?

$$3.41 \text{ Btu/hr} = 1 \text{ W}$$

(b) Because of the cool, Sierra Foothills location of the winery and vineyards, you initially feel that must or juice chilling will not be necessary. Therefore, when you go to size your refrigeration system, you decide to design the system based on the simultaneous maximum heat output of all fermentors and add 20% extra capacity for surge. What size refrigeration system should you specify?

(c) Your refrigeration contractor tells you that you must decide between a direct or indirect refrigeration system. Explain the difference and give one possible disadvantage of each system.

(d) After winemaking begins in the new facility, you discover that the days are hotter during harvest than you expected, and you need to install a tube-in-tube juice chiller between your press and your white wine fermentation tanks. If your juice is leaving the press pan at 85°F and 50 gpm, what is the coldest temperature (given your maximum refrigeration capacity) it can enter the fermentor assuming that 10 of the white fermentors and 12 of the red fermentors are fermenting at their maximum rate? Assume a $C_p = 0.91$ Btu/lb$_m$ °F and $\rho = 68$ lb$_m$/ft^3 for the juice.

(e) If you finally decide on a direct refrigeration system with a suction temperature of 40°F, how much area will you need for the heat exchanger and what length will this correspond to for a 2.5 in diameter pipe? Assume an overall heat transfer coefficient of 400 Btu/hr ft^2 °F.

16. **[Integrative Problem]** After completing the design of the Blue Note Winery, you decide to join the newly started Breakfast Winery. They are in the process of designing their new winery and therefore want to hire you for your previous design experience. This winery makes 80,000 cases of wine including a Cabernet Sauvignon, a Chardonnay, a Petit Syrah, and a sparkling wine. The main fermentors for this winery will be used for the red wines and for the base wines for the sparkling. The Chardonnay will be barrel fermented. Each of the stainless tanks are 12,000 gal (= 45,360 L) with a working volume of 9,000 gal (= 34,020 L). There are 15 of these fermentors.

(a) During the height of harvest, it is possible that all of these tanks will be fermenting at their maximum at the same time. This would include 10 tanks of Cabernet Sauvignon fermenting at 5 Brix/day and 5 tanks of Petit Syrah fermenting at 6.5 Brix/day. If no must chilling will be done during this period, what is the minimum refrigeration load for which we should plan?

$$3.41 \text{ Btu/hr} = 1 \text{ W}$$

(b) You decide that this winery will use an indirect refrigeration system. The refrigeration cycle will use Freon R22 and have an external chilled water loop that will circulate through the jackets of the fermentors to maintain temperature. Draw a diagram of the refrigeration system and **briefly** describe the function of each part.

(c) You need to size the heat exchanger to be used as the evaporator. It will be a shell-and-tube heat exchanger. Freon will enter the shell side as a liquid and leave as a gas at the same temperature. Water to be chilled will enter the tube side of the heat exchanger at 50°F and leave at 40°F. If the standard size tube has a 1 in diameter and is 8 ft long, how many tubes will you need for this heat transfer? Assume an overall heat transfer coefficient, U, of 250 Btu/hr ft^2 °F.

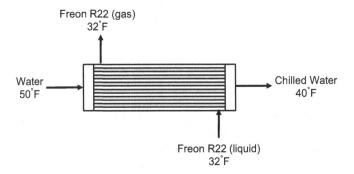

17. **[Integrative Problem]** After you leave UC Davis, you get a job with Arboreal Winery as their new Assistant Winemaker. As the winery is not yet built, your first task is to verify that you can do everything that you need to do with the refrigeration system that has been specified. Based on earlier discussions, the decision was made to purchase a 40 tonR system. This should be sufficient for controlling the temperature in both the Riesling and the Nebbiolo fermentations. However, as you think through the operations you need to perform, you realize that you will need to chill down your Rieslings at the end of fermentation to keep them from going dry. This is going to add to the refrigeration load.

(a) At peak refrigeration load during Riesling fermentations, you have calculated that you will need a total of 3 tonR for temperature control. That leaves 37 tonR for chilling the wine to stop a fermentation. You plan to accomplish this chilling by pumping the fermenting wine through a plate and frame heat exchanger using chilled glycol at 24°F as the coolant. The fermentations are at 55°F and you want to chill each fermentation to 40°F to stop it when it gets to the right sugar level. What is the fastest you can chill the 2,400 gal (= 320.9 ft^3) of fermenting wine without increasing the size of the refrigeration unit?

$$1 \text{ tonR} = 12,000 \text{ Btu/hr}$$

$$C_{p, \text{ wine}} = 0.92 \text{ Btu/lb}_m \, °F$$

$$\rho_{wine} = 63 \text{ lb}_m/\text{ft}^3$$

(b) For this heat exchanger, each plate has 6 ft² of area. How many plates will you need to accomplish this chilling if you expect the glycol to exit the heat exchanger at 32°F? Assume an overall heat transfer coefficient, U, of 350 Btu/hr ft² °F.

(c) Would this be a "direct" or "indirect" refrigeration system? How do you know? Explain briefly. What is one **disadvantage** of this choice?

(d) If you were to use a centrifugal pump to pump the fermenting wine to the heat exchanger for chilling, would you have to worry about your pump's NPSH more or less than if you waited for the fermentation to complete? Explain.

18. **[Integrative Problem]** Your white fermentation area at the Yi-Di Winery will produce 80,000 cases of wine annually and is designed to have its own crush pad, tanks, and refrigeration system. Even though the production in this winery is large, labor is relatively inexpensive, so all the fruit is hand-picked.

(a) How will your choice of press change if you decide to use whole-cluster pressing versus destemming and then pressing?

(b) The refrigeration system for this part of the winery was based on the rule of thumb of 1 tonR is needed for each 10,000 cases. Therefore, the refrigeration system was sized for 80 tonR. Based on your projected grape receiving pattern, you expect to need twenty (20) 10,000 gal (= 37,800 L) working volume fermentors. You are also expecting that, at the peak of crush, ten of these fermentations will be at the peak of their fermentation rate and the other ten will be at half their peak sugar utilization rate. What is the maximum sugar utilization rate for which this refrigeration system will work?

$$1 \text{ tonR} = 12,000 \text{ Btu/hr}$$

$$3.41 \text{ Btu/hr} = 1 \text{ W}$$

(c) If you are using an indirect refrigeration system and have a chilled glycol loop that runs through all 20 jackets in parallel, what flow rate will you need in the main glycol line entering the fermentation area to meet the peak load expected? The glycol loop is 35% propylene glycol in water and is maintained at 30°F. The return loop for the glycol is expected to be at 45°F.

$$\rho_{glycol} = 64.4 \text{ lb}_m/\text{ft}^3$$

$$Cp_{glycol} = 0.915 \text{ Btu/lb}_m \,°F$$

(d) What size pipe should this main glycol line be to assure turbulent flow and a reasonable pressure drop? You can assume smooth tubes.

$$\mu = 7 \text{ cp}$$

$$1 \text{ cp} = 0.000672 \text{ lb}_m/\text{ft s}$$

19. **[Integrative Problem]** At the first winery you work at after leaving UC Davis, Gordon Estate, you are put in charge of making excellent red wine from marginal grapes. The head winemaker tells you that he thinks that the best way of accomplishing this is to use a process called flash détente, in which whole must after destemming and crushing is heated up quickly and then rapidly cooled in a vacuum chamber to increase extraction.

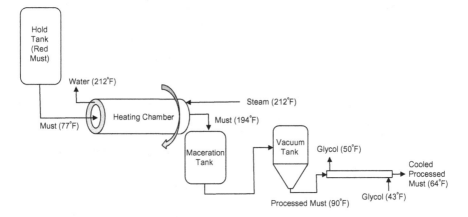

(a) The first stage of this process, pictured above, is to heat the must from 77°F to 194°F in what amounts to a large tube-in-tube heat exchanger that rotates to give good heat transfer. Steam enters the outer tube of this heat exchanger at 212°F and leaves as liquid at the same temperature. The inner tube that carries the must is 4 ft in diameter and the outer tube that carries the steam is 4.5 ft in diameter. If this system is to have a throughput of 20 tons/hr to meet your harvest demand, how long does the heating chamber need to be? You can assume that the must has a heat capacity of 0.95 Btu/lb$_m$ °F and the heat exchanger will have an overall heat transfer coefficient of 400 Btu/hr ft^2 °F.

$$1 \text{ ton} = 2000 \text{ lb}_m$$

(b) After the must goes through the vacuum chamber, it is still at 90°F and needs to be cooled down further to 64°F prior to the beginning of a cold soak and fermentation. This will be accomplished using a traditional tube-in-tube heat exchanger, in which the inner tube has a 3 in inner diameter. How long should this heat exchanger be for the same flow rate of must as above? Assume the glycol comes into this counter current heat exchanger at 43°F and leaves at 50°F. The overall heat transfer coefficient of this heat exchanger is 300 Btu/hr ft^2 °F.

(c) What volumetric flow rate of glycol will you need for this heat exchanger?

$$C_{p, \text{glycol}} = 0.24 \text{ Btu/lb}_m \text{ ° F}$$

$$\rho_{\text{glycol}} = 68.4 \text{ lb}_m/\text{ft}^3$$

$$7.48 \text{ gal} = 1 \text{ ft}^3$$

(d) Do you have a direct or indirect refrigeration system in this winery? Explain how you know. Name one disadvantage of this system.

HVAC

20. **[Integrative Problem]** The HVAC at Eagle's Pride was designed to provide 10 room changes of air per hour in the tank room when CO_2 needs to be evacuated. The tank room has a volume of 19,500,000 L. At first, there is a bank of twenty-five 15,000 gal fermentors, each with 12,000 gal (= 45,360 L) working volume. At the peak of crush, all of the fermentors are fermenting at an average rate of 4 Brix/day, which corresponds to 18,144 L/hr of CO_2 evolved per fermentor. What will the CO_2 level of the air in the tank room be at this point during crush? Is this a safe level? Assume an inlet air CO_2 level of 500 ppm (= 0.0005 vol fraction).

21. **[Integrative Problem]** At the Breakfast Winery, two separate rooms will be used for barrel storage. One will be used for Chardonnay barrel fermentation and subsequent aging, and the other will be used for aging the red wines.

(a) In the Chardonnay barrel fermentation room, you are concerned that the air flow may not be sufficient to reduce the CO_2 concentration to below what you consider to be a safe level. The level that you choose is 3,500 ppm. The architectural company designing the winery tells you that they have designed the room for 5 room changes per hour (i.e. 5 room volumes of fresh outside air pass through the room each hour). If the barrel room holds 800 barrels and is 40 ft wide by 50 ft long by 22 ft high, is this a high enough air flow rate to maintain safe levels of CO_2? At the peak of fermentation, you can expect that each barrel will generate 65 L/hr of pure CO_2 and outside air contains 500 ppm of CO_2.

$$1 \text{ ft}^3 = 28.3 \text{ L}$$

(b) For the red wine barrel aging room, you want to take the outside air in Cloverdale (where the winery is located), which will be 90°F and 25% RH (Relative Humidity) and condition it to 60°F and 85% RH. How much energy will be needed to accomplish this conditioning (in Btu/lb mol air).

(c) If you decide instead to keep your barrel room at 69°F and 85% RH, will you save energy? Show your work.

(d) Will the ethanol level of the Cabernet and Petit Syrah go up or down in these conditions? Explain your answer briefly.

(e) If your main goal in barrel aging your Cabernet is extraction of flavor components from the oak, should you plan to keep the wine in barrel for on the order of 6 weeks, 6 months, or 2 years? Explain your answer.

22. **[Integrative Problem]** The main fermentation hall of the new UC Davis Research and Teaching Winery will have all of the CO_2 from the fourteen 2,000 L working volume fermentors exhausted directly to the outside, as opposed to being vented directly into the room as in most wineries. However, as a backup to this system, the HVAC has to be designed to evacuate the CO_2 sufficiently to bring CO_2 levels below a danger level set to 4,000 ppm. The fermentation room will likely be most fully utilized during fall teaching with half of the fermentors (the red fermentations) generating 12 L CO_2/L must/day and the other half of the fermentors (the white fermentations) generating 7 L CO_2/L must/day maximum. What air flow rates will be necessary for this emergency system? Assume that outside air is typically 500 ppm CO_2.

12 Control and Information Management Systems

In the general field of process control and information management, there are really three types of systems that are in use in industry. These are process control systems, data acquisition systems (DASs), and data management systems. For systems that are commercially available, there may be some functional overlap that blurs this distinction (i.e. some process control systems may also acquire and save data). However, for the purpose of this discussion, these three types of systems will be discussed separately.

12.1 PROCESS CONTROL SYSTEMS

A process control system is a system used to control the action of equipment within some predetermined guidelines or setpoints. Examples of this control would be temperature control on a fermentor, flow rate control on a pumping system, or fill level control on a bottling machine. In general, for any control system (or individual "control loop"), there must be a signal coming from a sensor, a setpoint with which to compare it, a manipulated variable that can be used to return the value to the setpoint, and a controller to compare the actual value with the setpoint and manipulate the manipulated value. Using the temperature control in a fermentor as an example, the sensor would be some sort of temperature sensor (e.g. a thermocouple or resistance temperature detector [RTD]), the setpoint would be the desired fermentation temperature, and the manipulated variable would be the position of a flow control valve that allows coolant into the jacket of the tank. Similarly, the transmembrane pressure (TMP) in a cross-flow filter is often controlled. In this case, the sensor is a differential pressure sensor across the filter membrane (or the difference between pressure sensors on the feed/retentate side and the permeate side of the membrane). The setpoint would be set to maximize throughput while protecting the membrane, and the manipulated variable would be the speed of the feed pump and/or permeate pump. This concept is illustrated in Figure 12.1.

12.1.1 Control System Signals

There are four types of "signals" in a control system. These are analog input, analog output, digital input, and digital output (Figure 12.2). *Analog* means that the signal is a continuous value (e.g. temperature, pressure, pH, flow rate, etc.). Analog signals are commonly measured in the field using a 4–20 mA signal range, which is converted in a linear fashion to the measured parameter (e.g. pressure). Temperature is also an example of an analog signal. However, temperature devices like RTDs can have nonlinear outputs and may be handled separately or in a different manner than

DOI: 10.1201/9781003097495-12

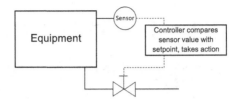

FIGURE 12.1 Schematic of a control loop. An input is converted to an action to be taken on the system to assure that operations fall within a limited range of a setpoint.

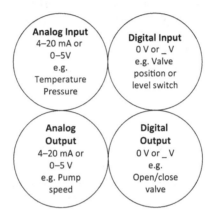

FIGURE 12.2 The types of signals found in control systems.

all other analog signals. *Digital* means that the signal is discontinuous (i.e. either on or off, but not in between). This is often handled electronically by having standard voltages for on/open or off/closed (e.g. 0 V or 5 V for "off" or "on"). An *input* is a signal that comes from a sensor or device in the field TO the controller. An *output* is a signal that comes FROM the controller to a device in the field.

12.1.2 Process Sensors

A key part of any control loop is the sensor. Sensors are available to measure a wide variety of variables, but the most commonly encountered process sensors are temperature sensors, pressure sensors, and flowmeters. More exotic sensors exist for dozens of applications, including pH, dissolved or atmospheric gas concentration (carbon dioxide, oxygen), color, conductivity, liquid height, weight, total organic carbon, or even cell concentrations. New types of sensors are being developed and adapted continuously.

Temperature sensors are typically thermocouples, where two different metals with different conductivity are joined. The electrical properties of the union change with temperature, which can be used to measure temperature. Thermocouples are almost never in direct contact with a process liquid, instead, an open-ended tube (called a "thermowell") is added to the equipment, with the open side facing the outer

TABLE 12.1
Types of Flow Meters

Meter Type	Pros	Cons
Orifice Plate/ Differential Pressure	Simple, cheap	Requires known density, poor solids handling, not sanitary
Magnetic Flow Meter ("Mag Meter")	Directly measures fluid velocity, non-invasive, handles solids well, can be strapped onto already existing pipe, sanitary	Cannot measure non-conductive liquids, intermediate price
Turbine Meter	Simple, extremely cheap, easy to install, good turndown	Terrible solids handling, easily jammed, not sanitary
Coriolis Flow Meter	Can directly measure density and mass flow in addition to volume flow, extremely accurate, handles solids well, no interruption of flow, sanitary	Extremely expensive, much larger than other flow meters

environment and the closed side poking into the process fluid. The thermocouple rests in the thermowell, which rapidly comes to thermal equilibrium with the process fluid. Other types of temperature sensors are also in use in the wine and allied industries, such as RTDs. RTDs typically use a metal wire or coating whose electrical resistance is a predictable function of temperature. RTDs are generally more accurate and have lower drift over time than thermocouples, though their response time (the time they take to read an accurate temperature when placed in a solution) can be slower. This latter point is only an issue if the dynamics of the system change on the order of seconds or less, which would not be typical for wine or beverage processing.

Pressure sensors detect pressure by measuring the force applied on a known area. Pressure sensors typically operate via the piezoelectric effect, where the strain on a piezoelectric material (such as a crystal) produces a current proportional to the strain. Pressure sensors can be used indirectly to measure tank height and even fluid density. Liquid height in a tank can be inferred from a single pressure sensor for a known fluid density, as the pressure exerted on the sensor is a function of fluid column height. Two pressure sensors, placed a known height apart in a tank, can be used to infer fluid density (by measuring the pressure difference between the height of each probe), and from there the liquid height. This is often used to track fermentation progression in wine and beer fermentors, as density drops over the course of fermentation as sugar is converted to alcohol and carbon dioxide.

There are numerous types of flow meters, with at least a half dozen finding common usage in wineries. Table 12.1 outlines the most common types, along with some pros and cons.

12.1.3 CONTROLLERS

The controller takes the signals from the sensors and calculates the action that needs to be taken to maintain a setpoint or desired process state. Three types of controllers can be found in the food and beverage processing industries. These are (1) local controllers; (2) programmable logic controllers (PLCs); and (3) distributed control systems (DCS). Local controllers are perhaps the most common type of controller in the wine industry. Typically, a local controller will be dedicated to one control loop. That is, the controller will get one signal from the field, compare the signal to the setpoint, and then send a signal to a device to do something to return the value to the setpoint. The advantages of local controllers are that they are relatively simple and low maintenance, and they are easy to configure and replace. Small to medium size wineries use local controllers extensively for temperature control on fermentors. The disadvantages of local controllers are that they are less flexible in the way that they control than other types and difficult to use if various controllers around the facility need to communicate with each other.

PLCs are the next step up in complexity. PLCs have a true microprocessor (computer) and can handle a large number of inputs and outputs (both analog and digital) simultaneously. Usually, one PLC is used for each major piece of equipment or groups of equipment. They were first developed to control production lines for consumer products and are fast and relatively inexpensive. The equipment can be ordered "off-the-shelf," and a large number of "system integrators" know how to configure and program the major brands of PLCs. The PLC consists of a power supply, the microprocessor, and a number of cards into which inputs and outputs are wired, usually organized by the categories described above (digital inputs, digital outputs, analog inputs, and analog outputs). Depending on the configuration, RTDs will sometimes have their own input cards. The programming language for PLCs is called "ladder logic," and is based on a series of IF/THEN statements such as IF the valve is open, THEN turn on the pump. If the left side of the "rung" is true, then the rung is complete or made, and the action on the right side of the rung is taken. An example of ladder logic is seen in Figure 12.3. It can be seen in this figure that rungs can rapidly become more complex with AND and OR statements on the left side. While a

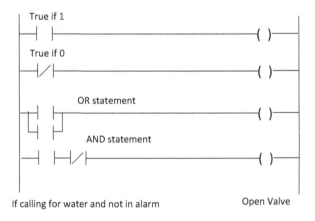

FIGURE 12.3   Examples of ladder logic "rungs."

former downside to the use of PLCs was difficulty in PLC/PLC communication, this is no longer a problem. PLCs can readily communicate using proprietary data highway protocols or by ethernet. PLCs do require some sort of separate operator interface as they are literally a black box (sometimes gray!). The operator interface can be as simple as buttons, knobs, and analog meters, or as complicated as a washable, touch-screen computer showing a picture of the piece of equipment and allowing monitoring and manipulation through interaction with the image on the screen. Operator interface computers need to be configured for each control system using a separate programming or operating system than the PLC itself.

DCS are less widely used in the wine industry currently but are widely used in other industries. In a DCS system, all signals are routed into a central computer that handles all of the process control. Therefore, control for a whole facility can be integrated easily, and usually, operator interfaces are part of the system (typically computers). The downside to DCS systems is that they are usually programmed in a proprietary language (so there are fewer people that can do it-making these systems potentially more expensive) and there is a single point-of-failure. That is, if the central computer crashes, you may not be able to run your facility. Given this possibility, companies with DCS systems generally plan for redundancy of computers and use special fault-tolerant hardware.

12.1.4 ACTUATION SYSTEMS

In addition to the sensors and controllers, a control loop needs a way to effect change in the process system. That is, it must be possible for the controller to prompt a state change in equipment to maintain process setpoints. While there are a number of actions a control system could take, the two most common means for effecting change in winery systems are through opening/closing valves and changing the speed of pumps. Automated actuation of valves can be accomplished through purely mechanical changes (e.g. motorized actuation), though the most common type of automated valve actuation is through the use of a solenoid-controlled pressurized air source. In this case, each valve is equipped with a spring-loaded device connected to the air source. When the control system calls for a change in valve state, it sends a signal to a solenoid that supplies air to the actuator that changes its state. These actuators can be purchased as "air to open" (also called "normally closed" or "fail closed") or "air to close" (also called "normally open" or "fail opened"). The type of actuator is chosen based on which position is safer in the event of a power loss to the system. Because the position of valves is often critical to operations, special sensors called limit switches can be purchased as part of the actuators in order to sense and report the state of the valves. These valves may also have some sort of mechanical/visual confirmation of position on the valve itself to aid with operator confirmation of correct operation or troubleshooting.

Most pumps can be purchased or modified to be turned on or off remotely. If pump speed is used for control, then the pump must be equipped with a variable frequency drive (VFDs). Depending on the size and application, VFDs can be local to the equipment they are controlling, such as a pump or impeller motor. In larger facilities, the VFDs are often placed remotely in banks in a clean, dry area close to the operating equipment. Each VFD is then wired to the motor it is controlling.

12.1.5 Control Logic, Feedback, and Feedforward Systems

Control systems are very rarely passive—while there are some systems that are purely monitoring, with humans reading measurements and responding, control systems typically include logic to measure and respond to process variations.

A common type of logic is "feedback control." In this type of logic, a controlled variable (for example, flow rate) is measured in real time, and a manipulated variable (% pump drive or % valve open) is changed to maintain the controlled variable at its setpoint. Table 12.2 shows some examples of manipulated and controlled variables.

Feedback control requires a pre-fixed "setpoint"—the target of the controlled variable. An example might be 50 gal/min, or 2,000 L, or 70°F. The rate at which a manipulated variable is changed to impact the controlled variable is a function of controller tuning, and it is related to the difference between the setpoint and actual value of the controlled variable (proportional), the integrated difference (integral), or the slope of the difference (differential). This type of control is commonly known as PID control. A complete description of PID control is outside the scope of this discussion, though it should be noted here that PID loops need to be "tuned" in order for them to work well for your system. This involves setting constants for the P, I, and D parts of the controller that give the desired system response, which is performed during system commissioning by a controls expert.

Another control scheme is "feedforward" control. In feedforward control, variables upstream are measured, and input variables are calculated from a pre-determined equation. An example might be controlling the outlet temperature of a heat exchanger by feedforward control. If the input flow rate and input temperature are varying, they can be measured, and the steam or coolant flow rate can be "trimmed" (adjusted) in real time to control for a target outlet temperature.

Feedforward systems can be extremely useful when rapid, fine control of a process is required, but have the drawback of requiring a fundamental understanding of a system. Complex systems which are hard to model (unlike a single heat exchanger) benefit from feedback control. Systems which are very simple (such as a flow rate controlled by a pump % drive) are also typically controlled via feedback control.

TABLE 12.2
Examples of Manipulated Variables Used to Control Controlled Variables

Controlled Variable	Manipulated Variables
Tank Volume	Input Flow Rate, Output Flow Rate
Heat Exchanger Outlet Temperature	Coolant Flow Rate, Steam Flow Rate, Process Fluid Flow Rate
Flow Rate	% Valve Open, % Pump Drive
Carbon Dioxide Concentration	Carbon Dioxide Pressure or Flow

12.2 DATA ACQUISITION SYSTEMS

A DAS is a system that will automatically and/or manually save process data in a usable form. In order to design a DAS, you need to know what data you want to collect, whether it is on-line or off-line, and whether the data are time-dependent or batch information. Sophisticated DASs can efficiently decide on the necessary frequency of data acquisition based on how rapidly a value is changing in time. This can serve to compress data storage. Data are often more easily analyzed by looking at visual representations (e.g. graphs) as opposed to simple tables of data. For this reason, it is common for a DAS to also create graphical reports of data, often over time. Simple DAS may preconfigure reports, while more complex systems may allow the operator to define which combination of variables to plot over a given time frame. If data are process critical or needed for regulatory purposes, then the DAS will require fault tolerant computers and redundant storage systems that automatically "mirror" each other.

Frequently, DASs are combined with supervisory control systems to form a supervisory control and data acquisition (SCADA) system. Supervisory control can provide control for a controller (e.g. go from three to two to one automated pumpovers per day at various time intervals). It can also manage alarms, setting off audible or visible alarms in a process area or on an operator interface and handle acknowledgement of alarms by process operators. These systems can also be used to call out to the right person to correct a problem or allow remote access to potentially help fix a problem from off-site.

12.3 DATA MANAGEMENT SYSTEMS

Two types of data management systems that are used to some degree in the wine industry are laboratory information management systems (LIMS) and management execution systems (MES). LIMS are data collection systems in which data are collected from several different pieces of analytical equipment and automatically stored in a common database that can be accessed remotely. For example, a LIMS system could be used to collect information from a spectrophotometer, autotitrator, and HPLC on the same sample. This information could then be used easily by winemakers, viticulturists, marketing or others at a winery that need this information. A management execution system is used to track information throughout an entire organization. For instance, information on retail sales could be fed back to vineyard and winery staff to adjust production. Inventory can be tracked automatically, so that the flow of product from the winery to the distributor to retail locations can be adjusted according to demand and product inventory. In addition, some of these systems link production with materials purchasing so that when raw materials such as DE or bentonite in inventory fall below some level, more is automatically purchased. A few specialized winery software systems for tracking some of these data are commercially available. Most are not at the level of complexity/automation found in other industries but may be sufficient for winery use depending on the nature and complexity of the specific winery business.

12.4   IMPLEMENTATION OF CONTROL AND DATA MANAGEMENT SYSTEMS

Implementation of modern control systems, automation, and information management systems can improve product quality and process reproducibility, while reducing reliance on personnel. However, adding automation to a process environment that has not historically utilized automation can be a difficult and stressful process. Therefore, a thoughtful, staged approach is important. This begins with identifying all of the current automation needs. It is important that this step is performed in conjunction with process operators that are familiar with all of the steps and processes performed during normal operations. Next, as many future automation needs as possible should be identified. Current management and operator comfort with automation should then be assessed. With all of this information, the new automation should be designed with current needs and current comfort level (or just above!) in mind using equipment capable of future automation needs. After implementation, as comfort level rises, automation can be increased with existing equipment. This approach can decrease the stress of implementation without having to wholesale change out the automation hardware as operators ask for increased automation (which they likely will in our experience!).

Notes on Designing for Operability, Sustainability, and Business Operations

To this point, this textbook has focused on individual unit operations within wineries, breweries, and distilleries. However, it is also important to consider design principles at a whole facility scale. Design principles are highly dependent on the resources, interests, and beliefs of those in charge of building the production facility at the time of the facility design. The impact of these principles, though, will persist throughout the useful life of the facility and thus are important to discuss here briefly.

One class of design principles is designing for operability. This includes designs that reduce capital costs or operating costs (sometime diametrically opposed), increase cleanability, increase functionality or how easy the facility is to use and how flexible, or increase the safety of operations. A second class of design principles is designing for environmental sustainability. In this case, design decisions could be made that would reduce water use, energy use, or carbon dioxide emissions, either in the building process itself or in the operations happening within the completed building. Finally, there are design principles related to business operations such as making an architectural statement and the aesthetics of the facility, integrating entertainment and hospitality spaces, and even considering how tours will be given without interrupting operations.

All of these design principles are viable and will need to be considered during the planning stages of a production facility. Some principles may be mutually exclusive, but most can co-exist with some forethought and creativity (and of course resources!).

In this chapter, we will examine some illustrations of the most important design principles and discuss some of the long-term implications of these choices. To do this, we will use some winery concepts developed as part of the UC Davis winery equipment course (as you have seen throughout this text), as well as some examples from our own Teaching and Research Winery at UC Davis. While this topic could take up an entire textbook itself, here we start the conversation and give highlights of the issues under consideration.

DESIGNING FOR OPERABILITY

Designing for operability may be the most important of the design principles. Afterall, if you cannot produce the products that you aim to produce, sustainability in any sense will not be as important and there will be no wine, beer, or spirits to sell. This is the area that should be given the most thought, and there should be major input from the winemaking, brewing, or distilling staff that will operate the facility (or trusted consultants if the staff has not yet been hired). There should be input on the layout of the facility and equipment and significant thought put into how operations will work during harvest and throughout the year. You should design for the style of product you are planning to make immediately, but with as much flexibility as possible to allow the evolution that will inevitably occur. Here, we will discuss four case-study wineries with various features that illustrate some of the choices that will need to be made.

In the first case study, Peak Winery (Figure N.1), we have a medium-sized winery located in western Colorado. The design here is a single-story building with an outdoors crush pad and tanks in rows running the length of the fermentation hall. The barrel room opens off of the fermentation hall and barrels are stacked here six high on four-barrel racks. There is indoor space between the crush pad and tanks that can be used for unit operations like filtration, barrel work, and bottling lines (mobile, for instance). This type of layout is very conducive to operations as it is all on one level. The UC Davis Teaching and Research Winery is also set up in a similar fashion. While a below-ground "cellar" for barrel storage or bottle storage could have its advantages—for instance, possible natural cooling to "cave" temperature—it also has its drawbacks which could be considerable. As examples, it is usually quite expensive to excavate flat land and build an actual cellar. In addition, having two levels will mean that people and equipment will need to be going between levels on a regular basis. If barrels are in the cellar, a specialized elevator that can handle the weight (even empty) will need to be installed at considerable expense. It will be difficult to get a forklift into the cellar, meaning that there will be considerable difficult

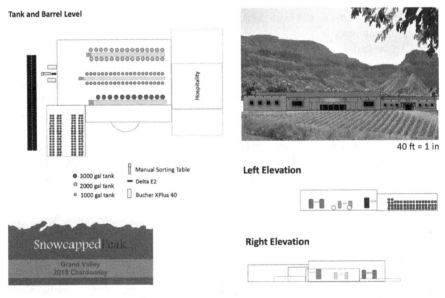

FIGURE N.1 Layout and elevations for the Peak Winery. This is a single level winery with an adjacent fermentation hall and barrel room with views into the fermentation hall from indoor and outdoor hospitality areas. It was meant to be a Colorado adobe style facility.

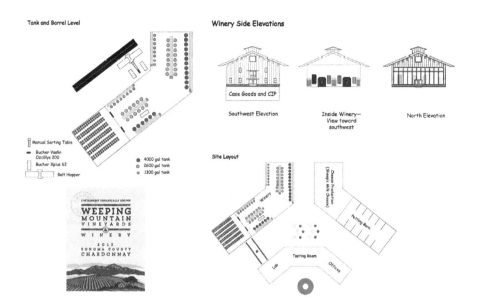

FIGURE N.2   Layout and elevations for Weeping Mountain Vineyards. This is a single level winery with central grape delivery and an adjacent barrel room. A small underground barrel hall (connecting the production facility with the tasting room) is reached by a large ramp on the production side. It is meant to have a modern farmhouse feel and has a mirror image building that holds a cheese making facility and petting farm for children. The space between these buildings and the tasting room is a patio that will be used for casual food service.

handwork necessary, and to be compliant with regulations like the ADA, a separate elevator for people will need to be installed as well.

The second case study, Weeping Mountain (Figure N.2), is a variation on this arrangement. One of the drawbacks of the Peak Winery design is that the distance from the crush pad to the far tanks will make for either long stretches of hose during harvest or expensive pipe manifolds. In the Weeping Mountain design, the fermentors are separated into two relatively equal pods of tanks, with the crush pad (and extra processing space) more central. This will lead to much shorter hose or pipe runs and likely less cleaning. The main barrel room is still on the same floor as the fermentation hall to ease movement of barrels back and forth, however, there is a small underground auxiliary barrel hall in between buildings. In order to make this possible, we have planned for a gradual, ADA-compliant ramp in the barrel room so that forklifts can drive barrels to the underground barrel hall—or at least a pallet jack can be employed. While this is one solution to allow some interesting underground spaces, it should be noted that the ramp will take up space that could otherwise be used for 15% expansion flexibility.

What if the siting of the winery is more conducive to underground facilities? This is the situation in the third case study, Traverse Winery (Figure N.3). Because this winery is built into the side of the hill, it affords outside access to each level of the winery. Here, we have chosen to place the fermentation hall over the barrel room. The crush pad is placed on the uphill side of the winery and there is direct access to

Front Elevation-Concept

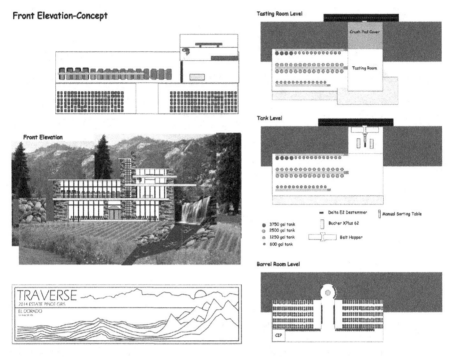

FIGURE N.3   Layout and elevations for Traverse Winery. This is a multi-level facility with production on two levels. It is built into the side of the hill and is meant to have a look similar to Falling Water, the home designed by Frank Lloyd Wright. There will be some energy savings from the barrel rooms being partially underground, but here most processing will be on a single level. Hospitality is on a mezzanine level above the tank tops with a view of production.

the barrel room area and cave from the downhill side. This will allow barrel deliveries to be made directly to the correct level, and forklifts, for instance, could drive from the crush pad level to the barrel level without issue. The tasting room area will be on a mezzanine overlooking the fermentation hall, also with access to the outside uphill from the winery.

While the Traverse design can employ some gravity flow, the multiple levels are really geared toward some energy savings from having the barrel storage areas partially underground. The last case study, Blue Note Winery (Figure N.4), is designed to be capable of nearly total gravity flow. Grapes arrive essentially at the roof of this facility and wine comes out the basement. This can be done, as this winery is also built into the side of the hill. Grapes are crushed on the top level and delivered to the fermentation hall on the middle level. White wines are delivered by gravity by to the barrel rooms on the lowest level. For red wines, conical retort tanks (like drainers) are used as fermentors and then skins are dropped through a slot in the floor to a press below on the lowest level. There are some tanks on this lowest level and the pressed red juice can be blended and/or barreled for aging on this level. Some pump work will be necessary at this point, though some alternatives, such a "bulldog pups" to push wine with pressure could also work. To have true and complete gravity flow, a

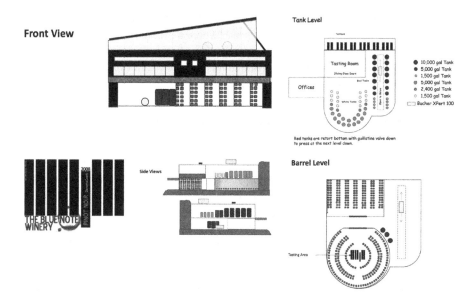

FIGURE N.4 Layout and elevations for the Blue Note Winery. This is designed to be a gravity flow facility with grape receiving on the top level, fermentation on the middle level, and red pressing (with a press on rails) and barrels on the bottom level. There are multiple entertaining spaces including an area in the middle of the white tanks that can be used for small concerts in the off season and in a circular barrel room.

fourth level would be necessary. This brings up the issue of the potential advantages of gravity flow in a modern winery. Gravity flow for wineries in the 19th century was a necessity. A lack of electricity meant that winemakers had to be creative about the flow of materials and product. Inevitably this meant that they figured out some way of driving or hoisting grapes to the top of the winery and the barrels were kept two levels below. Later, there were also arguments of energy savings for pumps and the potential damage that pumping would cause the product. In our experience, there are other major energy sinks in a winery setting and our own studies have shown little to no effect of pumping wine (we have not studied juice or must, and pumping must with the wrong pump could be an issue). There is also the question of whether pumpovers will be used for red wine, and whether this will negate some of the intentions of having a gravity flow winery. Finally, having a truly gravity flow winery may increase the required equipment. For instance, if the presses for white wines need to be at the top level and presses for reds need to be at the lowest level, this may require redundant presses as it will be difficult to impossible to move large presses between levels safely. For a winery all at one level, we can plan to purchase presses that would work for either red or white wines. This is not to say that a gravity flow or multi-level winery is not the best idea for your winery, but, again, all operations should be thought through, in addition to the cost, prior to making this decision.

In addition to the considerations discussed above, there are other overarching ideas that must be considered to create the safest facility possible. These include minimizing movement of large equipment, minimizing the need to climb on equipment or barrels, flooring that is smooth enough to be cleaned but rough enough to avoid slipping, carbon dioxide monitoring in any space where a primary or secondary fermentation might occur, and choosing equipment and facility design that minimizes entry into confined spaces (such as sending someone into a fermentor to shovel skins after a red fermentation). Safety should *always* take precedence over aesthetics.

DESIGNING FOR ENVIRONMENTAL SUSTAINABILITY

Designing a building and its operations for environmental sustainability is important. It allows us to contribute to reducing the use of natural resources that are important to the entire community. At UC Davis, we designed and built the most sustainable winery in the world. It was the first LEED Platinum certified winery in the world—it includes a major use of natural light, captures rainwater for use in irrigation and flushing toilets, relies on recycled construction materials, and employs night-time cooling and large amounts of insulation to cut down on cooling in the summer and heating in the winter, among other features. However, we also strive to maintain and promote sustainable operations within the facility. Here, we will focus on some of the ideas related to water, energy, and carbon dioxide that were originally inspired by Prof. Emeritus Roger Boulton.

As mentioned earlier in this text, it commonly takes 4–6 volumes of water to make 1 volume of wine (and 2–3 volumes for beer, including the product). Nearly all of this water is used in cleaning. Therefore, we are in the process of implementing a comprehensive strategy to save and recycle water at the UC Davis winery (with about half to two-thirds of the parts in place). First, we have installed a rainwater capture system off the larger buildings surrounding the winery in order to collect water to be used for processing. This water is rough filtered and sanitized prior to storage in 40,000 gal storage tanks. This rainwater is then RO filtered to potable levels slowly during the non-harvest season and placed in designated clean storage tanks for use in cleaning. We have designed a CIP system that will use this water to clean the winery equipment. As already explained, CIP systems save water from cleaning (among other benefits!) and facilitate the recycling of water during cleaning. All of the fermentors in the facility have been designed to be used with a CIP system. Finally, we are planning on taking the spent cleaning solutions from the CIP system and filter them to recover the water and chemistry with the idea of recycling as much as possible. As this system becomes a reality, the goal is to bring the water:wine ratio down to 1:1 or even less. Components of this system are already being utilized in other commercial wineries.

Energy use is also important in wineries, breweries, and distilleries, though the major uses of this energy are likely different. In a winery, chilling must, temperature control in fermentors and barrel rooms, and cleaning are likely some of the greatest uses. While many wineries and breweries, including ours, have installed solar panels on facilities or employed wind turbines or other even employed anaerobic digesters to generate energy, there are potential issues with this approach that need to be considered. For instance, will these installations take up space that could otherwise be used for other purposes such as agriculture? In addition, what if significant energy is still needed when these sources are not available (e.g. at night time or cloudy days for solar power)? For the later issue, UC Davis collaborators have taken the approach of using banks of second life electric car batteries. The batteries are charged during the day with solar power and then discharged at night or at peak usage periods to cut down draw that may determine overall utility rates. We are also cutting down on cooling needs for our winery by sending all carbon dioxide into a manifold that is venting outside instead of into the winery environment. As seen in the earlier chapter, this saves us on cooling huge amounts of outside air during harvest to maintain a safe and comfortable work environment. These power savings are harder to implement in breweries and distilleries, where material must be heated and boiled. However, application of heat exchanger networks (often called "Pinch Analysis") can be applied to dramatically reduce power costs.

Finally, there is carbon dioxide. While we are currently venting the carbon dioxide back into the atmosphere, researchers here at UC Davis are actively working on taking that manifolded carbon dioxide stream and converting it to useful products that will keep the greenhouse gas sequestered for as long as possible in building materials, as an example. This technology is particularly valuable to breweries, where nearly every end product is carbonated.

While the technology for environmental sustainability is evolving, there is one clear design feature that can be incorporated fairly inexpensively that will help make great strides in reducing water, energy, and carbon dioxide footprints. This is monitoring. It is really impossible to know where and how large your issues are without measuring your usage rates. The finer the detail of these measurements, the more you will be able to benchmark and monitor your progress in reducing your footprint. This is easiest when you are in design and planning phases, but still doable in a mature facility at reasonable costs.

DESIGNING FOR BUSINESS OPERATIONS

This book is really about the technical side of winemaking, brewing, distilling, and the equipment associated with these practices. However, sometimes making a product can be the easy part compared with selling the product—which, after all, is what makes this a business. Therefore, it is important to consider designing facilities for business operations as well. Even if a production facility is not open to the public, it is important to have an environment that facilitates interaction between all company functions and is conducive to worker morale and productivity. If the facility will be open to the public, there are a host of other considerations during the design. Many times, this means that an owner will want the facility to make a statement that can attract business. Perhaps this means making a bold architectural statement or emphasizing a comfortable space for customers to relax and soak in the ambiance. The customers will be more likely to buy product if they are buying the entire enjoyable experience. However, in creating this special environment, it is important not to degrade functionality of the winery or create spaces that are unsafe because of unplanned interactions between operators and the public. This is why it is important to consider the placement of business offices, hospitality and entertainment spaces, and even tours when designing the production part of the facility. The key is to make customers feel like they are getting an intimate look at the inner workings of the winery without actually getting in the way. This can be a lot of work and require a huge amount of creativity. It requires thinking about the adjacencies of all spaces, the view of the environment from production, hospitality, and offices, and the seasonality of processing. Again, starting this process early in the planning process is a necessity. In the end it will be worth it, as we are sure that all of us have had a quintessential wine or beer or spirits moment at a producer that we will never forget—that's the goal!

Index

Printed in the United States
by Baker & Taylor Publisher Services